DIGITAL ELECTRONICS

DIGITAL ELECTRONICS

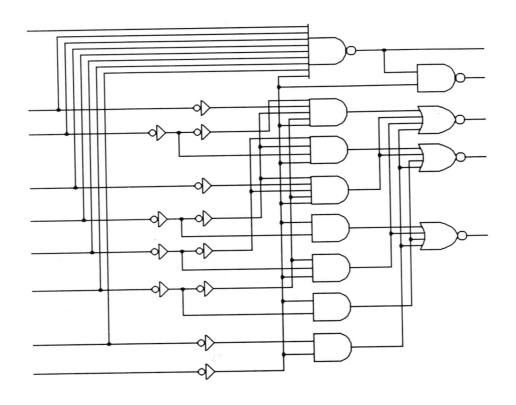

David L. Wagner

Department of Physics and Technology
Edinboro University
Edinboro, Pennsylvania

Harcourt Brace Jovanovich, Publishers

Technology Publications

San Diego   New York   Chicago   Austin   Washington, D.C.

London   Sydney   Toronto

ISBN 0-15-517636-6

Library of Congress Number 87-81903

Printed in the United States of America

Production supervision and interior design: WordCrafters Editorial Services, Inc.

CONTENTS

Chapter 2
THE CONSTRUCTION OF LOGIC GATES   36

Chapter 3
BOOLEAN ALGEBRA   68

Chapter 4
COMBINATIONAL LOGIC CIRCUITS

Chapter 5
FLIP-FLOPS

Chapter 6
FLIP-FLOP APPLICATIONS
156

Chapter 7
ASTABLE AND MONOSTABLE MULTIVIBRATORS 188

Chapter 8
KARNAUGH MAPPING AND DIGITAL DESIGN 212

Chapter 9
DAC, ADC, AND MICROPROCESSOR INTERFACING 252

Appendix
LABORATORY EXPERIMENTS   277

PREFACE

Historically, in most two-year associate of science programs and four-year engineering technology programs, digital electronics has been taught in a two-semester sequence, usually following a course in analog electronics. Recently, with the advent of microprocessor technology, the trend has been to compact the introduction to digital electronics into a one-semester course, often taught before or concurrently with a course in analog electronics. This practice leaves room in the curriculum for students to take one or more microprocessor courses, which draw heavily on a knowledge of digital electronics. It is also becoming increasingly common for four-year science majors in areas such as physics and chemistry to take a one-semester course in digital electronics—a subject that is playing an ever larger role in the science laboratory. This book addresses the need for a text suitable for a one-semester course.

This book is written for students in two-year associate of science or technology programs, four-year engineering technology programs, or four-year science programs. It introduces basic digital concepts and gives students a working knowledge of digital electronics. The material can be covered in one term of 15 to 16 weeks. In addition to the basic textbook, a complete set of laboratory exercises is included as an Appendix. The laboratory work can be done most conveniently with one of the many digital trainers on the market—those with +5-, +12-, and −12-volt supplies, logic indicators, debounced switches, and perhaps TTL-level square-wave outputs.

It is assumed that students using this book will have a command of basic algebra and a familiarity with the elements of dc and ac electricity. To make the book useful in a variety of curricula, however, there is a brief introduction to the operation of diodes, junction transistors, and MOS transistors; the material is not intended as a complete treatment but should prove sufficient to provide a basis for understanding the basic input and output characteristics of digital logic devices.

Chapter 9 includes a brief discussion of the 6502 microprocessor, which is intended as an illustration of the relative simplicity of interfacing digital circuits to microprocessor systems, not as a thorough discussion of the 6502 itself. The purpose of this material is largely to encourage students to go on to further study in the field of microprocessors; it can easily be omitted in curricula that include a course in microprocessor technology.

On a chapter by chapter basis, Chapter 1 introduces the basic logic gates NOT, AND, and OR. Other simple gates derived from these are also developed. Chapter 2 begins with a brief introduction to diodes, junction transistors, and MOS transistors and then shows how the basic logic gates can be implemented with these components. TTL and CMOS logic are explored in some detail. Chapter 3 develops the laws of Boolean algebra and introduces Boolean reduction techniques.

Chapter 4 introduces a number of useful combinational logic circuits such as decoders, multiplexers, code converters, and demultiplexers. The implementation of these devices is illustrated using both basic logic gates and MSI TTL circuits.

Chapter 5 covers flip-flop circuits. $R-S$ flip-flops are described first, followed by clocked $R-S$ flip-flops, D-type flip-flops, and $J-K$, master$-$slave flip-flops. Some basic uses of flip-flops are presented, such as switch debouncing and data latching. Chapter 6 shows how flip-flops can be used to implement counters, shift registers, and ring counters. Also included in this chapter are examples of MSI counter and shift registers.

In Chapter 7 a variety of clock circuits is developed. In particular, astable and monostable multivibrators are explained, as are multivibrators built around basic gates with Schmitt-trigger inputs and around the 555 timer.

Chapter 8 introduces the Karnaugh map reduction technique for combinational logic and gives several design examples using this technique. A Karnaugh map reduction technique is also presented for designing special counters using $J-K$ flip-flops, with design examples using the technique. The chapter concludes with an example involving the use of both combinational and sequential Karnaugh mapping to solve a design problem.

Chapter 9 begins with a discussion of digital-to-analog conversion using the $R-2R$ ladder technique. MSI DACs are then discussed, and an example is given of how to do analog-to-digital conversion using a DAC with suitable additional circuitry. Discussion of MSI ADCs follows, and finally some basic concepts of microprocessor interfacing are introduced. The 6502 microprocessor is used as an example because of its simplicity and because it forms the heart of many inexpensive computers (Apple II, Commodore 64, for example) likely to be available in electronics labs. Details are given for the construction of simple input and output ports and for accessing these from both BASIC and assembly language.

The emphasis throughout the book is on practical examples utilizing the most readily available digital circuits. Students who complete a course based on this textbook should be able to use digital electronics to solve many of the design problems they will encounter in the real world.

This book represents the efforts of many. For sharing their expertise and constructive criticism, I want to thank the following reviewers: Vincent K. Hall, ITT Technical Institute, Dayton, Ohio; Henry Hawkins, State University of New York–Oswego; Richard L. Koelker, Kent State University; William H. Mast, Appalachian State University; and Jack L. Waintraub, Middlesex County College.

David L. Wagner

DIGITAL ELECTRONICS

Chapter 1

FUNDAMENTAL CONCEPTS

Learning Objectives

After completing this chapter you should know:

- The basic digital functions NOT, AND, and OR and the logic symbols representing each.

- The derived functions NAND and NOR and the logic symbols representing each.

- The exclusive-OR (XOR) function and its logic symbol.

- The rudiments of Boolean algebra, including how to derive the logical function performed by a digital circuit consisting of simple logic gates.

- How to construct a digital circuit from a given Boolean function.

- What is meant by active inputs and outputs.

- How NAND gates or NOR gates can be used to construct any other gate.

- The binary number system, how to add and subtract binary numbers, and how to convert between the binary and decimal number systems.

- Several examples of alternative binary codes such as binary-coded decimal (BCD), excess three (XS3), and the Gray code.

- The hexadecimal number system and how to convert numbers between the hexadecimal, binary, and decimal systems.

In analog electronics, as opposed to digital electronics, we deal with quantities that can vary in a continuous fashion. Currents or voltages can take on a whole range of values between the limits imposed by the circuit under consideration. In digital electronics, on the other hand, we usually deal with voltages that are constrained to lie in one of two well-defined ranges. Each of these two ranges is referred to as a voltage *state*.

Perhaps the easiest way to understand the idea of two possible voltage states is to consider the simple switch arrangement shown in Figure 1.1. In the figure, the switch must be either open or closed. Thus there are two distinct states that the switch can exist in. If the switch is closed, the output of the circuit, V_{out}, will be 5 volts (V). On the other hand, the instant the switch opens the output will become 0 volts because of the resistor that ties the output directly to ground. Thus the output voltage can take on only the two values 5 volts and 0 volts.

Switches that must be either open or closed are perhaps the easiest type of digital circuit to visualize. There are, however, many ways electronic circuits can be constructed such that only two output voltages or "states" are possible. Transistors, for example, make excellent two-state devices. The important point is that, as long as we are dealing with two well-defined states, the circuit can be considered a digital circuit and can be analyzed according to the principles we will develop in this text.

In practice, the two voltage states of a digital circuit are two ranges of voltages that the circuit is restricted to. For example, the circuit may be restricted to a range of 0 to 0.8 volt or 2 to 5 volts. This means that the voltage of the circuit should never lie in the range of 0.8 to 2 volts. It is awkward to constantly refer to the actual voltage ranges of real circuits. Thus a shorthand terminology is used. We agree to call one of the two voltage ranges 0, and the other range 1. The most common convention is to refer to the system when it is in the lower voltage range as being in state 0, and when it is in the higher voltage range as being in state 1. This convention is referred to as *positive logic*. We could, of course, refer to the lower range as state 1 and the higher range as state 0. If we did this, we would be using the convention called *negative logic*. In this text we will restrict ourselves to positive logic.

Labeling the possible states of a digital circuit as 0 or 1 emphasizes the *binary* nature of digital circuits. Binary simply refers to a two-state device or system. A

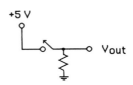

FIGURE 1.1 A simple digital circuit. The output is either 5 or 0 volts.

whole system of logic can be built around binary notation. This logic, represented by an abstract notation called *Boolean algebra*, involves the concepts of *true* and *false*, two distinct possibilities. By thinking of state 1 as true and state 0 as false, we see that we have an ideal notation for representing Boolean algebra. We will investigate the use of these concepts shortly. The symbols 0 and 1 also suggest numbers to us. It is entirely possible to do arithmetic in terms of just the numbers 0 and 1. These two numbers define the *binary number system*. Later in this chapter we will introduce the binary number system and show how arithmetic can be done in this system. The important point is that digital circuits, because of their binary nature, are well suited to do logical operations and, through the interpretation of the 0's and 1's as binary numbers, arithmetic operations as well.

Let us turn our attention first to the idea of binary logic and introduce three basic logical operations or functions in terms of which any complex logical function can be expressed.

1.2 THE BASIC DIGITAL FUNCTIONS

Real digital circuits usually carry out rather complex logical or arithmetic operations. Fortunately, these complex operations can be broken down and represented by only three quite simple logical operations. Each of these three simple operations is actually achieved by a small digital circuit, which we might refer to as *subcircuit*, since it refers to a building block of larger useful circuits. There are many different ways to actually build these subcircuits to achieve or *implement* the basic logical operations. At this stage, only the simplest of circuits will actually be examined. Our main objective is in defining and understanding the logical operations themselves.

The Inverter or NOT Gate

The simplest digital subcircuit is the *inverter*. This circuit takes a digital input and produces as its output the opposite state. That is, an input of 1 produces an output of 0, and an input of 0 produces an output of 1. Figure 1.2 shows the two basic symbols that are used to represent the inverter. The inversion is actually indicated by the small circle on either the output or the input side of the triangle. The triangle by itself with no circle is just the symbol for a *buffer amplifier*. A buffer amplifier is an electronic device that produces a voltage output that is the same as the input. However, it permits more current to be generated. We will see why this is important in Chapter 2.

Figure 1.2 contains some additional concepts. First, notice that the input side of the inverter has an input labeled A, and the output side has an output labeled $\overline{A}$. The label A is just a name given to the input so that we can discuss it in an abstract fashion. The actual input represented by A must be either 0 or 1. A symbol

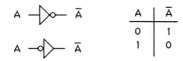

A	$\overline{A}$
0	1
1	0

FIGURE 1.2 The basic symbols and truth table for the inverter or NOT gate.

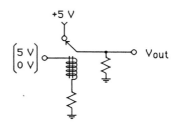

FIGURE 1.3 An inverter constructed from a relay.

such as $\overline{A}$ is called a *Boolean variable* and can be used in a type of algebra called Boolean algebra that is very useful in analyzing digital circuits (more about this shortly). The symbol $\overline{A}$ is one of the standard ways of indicating the *inverse* of A. The symbol $\overline{A}$ is read as *A not*. Whenever a bar appears over a logical (Boolean) symbol, this always means that whatever the value of the symbol itself is at the moment (1 or 0), it is to be inverted. Thus, if A is 1, then $\overline{A}$ or *A not* is 0.

The second thing to notice in Figure 1.2 is the truth table. A truth table is just a simple way of listing all the possible inputs to the circuit and the output that results in each case. In this case, the truth table is very simple. Under the column labeled A are listed the two inputs that are possible to the inverter. Under the column labeled $\overline{A}$ are listed the outputs that are produced for each input. It is clear from the truth table that, when A is 0 then $\overline{A}$ is 1, and when A is 1 then $\overline{A}$ is 0. A truth table makes a very handy way of summarizing the digital function of a circuit.

Figure 1.2 gives no clue to how an inverter might actually be constructed from electrical components. In fact, the logical idea of inversion does not even need to apply to electrical circuits at all. Digital logic was invented before there were any electrical circuits. Nevertheless, it would be nice to see how an inverter might be made. Figure 1.3 illustrates how an inverter might be constructed using a *relay*. A relay is just an electrical device that has a small coil of wire placed near a switch. When sufficient current flows through the coil, a magnetic field is produced that pulls on the switch. This causes the switch to be moved from its *normal* position. In the case of Figure 1.3, the switch is normally held closed by a small spring, so that V_{out} will be 5 volts when no current flows through the small wire coil. Thus 0 volts input produces 5 volts output. On the other hand, if the input to the relay becomes 5 volts, then current flows through the relay windings. This produces a magnetic field that pulls the switch open. Once this happens, the output drops to 0 volts. Thus an input of 5 volts results in an output of 0 volts. If we identify 5 volts as logical 1 and 0 volts as logical 0, when we have exactly the truth table shown in Figure 1.2. Relay logic was indeed used extensively in industry for many years and is in fact still in wide use for large power applications.

As a final comment on the inverter, we note that it is also referred to as the NOT gate. The term NOT simply means logical inversion. The use of the word gate is an historical artifact that arose with respect to AND gates and OR gates, which will be discussed next. Today the term is simply used as a synonym for circuit. This is, a NOT gate is simply a circuit that performs the NOT function (that is, inversion).

The AND Gate

The inverter has only a single input. Most logic gates, however, can have two or more inputs. The AND gate falls in this category. The logical idea of AND can be best understood by simple example.

Suppose that a sign says that only ticket holders over 16 years of age will be admitted. What this means is that, to be admitted, two things must be true. An individual must hold a ticket AND must be over 16 years of age. In this case there are two input states, each of which may be either true or false. If we identify logical 1 with true and logical 0 with false, then only when both inputs are 1 (both true) will the output be 1 (true; yes, you can enter). In digital logic, a 2-input AND gate performs this type of function. Figure 1.4 shows the basic logical symbol for a 2-input AND gate. Again notice the symbols A, B, and AB. These are simply a convenient way of representing the two inputs (A and B) and the output (AB). The combination of symbols AB (often written with a dot between A and B) is the standard way of indicating the AND function. It looks like the notation for normal multiplication, but *it does not mean multiplication*. The combination AB is read as "A and B." The truth table in Figure 1.4 describes the function of the AND gate. All possible inputs are listed in the truth table, along with the corresponding output in each case. We see that only when *both* inputs are 1 will the output be 1.

Figure 1.5 illustrates a simple two-relay implementation of an AND gate. The two relays include normally open switches, which are held open by small springs.

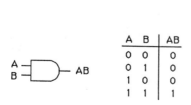

A	B	AB
0	0	0
0	1	0
1	0	0
1	1	1

FIGURE 1.4 The standard symbol and truth table for the AND gate.

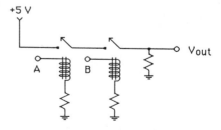

FIGURE 1.5 Implementation of an AND gate using two relays.

A and *B* are inputs to the two relays. Again let us assume that logical 1 corresponds to 5 volts and logical 0 corresponds to 0 volts. The output will clearly be 0 volts unless we can get both switches to close. To get both switches to close, current must flow in both relays to pull the switches closed. This requires that both *A* and *B* be 5 volts or logical 1. We see that this simple circuit does indeed produce the truth table in Figure 1.4.

The AND operation is not restricted to only two inputs. It is entirely possible to think of a situation where three conditions must all be true in order for the output to be true. Or we could have four conditions, or five, or any number. Figure 1.6 shows the logical symbol for a 3-input AND gate. However, a 3-input AND gate (or an AND gate with any number of inputs) can be implemented from a combination of 2-input AND gates. This is also shown in Figure 1.6. The 2-input AND gate is a fundamental logic unit. A larger-number input AND gate is a derived logic unit, since it can be constructed from a number of fundamental logic gates.

The OR Gate

The final basic logical function is the OR function. Like the AND gate, an OR gate can have two or more inputs. And, like the AND function, the idea of logical OR can best be understood by example.

At a local bank, free checking is extended to anyone who maintains a $100 balance in their account or who is at least 65 years old. Again we have two inputs, but now if *either one or both* are true the output is true (yes, you get free checking). If we identify logical 1 with true and logical 0 with false, then we have a situation where, if either or both inputs are 1, then the output will be 1. Figure 1.7 shows the logical symbol for a 2-input OR gate. The symbols *A* and *B* are used to conveniently represent the two inputs, and the output is represented by the com-

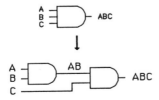

A	B	A+B
0	0	0
0	1	1
1	0	1
1	1	1

FIGURE 1.6 A 3-input AND gate and the equivalent circuit constructed from 2-input AND gates.

FIGURE 1.7 The symbol and truth table for the OR gate.

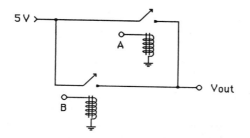

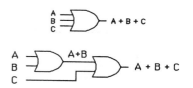

FIGURE 1.9 A 3-input OR gate and its implementation using 2-input OR gates.

FIGURE 1.8 An implementation of an OR gate using two relays.

bination of symbols $A + B$. The combination $A + B$ simply indicates the logical OR operation. Don't be deceived by the $+$ symbol. *It does not mean add*, it means A and B are to be logically ORed together. The combination of symbols $A + B$ is read as "A OR B." The truth table in Figure 1.7 summarizes what we have said about the 2-input OR function. If either or both inputs are 1, the output is 1. The OR function is sometimes referred to as the *inclusive* OR since the input state $A = 1$, $B = 1$ is included as one of the sets of inputs that produces an output of 1.

Figure 1.8 shows how a 2-input OR gate could be constructed from relays. Both switches are normally open. Clearly, if either switch (or both) is closed, the output will be 5 volts. Only if both switches remain open will the output be 0 volts. To close a switch, current must flow through the appropriate relay. This requires that the input to that relay be 5 volts. Thus if either input A or input B goes high (to 5 volts), the corresponding switch will close and the output will go to 5 volts. Again identifying logical 1 with 5 volts and logical 0 with 0 volts, we see that this circuit implements the OR function.

As with the AND gate, the OR gate can have more than two inputs. In such cases, if *any* of the inputs is 1 then the output is 1. Figure 1.9 illustrates the case of a 3-input OR gate. Also shown in the figure is an implementation of the 3-input OR gate using only 2-input OR gates. The 2-input OR gate is a fundamental logic gate, while the 3- (or higher) input OR gate is a derived logic gate since it can be constructed from fundamental logic gates.

We have now seen the three fundamental logic gates, the inverter, the AND gate, and the OR gate. From a strictly logical viewpoint, all the more complex logic circuits we will deal with could be implemented with various combinations of just these three basic gates. As a practical matter, however, it proves useful to define some additional logic gates that are very closely related to the basic three introduced. The reason for this is twofold. First, from a circuit implementation viewpoint, it proves easier to construct these additional gates than to construct some of the basic gates. Second, the flexibility gained by having these additional gates available greatly simplifies circuit design.

1.3   SOME ADDITIONAL LOGIC GATES

The NAND Gate

The NAND gate is simply an AND gate with its output inverted. The symbol for the 2-input NAND gate is shown in Figure 1.10. Shown directly below the symbol for the NAND gate is the logical equivalent of the NAND gate constructed from an AND gate and an inverter. Notice that the symbolic representation of the NAND function is $\overline{AB}$, which is read generally as "$(A$ and $B)$ NOT." It is important to keep in mind that the entire term AB is inverted. This is more easily seen by looking at the AND-inverter equivalent combination. The truth table in Figure 1.10 illustrates the NAND function. In the truth table, the AND function has been listed first as a matter of reference. It is clear that the NAND function is just the inversion of the AND function.

The NOR Gate

The NOR gate is an OR gate with its output inverted. Figure 1.11 shows the logical symbol for the 2-input NOR gate. Below this symbol is the logical equivalent of the NOR gate using an OR gate and an inverter. The symbolic representation of the NOR function is $\overline{A + B}$, which is usually read as "A OR B NOT." Again note that the entire term $A + B$ is inverted. The NOR function is spelled out more clearly in the truth table in Figure 1.11. In the truth table, the OR function has been included for handy reference, and it can be seen that the NOR function is just the inverse of the OR function.

The Exclusive-OR Gate

The exclusive-OR (XOR) gate is not as easily derivable from the basic gates as are the NAND and NOR gates. The easiest way to understand the XOR gate is to examine the truth table in Figure 1.12. This truth table shows that the XOR

FIGURE 1.10   The circuit symbol and truth table for a 2-input NAND gate. Also shown is the basic gate equivalent circuit. Note the bar over the AB expression, which indicates inversion of the entire expression.

A	B	AB	$\overline{AB}$
0	0	0	1
0	1	0	1
1	0	0	1
1	1	1	0

FIGURE 1.11 The circuit symbol and truth table for a NOR gate. Also shown is the basic gate equivalent circuit. Note the bar over the A + B expression denoting that the entire expression is to be inverted.

A	B	A + B	$\overline{A+B}$
0	0	0	1
0	1	1	0
1	0	1	0
1	1	1	0

function is 1 when one input is 1 and one input is zero, and is 0 otherwise. This is slightly different than the standard-OR (*inclusive*-OR) gate, which had an output of 1 when either *or both* inputs were 1. The standard symbol for the XOR operation is $\oplus$, and an expression such as $A \oplus B$ is usually read as "A ring-sum B."

An XOR gate is not one of the fundamental gates, so it must be constructed from a combination of these fundamental gates. The direct method of construction results from looking at what logic operation is actually being performed by the XOR gate. If we label the two inputs to the XOR gate as A and B, then the XOR gate goes high (to logical 1) whenever [A is high (1) AND B is low (0)] OR [A is low (0) AND B is high (1)]. In abstract notation, this can be written as

$$A \oplus B = A\overline{B} + \overline{A}B \tag{1.1}$$

Equation (1.1) is our first look at a Boolean equation. In Chapter 3, the rules of Boolean algebra will be developed in detail. For the moment, it is only necessary to introduce a few basic ideas

The idea behind an equation like Eq. (1.1) is that one logical operation, in this case $A \oplus B$, is defined in terms of previously defined operations. We already know the meaning of the individual terms on the right side of the equation. We only need to be clear about the order in which the operations on the right side of the equation are done. The rules of Boolean algebra state that the AND operation takes precedence over the OR operation. Thus, in Eq. (1.1), the terms $A\overline{B}$ and $\overline{A}B$ are evaluated first and then the results are ORed together. In more complex equations, it is necessary to introduce parentheses to keep terms properly grouped together. As in standard arithmetic, all terms within a pair of parentheses are to be evaluated before terms outside the parentheses are evaluated.

Returning to Eq. (1.1), its logical meaning is quite simple. The term on the left side of the equation is the notation for the XOR function. The terms on the

FIGURE 1.12 The basic symbol and truth table for the exclusive-OR (XOR) gate.

A	B	A⊕B
0	0	0
0	1	1
1	0	1
1	1	0

$$A \oplus B = A\overline{B} + \overline{A}B$$

FIGURE 1.13 The basic Boolean expression for the XOR function followed by a truth table demonstrating the correctness of the Boolean expression.

A	B	$\overline{A}$	$\overline{B}$	$A\overline{B}$	$\overline{A}B$	$A\overline{B} + \overline{A}B$
0	0	1	1	0	0	0
0	1	1	0	0	1	1
1	0	0	1	1	0	1
1	1	0	0	0	0	0

right side of the equation are the logical equivalent of $A \oplus B$. The right side of Eq. (1.1) will be 1 whenever either of the two terms on the right side equals 1. We can easily see when this occurs by constructing a simple truth table. This is shown in Figure 1.13. The truth table in Figure 1.13 is somewhat more elaborate than might ordinarily be the case in order to make a few important points. When we examine an expression such as that shown in Eq. (1.1), we are interested in what the value (1 or 0) of that expression will be for each of the possible combinations of A and B that can occur. Since A and B can each take on two values, four combinations are possible. These are listed at the left side of the truth table.

The next section of the truth table shows the corresponding values of $\overline{A}$ and $\overline{B}$. The third section of the truth table shows the values of $A\overline{B}$ and $\overline{A}B$ for each combination of A and B. Let us examine the values listed under $A\overline{B}$. This term is read as A AND (NOT B). For this term to equal 1, A must be 1 and $\overline{B}$ must also be 1 since A and $\overline{B}$ are being ANDed together. For $\overline{B}$ to be equal to 1, B must equal 0. Remember, $\overline{B}$ is the inverse of B. Thus A and $\overline{B}$ will be 1 for that unique combination of A and B where A is 1 and B is 0. The truth table shows that this is indeed the case. From what we have said, it is also clear that the $\overline{A}$ and B term will be 1 only when A is 0 (that is, $\overline{A}$ is 1) and B is 1. The truth table shows this to be the case also.

Finally, the last section of the truth table shows the value of $A\overline{B} + \overline{A}B$ for each possible input combination of A and B. To ascertain the value of $A\overline{B} + \overline{A}B$, we need only look to the left in the truth table and OR the two sets of values under $A\overline{B}$ and $\overline{A}B$. We see finally that $A\overline{B} + \overline{A}B$ is 1 only for the two input conditions $A = 1$, $B = 0$, and $A = 0$, $B = 1$. This is, of course, exactly our original definition of the XOR function.

Once we have an equation like Eq. (1.1), it can be used to guide us in constructing an XOR gate from the three fundamental logic gates. Figure 1.14 shows the direct circuit implementation of the right side of Eq. (1.1). The input(s) to and output from each gate are carefully labeled to allow you to see the final development of $A\overline{B} + \overline{A}B$. Notice how inverters are used to produce $\overline{A}$ and $\overline{B}$, and then AND gates are used to produce the $A\overline{B}$ and $\overline{A}B$ terms. Finally, an OR gate produces the end result.

Although the circuit shown in Figure 1.14 is certainly correct, it would probably not be the method of choice of most designers for constructing an XOR gate

FIGURE 1.14 The direct circuit implementation of the XOR gate that follows from the Boolean expression for the XOR function. Notice that all three kinds of basic gates are required.

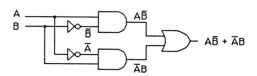

from other more fundamental gates. The reason for this is practical. Modern technology provides logic gates in the form of small *integrated circuits* (ICs), which usually have several identical logic gates in a single package. An integrated circuit is the fabrication of a complete electrical circuit, including several transistors, diodes, resistors, and capacitors, on a single piece of silicon. The circuit shown in Figure 1.14 uses three different kinds of logic gates, which would normally come in three different IC packages.

Another way to construct an XOR gate is shown in Figure 1.15. We will not do a Boolean analysis of this circuit at the present time. However, you should be able to convince yourself, by looking at the results of the four possible input combinations, that this circuit will indeed function as an XOR gate. It is interesting that this design does not even use any of the "fundamental" gates, but rather uses four NAND gates. From the designer's viewpoint, this is no handicap at all. In fact, NAND gates are actually easier to fabricate in IC form than are AND gates. In addition, one standard type of digital IC contains four NAND gates in a single package. Thus the circuit in Figure 1.15 requires only a single IC package to construct, while the circuit in Figure 1.14 requires three such packages. Apparent simplicity on paper does not always translate into actual simplicity in the final circuit from the designer's viewpoint.

The XOR gate proves to be quite useful, but the inverse of the XOR function is also very useful. An XOR gate with the output inverted is called an XNOR gate. The symbol for the XNOR gate is the XOR gate symbol with a circle on the output to indicate inversion. The symbolic representation of the XNOR function is $\overline{A \oplus B}$, which is usually read as "A ring-sum B NOT." The basic symbol and truth table for the XNOR function are shown in Figure 1.16.

FIGURE 1.15 An alternative implementation of an XOR gate using four NAND gates. This approach requires only one type of gate, four of which come in a single integrated-circuit package.

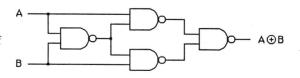

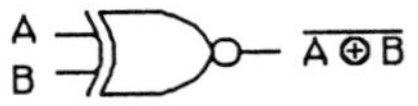

FIGURE 1.16 The basic symbol and truth table for the exclusive-NOR (XNOR) gate.

A	B	$\overline{A \oplus B}$
0	0	1
0	1	0
1	0	0
1	1	1

1.4 MORE ON COMBINATIONS OF GATES

In Section 1.3, we began to see how logic gates can be combined in various ways to produce more complex logical functions. In this section, we will investigate gate combinations in more detail. Basically, two general approaches are of interest. First, it is often necessary to examine an existing circuit diagram and to infer from the diagram what the Boolean function of the circuit is. One reason for doing this is obviously to learn the basic function of the circuit. A second reason is to obtain the Boolean function so that it can be manipulated according to the rules of Boolean algebra, perhaps leading to circuit simplification or to a less expensive circuit.

The second approach begins with a known Boolean function. This function then serves as a guide for the construction of a circuit that implements the function.

Examples of Deriving the Function from the Circuit

■ EXAMPLE 1.4—1

Figure 1.17 shows a simple logic circuit having two inputs, A and B, and four logic gates. Let us follow the two inputs as they are operated on by the various logic gates. First, we see that both A and B are input directly to a 2-input NAND gate. The output of this gate must then be $\overline{AB}$. This output will be one of the inputs to the final OR gate. Returning to the two inputs, A and B, we see that A is input to an inverter, the output of which goes to an AND gate. If A goes into the inverter, then $\overline{A}$ is the output from the inverter. This output, $\overline{A}$, is one of the inputs to the lower AND gate. The other input to this gate is B. The output is therefore $\overline{A}B$.

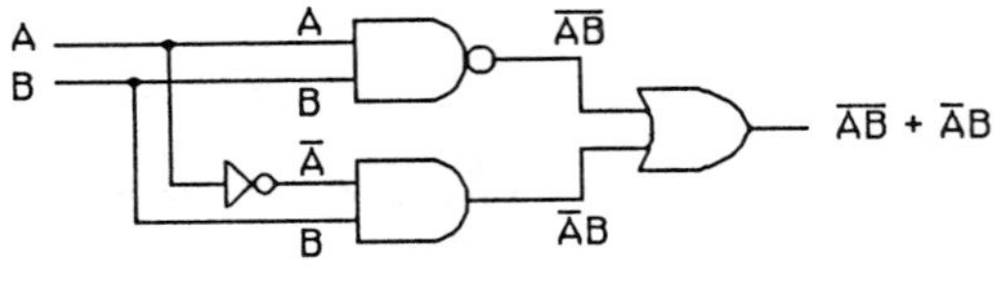

FIGURE 1.17 Circuit for Example 1.4-1 showing how the Boolean expression is arrived at by following the inputs carefully through each gate in the circuit.

FIGURE 1.18 Circuit for Example 1.4-2 showing how the Boolean expression for the circuit is derived by following the inputs through each gate in the circuit.

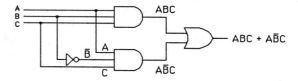

Notice how this is written on Figure 1.17. Finally, the two inputs to the OR gate are $\overline{AB}$ and $\overline{A}B$. These two inputs are ORed together to produce the final output, which is shown on the figure. ∎

■ **EXAMPLE 1.4—2**

It is of course possible to have more than two inputs, and we also need not restrict ourselves to 2-input gates. Figure 1.18 illustrates a 3-input example that also uses two 3-input AND gates. Again, the input(s) to and the output from each gate are labeled on the figure so that you can see how the final result is arrived at. ∎

■ **EXAMPLE 1.4—3**

The final example shown in Figure 1.19 is a bit more complicated. This is the type of situation where errors often creep in. Look carefully at how each input and output is labeled in Figure 1.19. Notice particularly that the output from the 2-input OR gate has parentheses around the entire expression. These parentheses, although they are not really necessary in the figure, emphasize that the expression is a *single* entity. Notice also that the output from the 2-input OR gate must be preserved as a single entity as it is operated on by the final AND gate. This is done by enclosing this term in parentheses so that it is kept as one term.

By carefully labeling inputs and outputs as we have done in our three examples, it is possible to determine the Boolean expression for any circuit constructed

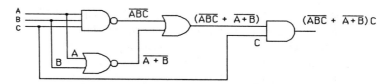

FIGURE 1.19 Circuit for Example 1.4-3 showing how the Boolean expression for the circuit is derived by following the inputs through each gate in the circuit.

from digital logic gates. In complex cases, the expression thus derived may seem hopelessly messy to deal with. However, we will soon develop methods for *reducing* complex expressions to much simpler form.

Examples of Constructing the Circuit from a Boolean Expression

Generally, the circuit designer has a logical function in mind before a circuit is actually implemented. That is, the designer has arrived at some Boolean expression that will achieve the desired result, and now must design a circuit to implement the Boolean function. In Chapter 3, we will see that there are actually many circuit designs that implement the same Boolean function. However, we can always implement the function directly as it stands, so to speak. This can be done by working from the output of the function back toward the inputs to the function or from the inputs to the function toward the output of the function. The following examples illustrate the two methods.

■ **EXAMPLE 1.4—4**

Figure 1.20 illustrates the output-to-input approach. We begin with a basic Boolean expression to be implemented. In this case the expression is $(A + B)C + \overline{AB}$. It is clear that the final gate in the circuit must be an OR gate that ORs together the two terms $(A + B)C$ and $\overline{AB}$. Thus we construct this gate first and label the two inputs to it. This is shown in Figure 1.20(a). Next we think about how the $(A + B)C$ and $\overline{AB}$ terms must be produced. Clearly, the $(A + B)C$ term comes from ANDing $(A + B)$ with C. This requires an AND gate. The $\overline{AB}$ term is just A NANDed wth B. Figure 1.20(b) shows this new information. Finally, the $(A + B)$ must still be produced. This requires an OR gate to OR A with B. Our task is now complete, and the result is shown in Figure 1.20(c). ■

In implementing the Boolean expression of Figure 1.20, we could just as well have used an input-to-output approach. This method is illustrated in Figure 1.21. We begin by identifying the basic input variables, in this case A, B, and C. Next we look at what must be done to each of these variables to build up the final expression. Clearly, A and B must be NANDed, so we do this using a NAND gate. A and B must also be ORed, so we do this with an OR gate. This leaves us with the circuit shown in Figure 1.21 (a). Now we look at the outputs that we have created to see what must be done with them. We see that the $(A + B)$ term must

Boolean function to be implemented: $(A+B)C + \overline{AB}$

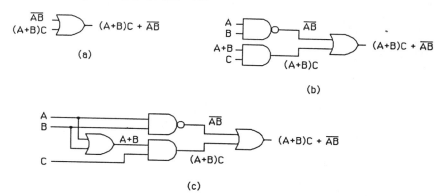

(a)

(b)

(c)

FIGURE 1.20 The circuit for Example 1.4-4 showing the implementation of a Boolean function from an output-to-input approach.

be ANDed with *C*, so we do this with an AND gate. This step is shown in Figure 1.21(b). Finally, it is only necessary to OR the two outputs shown in Figure 1.21(b) to get the desired result. This is shown in Figure 1.21(c), which is identical to Figure 1.20(c).

Boolean function to be implemented: $(A+B)C + \overline{AB}$

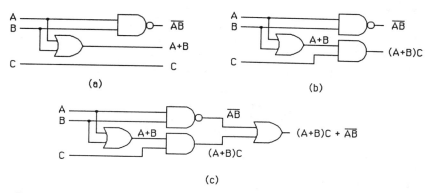

(a)

(b)

(c)

FIGURE 1.21 The same circuit as in Figure 1.20, but developed from an input-to-output approach.

Boolean function to be implemented: $\overline{(\overline{A}B + C)B}$

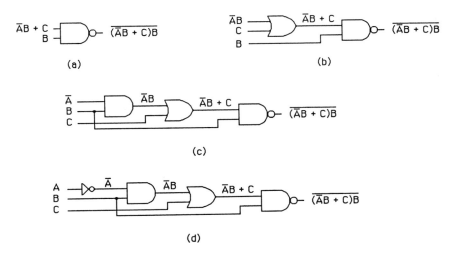

(a)

(b)

(c)

(d)

FIGURE 1.22 Circuit for Example 1.4-5 showing the development of the circuit implementation of a Boolean function from an output-to-input approach.

■ EXAMPLE 1.4–5

As a final example, let us examine the Boolean expression shown at the top of Figure 1.22. Perhaps one confusing thing about this expression is the bar over the entire expression. Under the bar, there is a product of two distinct terms, $(\overline{A}B + C)$ and B. The operation being performed is thus a NAND operation between the terms $(\overline{A}B + C)$ and B. If we adopt an output-to-input approach, then the final gate must be a NAND gate. This is shown in Figure 1.22(a). One input to this NAND gate is just B, which requires no further attention. The other term, $(\overline{A}B + C)$, must be produced by additional gates. We clearly need an OR gate. This is shown in Figure 1.22(b). One input to this OR gate is just C, which needs no further attention. The other input, $\overline{A}B$, requires an AND gate. This is shown in Figure 1.22(c). One input to the AND gate is just B, which requires no further attention. The other input is $\overline{A}$. This input is produced by putting A through an inverter or NOT gate. Thus the final circuit is shown in Figure 1.22(d). ■

These few examples hardly exhaust the complexities of constructing a circuit to implement a given Boolean function. However, they should provide sufficient guidance to allow you to design circuits for fairly straightforward Boolean functions.

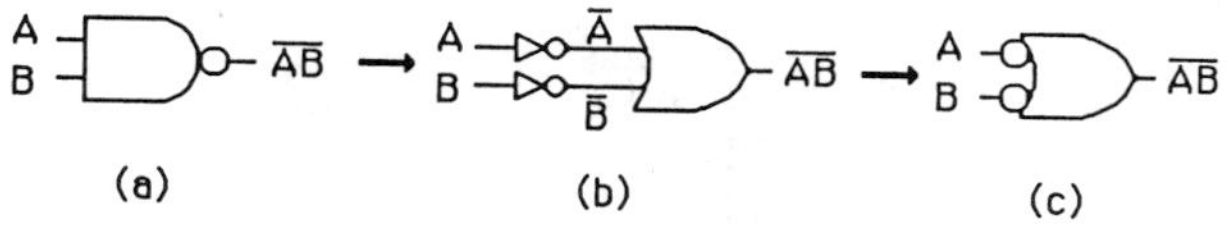

FIGURE 1.23   (a) The basic NAND gate symbol. (b) The low-level inputs shown explicitly as inverters. (c) The equivalent low-level input OR gate symbol for a NAND gate.

1.5   THE IDEA OF ACTIVE INPUTS AND OUTPUTS

In Section 1.3 we introduced the NAND gate and the NOR gate and the symbols used to represent them. Figures 1.10 and 1.11 show these symbols and the truth tables that define the NAND and NOR operations. Let us examine the NAND truth table in Figure 1.10 a little more closely.

The standard interpretation of this truth table is to say that, if both A and B are high, then the output is low; otherwise, the output is high. However, you might notice that another way of viewing the NAND truth table is to realize that, if either input A or input B (or both) is LOW, then the output is high. This implies an OR operation. That is, the Boolean operation $\overline{AB}$ is entirely equivalent to the expression $\overline{A} + \overline{B}$. Figure 1.23(a) shows the normal high-level-logic NAND symbol. Figure 1.23(b) shows the implementation of the expression $\overline{A} + \overline{B}$ using inverters and an OR gate. In Figure 1.23(c), the inverters have been reduced to small circles at the inputs to the OR gate. The symbol in Figure 1.23(c) is entirely equivalent to, and can be used as an alternative to, the standard NAND gate symbol. The symbol in Figure 1.23(c) is called the low-level input OR gate equivalent symbol for the high-level input NAND gate. Further verification of the equivalence of $\overline{AB}$ and $\overline{A} + \overline{B}$ is done by truth table in Figure 1.24.

Given that the two symbols in Figure 1.22(a) and (c) are equivalent, you still might wonder what is to be gained by having two separate symbols that mean exactly the same thing logically. The answer is that judicious use of one or the other of these symbols can make reading a circuit diagram significantly easier.

A	B	$\overline{A}$	$\overline{B}$	$\overline{A} + \overline{B}$	AB	$\overline{AB}$
0	0	1	1	1	0	1
0	1	1	0	1	0	1
1	0	0	1	1	0	1
1	1	0	0	0	1	0

Note equivalence

FIGURE 1.24   Truth table demonstration that $\overline{A} + \overline{B}$ and is equivalent to $\overline{AB}$.

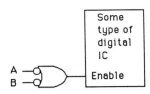

FIGURE 1.25 Either *A* low OR *B* low enables the digital IC. Note that the enable is active high since there is no low-level indicator (small circle) at the enable input, and there is no line over the word enable.

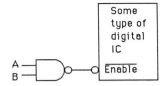

FIGURE 1.26 Both *A* AND *B* must be high to enable the digital IC. Note that the enable input is active low as indicated by the low-level indicator (small circle) and by the line over the word enable.

Suppose, for example, that we had some large digital IC that had a certain function enabled (turned on) when a particular input pin went high. Figure 1.25 shows such a hypothetical situation. Suppose further that we wish the enable input to go high whenever either or both of two control lines (*A* and *B*) go low. We need a NAND gate to do this; but if we use the low-level input OR equivalent symbol, it is much easier to visualize that the OR gate's output goes high whenever either of its inputs goes low. The logic of the diagram is easy to follow.

We could just as easily encounter the reverse situation. Suppose that we had a digital IC that was enabled by a low input to a certain pin. And suppose further that we wished to have this pin be low anytime both of two control lines (*A* and *B*) simultaneously go high. Figure 1.26 shows this situation. The NAND gate in its normal form makes it easy to understand the logic of the figure.

The Generalized OR Gate

Digital designers often make use of equivalent circuit symbols to enhance the readability of their diagrams. Thus you might encounter any of the symbols in Figure 1.27, all of which implement a different logical function. It is actually very easy to keep the meaning of these symbols straight, however. Let us see one way this might be done.

We begin by introducing the idea of active inputs and active outputs. We have now seen small circles introduced on both the inputs of gates, such as in Figure 1.23(c), and on the output of gates, such as the standard NAND gate symbol. These small circles are called low-level indicators. They indicate that a particular input or output is *active low*. When an input or output has no low-level indicator, the input or output is said to be *active high*. The terms active low and active high can best be explained in a specific context.

Let us consider the basic OR function (and later the AND function) in terms of active-low and active-high inputs and outputs. We use the truth table in Figure

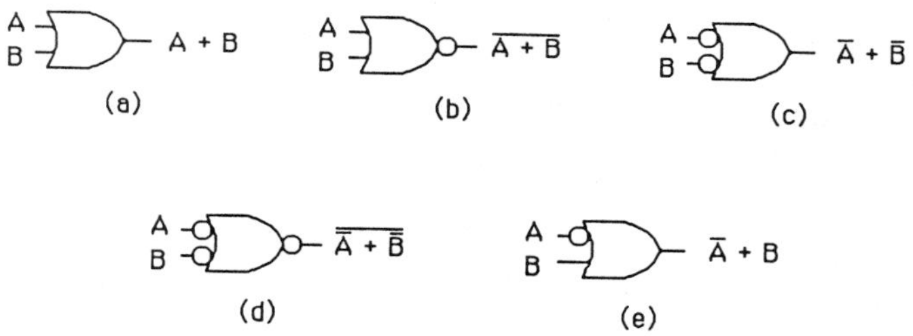

FIGURE 1.27 Basic forms of the OR gate showing different combinations of active high and low inputs and outputs.

1.28 as the basic OR truth table. The words inactive and active have replaced the states 0 and 1, respectively. The basic idea of an OR gate now is that, when either or both of the inputs is ACTIVE, the output is ACTIVE. The particular version of the circuit symbol tells us by inspection which state (high or low) is active for the inputs and output. Consider the following examples.

■ **EXAMPLE 1.5—1**

The symbol in Figure 1.27(a) is just the standard OR gate. The symbol shows that there are no low-level indicators on either input, so both inputs are active high. Likewise, there is no low-level indicator on the output, so the output is also active high. Looking at the basic OR truth table in Figure 1.28, we see that this means that whenever either input is high (active) the output will be high (active).   ■

■ **EXAMPLE 1.5—2**

Consider Figure 1.27(c). Here we see that the circles indicate that both inputs are active low, but the lack of a circle on the output indicates that it is active high. Again, looking at the basic OR truth in Figure 1.28, we see that this means that whenever either input is low (active) the output will be high (active).   ■

FIGURE 1.28 The truth table for the basic OR gate in terms of active and inactive.

A	B	A + B
inactive	inactive	inactive
inactive	active	active
active	inactive	active
active	active	active

■ EXAMPLE 1.5–3

Consider Figure 1.27(e). Here we see that input A is active low while input B is active high. The output is active high. Thus the output will go high (active) whenever either input A goes low (active) or input B goes high (active). ■

We need not memorize different truth tables for each symbol or for that matter even remember that Figure 1.27(c) is really the same as a NAND gate. The symbol itself immediately tells us what logical function is being performed.

The Generalized AND Gate

Figure 1.27 illustrated how the basic OR gate could have different combinations of active high or low inputs or output. The same thing is true of the basic AND gate.

If we examine the truth table for a NOR gate in Figure 1.11, we see that the NOR gate could be thought of as performing an AND function. In particular, the output of the 2-input NOR gate goes high when both inputs A AND B are *low*. That is, the logical function $\overline{A + B}$ is equivalent to $\overline{A}\ \overline{B}$. Figure 1.29 illustrates the equivalence. Just as we introduced a basic OR truth table in terms of active and inactive, so we are now led to introduce a basic AND truth table as well. This is shown in Figure 1.30. Figure 1.31 illustrates some of the symbols you are likely to see on digital circuit diagrams. Again, it is easy to understand what each of these symbols actually means by looking at each symbol and at Figure 1.30.

■ EXAMPLE 1.5–4

Consider the symbol shown in Figure 1.31(b), which is just the basic NAND gate. We see from the basic AND truth table that the output is active only when both

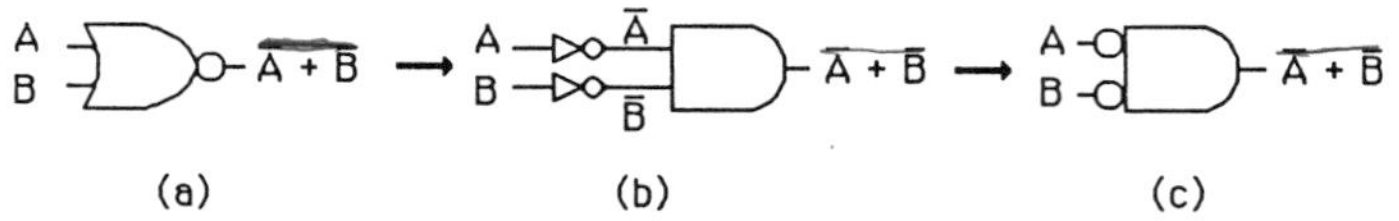

FIGURE 1.29 (a) The basic NOR symbol. (b) The low-level inputs shown explicitly as inverters. (c) The equivalent low-level input AND gate symbol for a NOR gate.

A	B	A B
inactive	inactive	inactive
inactive	active	inactive
active	inactive	inactive
active	active	active

FIGURE 1.30 Truth table for the basic AND gate in terms of active and inactive.

inputs are active. Figure 1.31(b) shows us that both inputs are active high and the output is active low. Thus, when both inputs are high (active), the output will be low (active). ∎

∎ EXAMPLE 1.5—5

Consider Figure 1.31(c). Here both inputs are active low and the output is active high. Thus only when both inputs are low (active) will the output be high (active). This is, of course, just the NOR function. ∎

∎ EXAMPLE 1.5—6

Consider Figure 1.31(d). Here both the inputs and output are active low. Thus only when both inputs are low (active) will the output be low (active). This is a rather awkward way of saying that we have a standard OR gate. Even so, you may see the symbol in Figure 1.31(d) actually used in a circuit diagram if the low output is the state of interest, and it should occur only when the two input lines go low. ∎

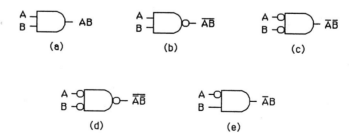

FIGURE 1.31 Basic form of the AND gate showing various combinations of active high and low inputs and outputs.

1.6 GATE CONVERSION

As a practical matter, several simple logic gates are packaged in one IC. Thus four 2-input NAND gates come in a single IC package. Often a circuit designer will not have a direct need for every gate in a given package. For example, the design may call for three 2-input NAND gates, leaving one left unused in the standard package. It therefore becomes useful to know that one kind of gate can sometimes be transformed into a different type of gate.

Using NAND Gates to Build Other Gates

Imagine that we have a NAND gate with one input permanently tied high. This is shown in Figure 1.32(a), where one input is tied directly to 5 volts and the other is called input B. If we examine the NAND truth table in Figure 1.10 and think of the 5-volt input as input A, then only the last two lines of the truth table are possible. Examination of the truth table then shows that the output is always the opposite of input B. The NAND gate has become an inverter for the single input B. Figure 1.32(b) shows a second way of making a NAND gate into an inverter. In this case, the two inputs are tied together to form a single input. The NAND truth table again shows us that, for the two states where both inputs are equal, the output is always the inverse of the input states. Circuit designers often use an unneeded NAND gate as an inverter.

Once we know how to construct an inverter from a NAND gate, it becomes possible to use a number of NAND gates to construct any other basic gate. For example, if an ordinary NAND gate is followed by an inverter (constructed from a NAND gate), the result is just an AND gate. If both inputs of the AND gate so constructed are preceded by inverters, the result is a NOR gate. And if the inputs of a NAND gate are inverted, the result is an OR gate. All this is shown in Figure 1.33.

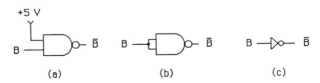

(a)   (b)   (c)

FIGURE 1.32 (a) and (b) Two ways of wiring a NAND gate to produce an inverter (c).

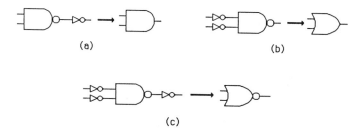

FIGURE 1.33 Constructing various gates from a NAND gate and inverters (which can be made from a NAND gate). (a) An AND gate. (b) An OR gate. (c) A NOR gate.

Using NOR Gates to Build Other Gates

NOR gates can also be used to construct other gates. Figure 1.34 shows two ways that a NOR gate can be wired as an inverter. The first method is analogous to the method shown in Figure 1.32(a) for constructing an inverter from a NAND gate, except that one input is permanently tied low instead of high. This method is shown in Figure 1.34(a). Examination of the NOR gate truth table in Figure 1.11 shows that, with one input tied low, the output of the NOR gate will be the inverse of the other input.

The second method of constructing an inverter from a NOR gate is shown in Figure 1.34(b). Like the case of the NAND gate, examination of the NOR gate truth table in Figure 1.11 shows that, when both inputs are constrained to be equal, the output will be the inverse of the inputs.

Once we know how to construct an inverter from a NOR gate, we can use a number of NOR gates to construct any other basic gate. This is shown explicitly in Figure 1.35. Knowledge of how to transform NAND or NOR gates into other basic gates is very useful to the designer who is faced with making the most efficient use of the commercial packaging of digital logic gates.

FIGURE 1.34 Two ways of constructing an inverter from a NOR gate.

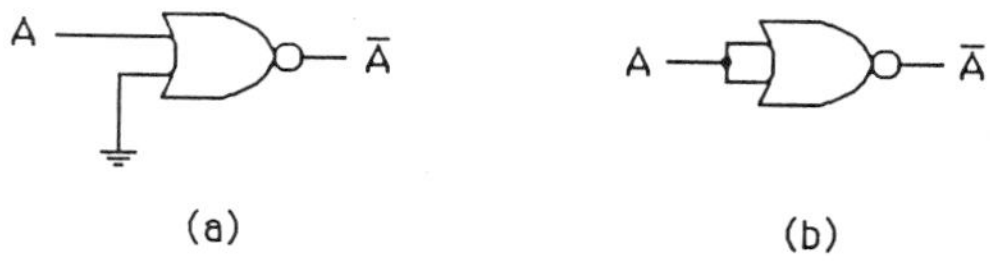

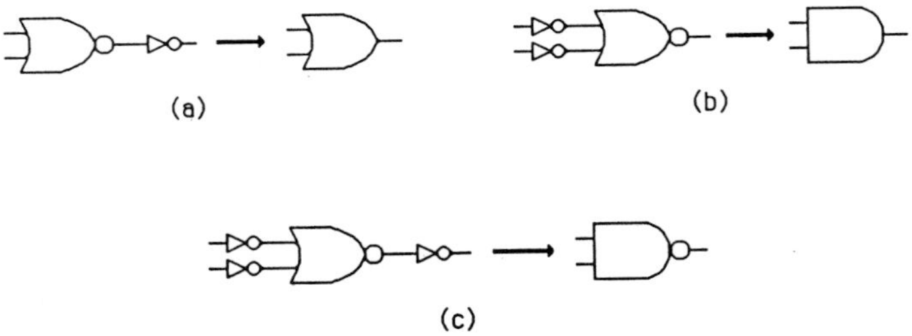

FIGURE 1.35 Constructing various gates from a NOR gate and inverters. (a) An OR gate. (b) An AND gate. (c) A NAND gate.

1.7 BINARY CODES

As we mentioned earlier, because digital logic is usually represented in terms of 1's and 0's, it is natural to think about digital inputs and outputs in terms of binary numbers. This is particularly true when several inputs or several outputs are involved. For example, suppose a digital circuit has three inputs labeled A, B, and C. Each of these inputs can take on the value 0 or 1. The possibilities are shown in the table in Figure 1.36. The entries in Figure 1.36 can be thought of as simply a list of input combinations. However, it is also possible to associate a binary number with each combination of 1's and 0's. One way of doing this is the standard binary number system.

Binary Numbers

In the decimal number system there are 10 distinct digits, 0 to 9, which can be used to represent various numbers. These digits can be combined according to simple rules to represent any desired number. The basic rule is one of attaching a *weight* to each digit in a number according to the position that the digit occupies within the number. If we restrict ourselves to integer numbers for the time being,

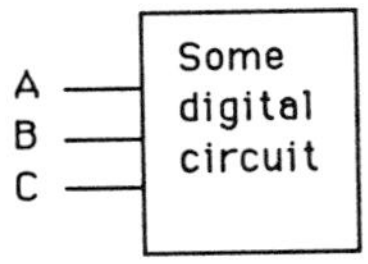

A	B	C
0	0	0
0	0	1
0	1	0
0	1	1
1	0	0
1	0	1
1	1	0
1	1	1

Possible input states

FIGURE 1.36 A 3-input digital circuit and a list of all possible input states.

then the rule is simple. The rightmost digit is assigned a weight of 1, and each succeeding digit to the left is assigned a weight that increases by a factor of 10 for each digit. An example will make this clearer.

■ EXAMPLE 1.7—1

Consider the number 835. The rightmost digit has a weight of 1, so the digit 5 is worth $5 \times 1 = 5$. The next digit has a weight of 10, so the 3 is worth $3 \times 10 = 30$. The next digit has a weight of 100, so the 8 is worth $8 \times 100 = 800$. The total value of the number is the sum of values contributed by each digit (with its corresponding weight). Thus we have

$$835 = (8 \times 100) + (3 \times 10) + (5 \times 1)$$ ■

The weights used for each digit are just determined by the *base* of the number system (10 in the decimal system) raised to succeedingly higher powers. Thus the weight of the first (rightmost, least significant) digit is 10 raised to the zero power. The next digit has a weight of 10 raised to the first power, and so on. This leads of weights of 1, 10, 100, 1000, and so on.

Now let us imagine a system that has only two digits, 0 and 1. We can count in this system in a manner analogous to the decimal system except now each digit within a number can only be 0 or 1. Again we will assign a weight to each digit within a number according to its position. Since our new system has only two digits, we will assign weights to digits based on powers of 2. Thus the rightmost (least significant) digit will be given a weight of 2 to the zero power, or just 1. The next digit will be given a weight of 2 to the first power, or 2. The table in Figure 1.37 summarizes the weights for each of the first eight digits in a binary number. In the table the position of each digit is listed relative to the least significant digit. Thus the least significant digit itself is in position 0. In terms of the weights in Figure 1.37, we can easily determine the value (expressed for convenience in decimal form) of any binary number. Consider the following example.

Position relative to least sig. digit	weight
0	$2^0 = 1$
1	$2^1 = 2$
2	$2^2 = 4$
3	$2^3 = 8$
4	$2^4 = 16$
5	$2^5 = 32$
6	$2^6 = 64$
7	$2^7 = 128$

FIGURE 1.37 The binary digit positions and the weight attached to each.

■ EXAMPLE 1.7–2

Consider the number 10011. Starting at the right with the least significant digit, we have a value given by the digit (1) times the weight of the digit (1), or $1 \times 1 = 1$. The value for the next digit is the digit (1) times the weight of the digit (2), or $1 \times 2 = 2$. We can continue this way through the entire number. Thus

$$10011 = (1 \times 16) + (0 \times 8) + (0 \times 4) + (1 \times 2) + (1 \times 1)$$

$$= 16 + 0 + 0 + 2 + 1 = 19 \text{ decimal} \qquad ■$$

Figure 1.38 lists the first 16 binary numbers and their decimal equivalents. If we return now to the inputs listed in Figure 1.36 for our digital circuit, we see that they could be viewed as binary numbers equivalent to 0 to 7, provided that A is regarded as the most significant digit and C as the least. This is often a convenient way of thinking about binary inputs or outputs.

Adding Binary Numbers

Just as decimal numbers can be added, so also can binary numbers. The rules are basically the same. When two digits are added, these two digits produce a sum and possibly a carry. The carry must then be added to the next sum, and so on. The following examples illustrate the result of adding two digits.

$$
\begin{array}{cccc}
0 & 0 & 1 & 1 \\
+0 & +1 & +0 & +1 \\
\hline
0 & 1 & 1 & 10
\end{array}
$$

The last example shows that a carry has been produced. The carry is placed in the column with the next significant digit. An example will make this clearer.

Binary Number	Decimal Equivalent	Binary Number	Decimal Equivalent
0000	0	1000	8
0001	1	1001	9
0010	2	1010	10
0011	3	1011	11
0100	4	1100	12
0101	5	1101	13
0110	6	1110	14
0111	7	1111	15

FIGURE 1.38 The first 16 binary numbers and their decimal equivalents.

■ EXAMPLE 1.7—3

Suppose we add the two numbers 10110 and 01100.

$$
\begin{array}{r}
\text{carries} \quad 1 \ \ 1\ 1 \ \ \ \ \ \ \\
1\ 0\ 1\ 1\ 0 \\
+\ \ 0\ 1\ 1\ 0\ 0 \\
\hline
1\ \ \ 0\ 0\ 0\ 1\ 0
\end{array}
$$

Let us consider this problem step by step. First, we add the two least significant digits $(0 + 0)$. The result is 0 with no carry. We then proceed with the next two digits $(1 + 0)$. The result is 1 with no carry. Next we have $(1 + 1)$. The result is 0 *with* a carry of 1. This carry is placed over the next most significant digit column. Thus the next sum involves this carry as well as the two digits of the original number. We have, therefore, $(1 + 0 + 1)$, which is 0 with a carry of 1. This carry is placed over the next most significant digit column. The next sum is therefore $(1 + 1 + 0)$, which is 0 with a carry of 1. This last carry is placed over the column where the next most significant digits would be and then added to the two implied zeros to produce $(1 + 0 + 0)$, which is 1 with no carry. The addition is now complete.

■

Binary arithmetic takes on great importance in any digital circuits that are designed to carry out mathematical functions. Since it is not possible to deal with decimal arithmetic in a direct electronic manner, all decimal numbers are first converted to binary numbers, and then the desired operations are carried out. The results are then converted back to decimal form for human convenience. In a future chapter we will see how digital circuits can be used to directly add binary numbers to each other.

Other Binary Codes

There are times when it is convenient to represent numbers using a binary system other than the conventional binary number system discussed previously. One common system of this sort uses the standard binary system to represent *each digit* of a decimal number. This system is called the binary-coded decimal (BCD) system. It is very simple and works as follows. Suppose we wish to represent the decimal number 938. We simply convert each decimal digit to its binary equivalent and then write down the results. Thus

$$9 = 1001, \quad 3 = 0011, \quad 8 = 1000$$

Thus $938 = 1001 \quad 0011 \quad 1000$ in BCD.

Number	BCD	Excess 3	Gray
0	0000	0011	0000
1	0001	0100	0001
2	0010	0101	0011
3	0011	0110	0010
4	0100	0111	0110
5	0101	1000	0111
6	0110	1001	0101
7	0111	1010	0100
8	1000	1011	1100
9	1001	1100	1101

FIGURE 1.39 The BCD, excess 3, and Gray binary codes.

Another code that is sometimes used is called the excess 3 code. This code is also used to represent each digit of a decimal number by a four-digit binary number. However, the binary number used is equal to the binary equivalent of the decimal number *plus three*. For example, the decimal number 6 would actually be represented by the binary equivalent of 6 + 3, or 9 (1001 binary). Thus in excess 3 code with the number 938 would become

$$9 = 1100, \quad 3 = 0110, \quad 8 = 1011$$

And 938 = 1100 0110 1011   in excess three code.

A final code that is occasionally used is the Gray code. The Gray code is also used to represent decimal digits. The BCD and excess 3 codes are such that, as we go from one number to the next, often more than one binary digit (called a *bit*) will change in consecutive numbers. For example, in the BCD system as we go from 3 (0011) to 4 (0100), the last 3 bits all change. The Gray code is set up so that as we go from one number to the next only 1 bit changes. The table in Figure 1.39 compares the BCD, excess 3, and Gray codes.

Hexadecimal Code

Large, standard binary numbers such as 110110011000 are not particularly easy to mentally convert to an understandable decimal number. It would be most convenient if long binary numbers could be represented in some easily understandable, succinct fashion. Hexadecimal notation is ideally suited for this. The hexadecimal number system uses the base 16 just as the binary system uses the base 2 and the decimal system uses the base 10. In a base 16 system, there must be 16 distinct digit symbols (just as there are 10 symbols, 0 to 9, in the decimal system). The symbols used are the numbers 0 to 9 for the first 10 digits and the letters A to F for the next 6 digits. A comparison of the binary, decimal, and hexadecimal systems for the first 17 numbers will make the system somewhat clearer.

Binary	Decimal	Hexadecimal
0000	0	0
0001	1	1
0010	2	2
0011	3	3
0100	4	4
0101	5	5
0110	6	6
0111	7	7
1000	8	8
1001	9	9
1010	10	A
1011	11	B
1100	12	C
1101	13	D
1110	14	E
1111	15	F
10000	16	10

Notice that each hexadecimal digit represents a group of four binary digits. This means that any hexadecimal number can easily be converted to its binary equivalent.

■ EXAMPLE 1.7—4

Consider the hex number 5A. The table shows that the five becomes 0101 and the A becomes 1010. Thus 5A is equivalent to 01011010.   ■

Conversion from binary to hex is likewise very easy.

■ EXAMPLE 1.7—5

Consider the binary number 110011. This number is six digits in length. To make the conversion easier to visualize, it is convenient to expand the length of the binary number to a multiple of four digits. This can be done by adding a sufficient number of *leading* zeros. Thus our number 101011 can be written as 00101011. To make things even easier, let us segregate the number visually into groups of four binary digits. Our number then becomes 0010 1011. This number can be directly converted to two hexadecimal digits using the preceding table. Thus the number is seen to be 2B.   ■

It is easy to confuse hexadecimal notation with decimal notation, especially for numbers including the digits 0 to 9. To keep the distinction clear, the symbol $ is used to precede hexadecimal notation. Thus the number 5A is usually written as $5A. Since the example includes the digit A, it is obvious that we are dealing with a hex number. However, if the number had been 23, it would be necessary to include the $ to make the distinction clear. Thus $23 is quite different from 23.

In the decimal system, the number 23 would have a value of 3 from the 1's place plus a value of 2×10 from the 10's place. That is, the least significant digit has a weight multiplier equal to the base (in this case 10) raised to the zero power. Since *any* number raised to the zero power is 1, the least significant digit in all number systems has a weight multiplier of 1. The second most significant digit has a weight multiplier equal to the base to the first power. This would be 10 in the decimal system. The third most significant digit has a weight multiplier equal to the base to the second power, which would be 100 in the decimal system. The weight of each digit increases by a factor equal to the base of the number system. Thus the weight multipliers in the decimal system are 1, 10, 100, 1000, and so on. In the hexadecimal system, the base is 16. Thus the weight multipliers are 1, 16, 256, 4096, and so on. With this in mind, it is easy to convert a hex number into its decimal equivalent.

■ EXAMPLE 1.7—6

Consider the number $23. The least significant digit, 3, is just worth 3. The second most significant digit, 2, has a weight multiplier of 16, so it is worth 2×16 or 32. Thus the value of the number $23 is just $32 + 3$, or 35 in decimal notation. ■

Hexadecimal notation is a very natural way of dealing with large binary numbers. For example, microprocessors have digital data buses that consist of 8 or 16 wires. The state of each wire can be easily visualized as a 1 or 0. Thus a 16-bit data bus might have on its wires at a given moment the binary number 1001001101101110. This is very difficult to make sense of in its binary form. However, converted to hexadecimal (by grouping the digits in groups of four), the number becomes $935E, which is much easier to deal with. Microprocessors also have address buses, which likewise consist of 16 or more wires. Thus the state of an address bus can also be visualized as a long series of 1's and 0's. And, as in the case of the data bus, it is much easier to convert the long string of binary digits into the equivalent hexadecimal number. A 16-bit address bus could then take on any value in the range $0000 to $FFFF. You will find that thinking of microprocessor addresses in hexadecimal form is much more natural than using binary or decimal notation.

PROBLEMS

1. Suppose that three single-pole, single-throw (SPST) switches (labeled A, B, and C) are wired to a light in such a fashion that the light will only turn on if all three switches are closed.

   a. What logical function is being performed?

   b. Draw a circuit diagram to produce the desired result.

   c. Draw a truth table for the circuit in part b, where the logic 1 state corresponds to a closed switch.

2. Show how three 2-input AND gates can be combined to produce a 4-input AND gate.

3. Suppose Figure 1.40 represents the inputs to a 2-input AND gate over a period of time. Draw a similar diagram showing the output of the AND gate as a function of time.

4. Make a truth table for a 3-input OR gate.

5. It is desired to have a light turn on when either switch A and B are both closed or switch C is closed.

   a. Write the Boolean function for the light on condition.

   b. Design the circuit that will achieve the desired result.

6. Make a truth table for the circuit shown in Figure 1.41.

7. The logical statement "A AND B OR C" can be interpreted in two ways unless parentheses are added for clarification. Using standard logic gates, illustrate the two interpretations, and label the output of each in an unambiguous way.

8. The seat-belt alarm in an automobile is to sound if the ignition key is on and either the driver's seat is occupied without the seat belt being fastened or the passenger's seat is occupied without the seat belt being fastened. Using appropriate symbols (such as A, B, and so on) for each condition, write the Boolean expression for the alarm sounding. Implement the Boolean expression using basic logic gates.

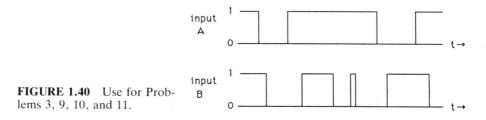

FIGURE 1.40 Use for Problems 3, 9, 10, and 11.

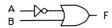

FIGURE 1.41 Use with Problem 6.

FIGURE 1.42 Use with Problem 13.

9. Suppose that Figure 1.40 represents the inputs to a 2-input NAND gate over a period of time. Graph the output of the NAND gate as a function of time.

10. Repeat Problem 9 assuming that the inputs are to a 2-input NOR gate.

11. Repeat Problem 9 assuming that the inputs are to a 2-input XOR gate.

12. Construct the truth table for Figure 1.15. Does this circuit indeed perform the XOR function?

13. Consider the circuit shown in Figure 1.42. First, determine the Boolean expressions for L, M, N, and Y. Then answer the following questions.

   a. If point M is 0, then point L must be _____ ?

   b. If A is 1 and L is 0, then B must be _____ ?

   c. If M is 1 and Y is 0, then D must be _____ ?

14. For the circuit shown in Figure 1.43, determine the Boolean expressions for L, M, and Y.

15. Design a circuit that will implement the Boolean expression

$$F = AB + \overline{(A + B)}$$

16. Design a circuit that will implement the Boolean expression

$$F = \overline{AB\overline{C} + BC}$$

17. Draw the circuit symbol for a 2-input OR gate with active-low inputs and an active-high output. Construct the truth table for this circuit. Compare this truth table to that of a standard 2-input NAND gate.

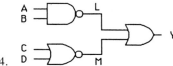

FIGURE 1.43 Use with Problem 14.

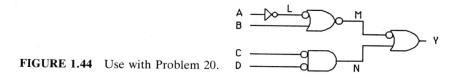

FIGURE 1.44 Use with Problem 20.

18. An OR gate has one active-low input and one active-high input. It has an active-low output. Draw the circuit symbol for this gate and construct the truth table.

19. Draw the circuit symbol for a 2-input AND gate with active-low inputs and an active-high output. Construct the truth table for this circuit, and compare this truth table to that of a standard NOR gate.

20. Examine the circuit shown in Figure 1.44.

    a. What state of M will force Y high?

    b. What state of N will force Y high?

    c. What has to occur for N to be high?

    d. What state of L will force M low?

    e. What state of B will force M low?

    f. What state of A will make L low?

    g. Assuming A is high, what is the state of Y?

21. Write the decimal numbers 10, 20, and 30 in binary form.

22. Write the binary numbers 10011, 011010, and 111011 in decimal form.

23. Using the BCD code, write the decimal numbers 935, 23, and 482.

24. A digital circuit has five inputs. How many unique input combinations can occur?

25. Add the two binary numbers 10010 and 01110.

Chapter 2

THE CONSTRUCTION OF LOGIC GATES

Learning Objectives

After completing this chapter you should know:

- How a junction diode works and how it can be used to construct a basic logic gate.

- How a junction transistor works and how it can be used as an electronic switch.

- How logic gates can be constructed from transistor circuits.

- The basic nature of the standard TTL logic gate.

- The input and output voltage characteristics of standard TTL logic gates.

- The sourcing and sinking current requirements of standard TTL logic.

- What is meant by gate delay.

- What is meant by fan-out.

- The important characteristics of other TTL logic families and how gates from various TTL families can be interconnected.

- What an open-collector output is.

- What a tristate output is.

- What is meant by metal oxide semiconducting (MOS) technology.

- What an enhancement-mode MOSFET is.

- What complementary MOS (CMOS) logic is and how it differs from TTL logic.

- The characteristics of newer CMOS logic families.

- How TTL and CMOS logic gates can be interconnected.

2.1   INTRODUCTION

In Chapter 1 we introduced the basic digital logic gates and some other simple gates that could be derived from them. We did not, however, look into how these gates could be produced in a useful electronic form. In this chapter we will remedy that situation. First, we will investigate two electronic devices that are very useful in constructing digital circuits, the diode and the transistor. Next, the basic properties of these devices will be examined to see how they can be used to construct logic gates. Finally, two common digital logic families, the TTL family and the CMOS family, will be introduced. We will see how gate construction is accomplished in these two families and examine many characteristics of integrated circuits belonging to these families. Let us begin with diodes and transistors.

2.2   DIODES AND TRANSISTORS

When we introduced the basic logic gates in Section 1.2, it was shown that these gates could be constructed from relays. A relay is basically a two-state device; it is either open or it is closed. Unfortunately, relays are large, relatively expensive devices. Modern digital integrated circuits often perform complex logic functions, and yet these integrated circuits are physically quite small. A circuit consisting of relays that performed the same function as even a fairly simple digital integrated circuit would be very large. Therefore, digital integrated circuits do not use relays, but instead use electronic components that can act as two-state devices. These devices are the transistor and the diode. Both of these devices have two well-defined states, both can be made very small, and both can change state very rapidly. Before we see how diodes and transistors can be used to construct logic gates, it is of value to gain an understanding of the basic electronic properties of these devices.

The Diode

Most diodes are made from either silicon or germanium. We will focus our attention on those made from silicon. Pure silicon is known as a semiconductor because it conducts electricity (thus it is not an insulator), but does not conduct as well as materials known as conductors (such as copper and silver). Pure silicon can have its properties changed in very useful ways by adding certain kinds of impurities or *dopents* to it. One type of dopent tends to add excess electrons (negative charges) to the silicon. The resulting material, called *N-type* silicon, is a better conductor than pure silicon. A second type of dopent tends to remove electrons from the silicon, leaving behind positive "holes," which are free to move about. The material

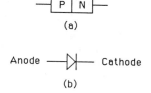

FIGURE 2.1 (a) The construction of a junction diode from *p*-type and *n*-type semiconducting material. (b) The circuit symbol for a diode, oriented in the same sense as (a).

that results from this second kind of doping is called *P-type* silicon, and it also is a better conductor than pure silicon.

A *junction diode* results when a piece of *N*-type and a piece of *P*-type silicon are brought together as in Figure 2.1(a). The critical operation occurs because of the junction between the two types of silicon, the *P–N* junction; thus the name junction diode. The electrical symbol for the diode is shown in Figure 2.1(b), where the orientation of the symbol is consistent with the orientation of the *P*- and *N*-type materials in Figure 2.1(a). Notice the names attached to the two ends of the diode, the anode and the cathode. These names allow us to refer unambiguously to either end of a diode.

Let us turn our attention to how a diode functions in an electrical circuit. We begin with the simple circuit shown in Figure 2.2(a). In this circuit, let us imagine that the voltage of the battery is continuously adjustable from zero up to a volt or so. As the voltage of the battery is increased above zero, a small current will begin to flow through the diode. In addition, the anode of the diode will have a positive voltage with respect to the cathode. A diode in this condition is said to be *forward biased* or *biased on*. As the voltage is increased, the current flow increases rapidly. This is shown in Figure 2.2(c) on the right half of the figure. For a forward-biased silicon diode, as the voltage approaches about 0.6 volt, the current flow begins to increase substantially. Once the knee of the current–voltage curve has been reached, any further increase in voltage results in a very large increase in current. Once the

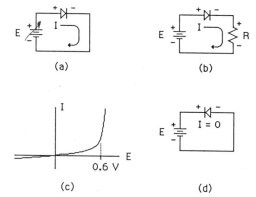

FIGURE 2.2 (a) A forward-biased diode. (b) A forward-biased diode with current-limiting resistor. (c) The current–voltage curve for a diode. (d) A reverse-biased diode.

applied voltage has reached 0.6 volt, the diode has become nearly a perfect conductor *with respect to further increases in voltage*. The current–voltage curve rises so rapidly at this point that a very large range of current is possible for *almost* the same voltage (about 0.6 volt). If we add a resistor to the circuit of Figure 2.2(a), the result is Figure 2.2(b). For battery voltages greater than 0.6 volt, the diode will, for all practical purposes, have a fixed 0.6-volt drop across it. The remainder of the battery voltage drop will be across the resistor.

■ EXAMPLE 2.2—1

In Figure 2.2(b), suppose that the battery voltage is 5 volts and the resistor value is 2 kilohms (kΩ). Let us calculate the current in the circuit.

The voltage drop across the diode and the resistor must together equal the voltage of the battery. Since the diode drops essentially 0.6 volt, then the resistor must drop 5.0 − 0.6 volt, or 4.4 volts. This gives a current through the resistor of

$$I = \frac{4.4 \text{ volts}}{2 \text{ kilohms}} = 2.2 \text{ milliamperes} \quad (\text{mA}) \qquad ■$$

In this example, if the value of the resistor were changed, we would still assume a voltage drop across the diode of 0.6 volt. Strictly speaking, if the current gets small enough, then the voltage across the diode will be less than 0.6 volt. This does not begin to happen, however, until the current gets well below 1 milliampere.

The essentially fixed 0.6-volt drop across a forward-biased silicon diode is referred to as *one diode drop*. That is, when a significant current (more than 0.1 milliampere or so) is flowing through a forward-biased silicon diode, we assume a voltage drop of 0.6 volt, "one diode drop."

Now let us consider the circuit shown in Figure 2.2(d). In this figure the cathode of the diode is at a higher voltage than the anode. A diode in this condition is said to be *reverse-biased* or *biased off*. Almost no current will flow under these circumstances (typically a few nanoamperes). A reverse-biased diode is very nearly an open circuit.

Diodes do have some operating limitations. In the forward direction, the diode will have a 0.6-volt drop across it for any appreciable current. Since the power consumed by an electrical device is just the product of current and voltage (*IV*), the power dissipated in the diode will steadily rise as the current is increased. This will produce heat in the diode. Depending on the actual size of the diode, there is a limit to how much heating can be tolerated. Thus specific diodes have a safe current rating that should not be exceeded.

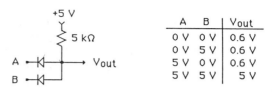

FIGURE 2.3 Two diodes and a resistor used to construct an AND gate. The voltage truth table for this gate is shown on the right.

A	B	Vout
0 V	0 V	0.6 V
0 V	5 V	0.6 V
5 V	0 V	0.6 V
5 V	5 V	5 V

When a diode is reverse biased, it is nearly an open circuit. Even if large reverse voltages are applied, the diode will not conduct. However, as the reverse voltage is increased, the internal electric field in the diode grows proportionately. At some critical field, the diode breaks down and begins to conduct very well. For most diodes, if this happens the diode will be destroyed. Thus diodes are also rated according to reverse breakdown voltage, which should not be exceeded.

Let us now see how diodes might be used to construct some basic logic gates. Figure 2.3 shows one possibility, where we assume that silicon diodes have been used. In this figure, A and B are input voltages, which we assume must be either 0 or 5 volts. Let us imagine that input A is made 0 volts (connected to ground), while input B is held at 5 volts. Current can now flow from the 5-volt supply through the 5-kilohm resistor, through diode A to ground. The diode becomes forward biased and in this condition has a voltage drop of 0.6 volt across it (one diode drop). Thus the anode of diode A, and thus the output of the circuit, will become 0.6 volt. Notice that diode B in this case is reverse biased.

This analysis applies equally well if input B is made 0 volts and input A is made 5 volts. If both inputs are made 0 volts, then current will flow through both diodes to ground. Both diodes will be forward biased, but the output voltage will still be 0.6 volt.

If both A and B are made 5 volts, then the input voltages will be at the same voltage level as the supply. There will be no voltage difference across the diodes or the 5-kilohm resistor. With no voltage across the 5-kilohm resistor, the output voltage must be 5 volts. The voltage truth table in Figure 2.3 shows the complete story. If we identify the high-voltage state as logical 1 and the low-voltage state as logical 0, then this circuit is obviously an AND gate.

The circuit in Figure 2.3 has a serious flaw as a practical logic gate. Imagine what would happen if several such gates were connected together, with the output of one going to the input of the next. Let us suppose that one of the inputs to the first gate is 0 volts so that the output of this gate is 0.6 volt. This output becomes an input to the next gate. However, this input is now 0.6 volt, not 0 volts. Thus the output of the next gate will be one diode drop higher, or 1.2 volts. If this output is used as an input to the next gate, the output of that gate will be one diode drop higher yet, or 1.8 volts. You can see that the low-voltage state slowly creeps up as it propagates through a series of gates. This would be intolerable in a practical system. For this reason, modern logic does not rely just on diodes.

Junction Transistors

The junction transistor is somewhat similar in construction to a junction diode, except that the transistor has three sections instead of two. Like the junction diode, the junction transistor consists of sections of *P*- and *N*-type material. However, because there are three sections in the transistor, it is possible to make two different kinds of transistors.

The two varieties of junction transistors are shown in Figure 2.4. The *NPN* transistor, shown in Figure 2.4(a), has *N*-type material sandwiched around *P*-type material. The circuit symbol for the *NPN* transistor is shown directly below the figure illustrating the transistor's construction. Figure 2.4(b) illustrates the construction and circuit symbol for a *PNP* transistor. We should note in passing that it appears from the upper figures that the transistor is a symmetric device. That is, it looks like the emitter and collector ends of the transistor have been arbitrarily labeled and could be interchanged with no effect. Actually, the geometry of real transistors is not symmetric, and interchanging the emitter and collector causes drastic changes in the transistor's performance.

To understand the operation of a transistor, let us focus our attention on an *NPN* transistor in a typical circuit. Figure 2.5 shows one arrangement, the common-emitter configuration. In this circuit, the voltage E is assumed to be a variable voltage that can take on any value from 0 to 5 volts. Let us focus our attention on the lower *loop* of the circuit consisting of the variable battery, the 2-kilohm resistor, and the base–emitter junction. The base–emitter junction behaves like a diode, where the base plays the role of anode and the emitter plays the role of cathode. In the loop we are examining, it is clear that the base–emitter diode will be forward biased if the voltage E is adjusted above 0 volts.

Let us begin by assuming that the voltage E is adjusted to 0 volts. Then essentially no current will flow through the diode. As the voltage E is increased, current will begin to flow in the lower loop, across the base–emitter junction. As the voltage of E is further increased, the current flow will also increase. When E reaches 0.6 volt, the base–emitter diode will be biased fully on. That is, the voltage

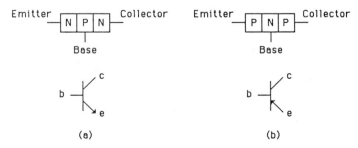

FIGURE 2.4   Basic construction and circuit symbol for (a) an *NPN* transistor and (b) a *PNP* transistor.

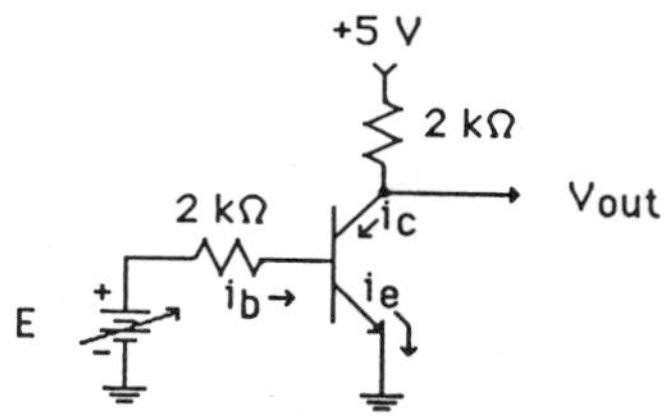

FIGURE 2.5 The basic transistor common-emitter circuit. The output is controlled by varying the input voltage E.

across the base–emitter junction will have reached nearly 0.6 volt (there will be a small voltage drop across the resistor). Any further increase in E will not change the base–emitter voltage, which will hold essentially fixed at 0.6 volt. Instead, the increased voltage drop will show up across the 2-kilohm resistor, and more current will flow in the lower circuit loop. This current must flow into the base and across the base–emitter junction and is called the *base current* of the transistor. It is identified as i_b on Figure 2.5.

Now let us focus our attention on the part of the circuit consisting of the 5-volt supply, the upper 2-kilohm resistor, and the main body of the transistor (from collector to emitter). The main body of the transistor presents an impediment to current flow that *is not fixed*, but that is controlled by the amount of base current flowing. Under normal conditions, the majority of current flowing through a transistor is from the collector, across the very thin base region, to the emitter. The current flowing into the collector is called the *collector current* and is indicated by i_c in Figure 2.5. The current flowing out of the emitter is called the *emitter current* and is indicated by i_e in Figure 2.5. As a practical matter, the collector and emitter currents are nearly equal, both being much larger than the base current. Thus the main current flow through a transistor can be referred to either the collector current or the emitter current, interchangeably.

In operation, the base current controls how easily the main body of the transistor conducts current. The more base current that flows, the better the main body of the transistor conducts. The operation of the collector–base–emitter combination is such that, when any current flows from the base to the emitter, a much larger current, perhaps 100 times larger, will flow from the collector to the emitter. This amplification of current is what makes the transistor so useful. In the present case, if E is set to zero volts, we have no base–emitter current. This means that we should have no collector–emitter (which we will call simply collector) current either. With no collector current, the transistor is said to be *cut off*. Also, since no current flows from the collector, then no current flows through the 2-kilohm collector resistor. If no current flows through the collector resistor, then there is no voltage *drop* across this resistor. Thus V_{out} will be 5 volts.

Now let us slowly begin to increase the voltage E. Now a small current begins to flow through the base–emitter diode (a small *base* current begins to flow). Because of the current-amplifying nature of the transistor, a much larger collector current flows. For example, if the base current has risen to about 1 microampere

(μA), then the collector current will be about 100 times larger, or about 0.1 milliampere. A current of 0.1 millampere flowing through the 2-kilohm collector resistor will produce a voltage drop of 0.2 volts. Thus V_{out} will be (5 − 0.2 volts), or 4.8 volts. If we increase the input voltage E still more, then the base current will continue to increase. This will cause the collector current to rise proportionately. A transistor operating such that the collector current increases proportionally to the base current is said to be *operating in the linear region*. The increased collector current flowing through the collector resistor will cause V_{out} to drop still more. Eventually, if we continue to slowly increase the input voltage E, there will come a time when the collector current is just big enough to produce nearly a 5-volt drop across the collector resistor. At this point, V_{out} will have dropped to nearly 0 volts. When this occurs, we say that the transistor is *saturated*. Any further increase in the input voltage will increase the base current, but not the collector current. This is because once the transistor has become saturated it cannot become a better conductor. One way of thinking about the action of a transistor is to visualize the collector–emitter section of the transistor as just a variable resistance under the control of the base current. When the base current is zero, this resistance is very high. Then voltage-divider action makes V_{out} nearly 5 volts. As the base current increases, the *resistance* of the transistor decreases and this reduces V_{out}. When the base current gets large enough to reduce the collector–emitter voltage to nearly zero, the transistor has become nearly a short circuit. Increasing the base current further cannot reduce this resistance below zero.

Reviewing what we have said, we see that the transistor can be cut off if no base current flows; it can be in the linear region where collector current is proportional to base current; or it can be saturated if the base current causes the emitter–collector voltage to drop to nearly zero. To use the transistor as the basis of a digital logic device, it is necessary to set up a situation where the transistor will always be either cut off or saturated, but never in the linear region. The circuit in Figure 2.5 would accomplish this if the input voltage were restricted to the two values of 5 and 0 volts. This is shown more clearly in Figure 2.6. In Figure 2.6(a), the input voltage is 0 volts, so the transistor is cut off and thus the output is 5 volts. In Figure 2.6(b), the input voltage is 5 volts. Because the base–emitter junction is essentially a diode, there will be one diode drop in voltage from the base to the emitter, that is 0.6 volt. This leaves 4.4 volts across the 2-kilohm base resistor. A voltage of 4.4 volts across a 2-kΩ resistor produces a current of 2.2 millamperes. For the transistor to avoid being saturated, the collector current would need to rise to about 100 times the base current. This would require a collector current of 0.22 amperes (A). But even if the transistor becomes a perfect short circuit, the 2-kilohm collector resistor alone would limit the collector current to 2.5 millamperes. Thus the transistor is well into saturation when the base current is 2.2 millamperes. A small-signal (low-power) transistor usually has a collector–emitter saturation voltage of about 0.1 to 0.2 volt. In Figure 2.6(b), we show an output voltage of 0.1 volt.

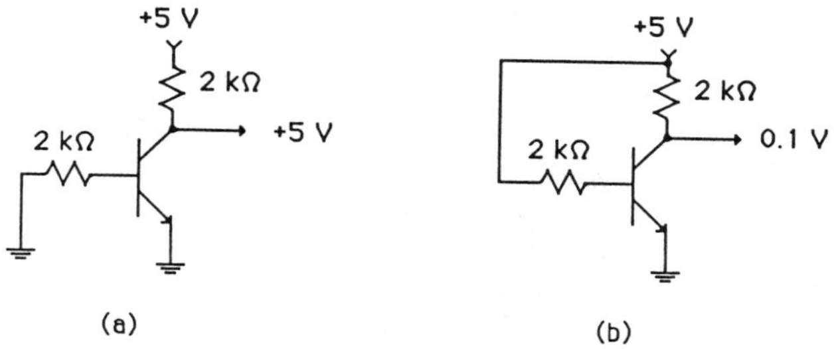

FIGURE 2.6 (a) The cutoff transistor state. (b) The saturated transistor state. Note the inverting action of this circuit. The transistor essentially acts like a switch in the circuit.

Notice that we have a fair amount of latitude in our input voltage. If the input voltage is kept below 0.6 volt (one diode drop), this will be sufficient to keep appreciable base current from flowing and thus will keep the transistor cut off. Likewise, an input voltage of perhaps as little as 2 volts would be sufficient to saturate the transistor. Thus, if we use transistors to produce logic gates, there is no reason for the signal to degrade as it does with diode logic gates. As a matter of fact, the circuit shown in Figure 2.6 is really just an inverter. When the input is high (1), the output is low (0), and vice versa.

As long as our circuit forces the transistor to exist in either the cutoff or saturated states, the transistor behaves very much like a switch. That is, it is either virtually a short circuit (saturated) or an open circuit (cut off). In digital electronics, transistors are often referred to as electronic switches.

A Junction Transistor NOR Gate

Figure 2.7 shows how a NOR gate might be constructed from two junction transistors. Note that the two transistors are essentially in parallel with each other. Thus, if *either* transistor becomes saturated, the other transistor will be effectively shorted out, and the output will go to nearly 0 volts. The truth table in Figure 2.7 summarizes the circuit action. We see that only when both inputs are at (or very near) 0 volts will both transistors remain cut off, producing an output of 5 volts. If either or both inputs become high (anything over about 2 volts), then the output will drop to nearly 0 volts. If we identify the high-voltage state with logical 1 and the low-voltage state with logical 0, then the truth table is that of a NOR gate.

The circuit in Figure 2.7 gives us a good opportunity to discuss an important point about logic gates. This is the concept of *fan out*.

The fan-out of a digital circuit is the number of digital inputs that can be successfully driven by the output of the digital circuit.

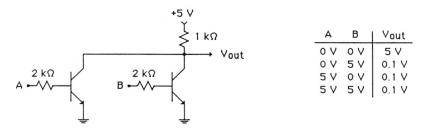

A	B	Vout
0 V	0 V	5 V
0 V	5 V	0.1 V
5 V	0 V	0.1 V
5 V	5 V	0.1 V

FIGURE 2.7 A resistor–transistor NOR gate. At right is the voltage truth table.

The output voltage from one circuit will usually be an input to one or more other, similar digital circuits. In terms of Figure 2.7, we could visualize V_{out} as an input to another digital circuit, that is, as being connected to the equivalent of input A or B. In the case where V_{out} is low, there should be no problem because a low *input* does not need to provide any current. However, suppose V_{out} is high. Figure 2.8 shows the output from one gate connected to the input of another gate. Since the transistor in the first gate is cut off (this is why the output is high), that transistor is like an open switch and can be removed from the picture. Likewise, since the base–emitter junction at the input of the second gate is like a diode, it might as well be pictured that way. Thus the circuit shown in Figure 2.8(a) is equivalent, for purposes of determining V_{out}, to the circuit in Figure 2.8(b). Since the diode drops 0.6 volt, this leaves 4.4 volts across the two resistors in series. The two resistors act like a voltage divider with 4.4 volts across them. The 2-kilohm resistor gets two-thirds of the 4.4 volts, or 2.93 volts. Adding the diode drop to this gives us a value of 3.53 volts for V_{out}. This is quite a bit less than we got when the output was not connected to anything.

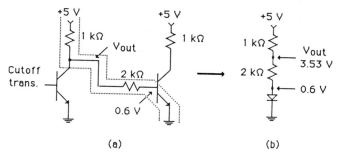

(a)   (b)

FIGURE 2.8 (a) The output of one gate driving the input of another gate. (b) The equivalent circuit used to compute V_{out}. Note that the emphasized portion (dotted lines) of figure (a) corresponds to figure (b).

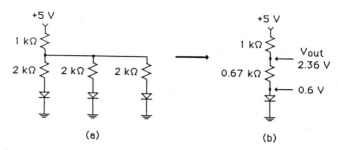

FIGURE 2.9 (a) The equivalent circuit showing the effect of one output driving three inputs. (b) Further reduction of the equivalent circuit shown in (a).

The situation gets rapidly worse if the output is connected to several other inputs. Figure 2.9(a) shows the effective circuit for the output connected to three inputs. Each input appears as a 2-kilohm resistor and diode in series. As long as enough current can be provided to keep each diode conducting, the three 2-kilohm resistors are effectively in parallel between the 1-kilohm resistor and 0.6 volt. Thus the circuit in Figure 2.9(a) can be reduced to that in Figure 2.9(b). We can easily calculate V_{out} in the same manner as for Figure 2.8(b). The result is 2.36 volts. Thus, as the output is connected to (or forced to *drive*) more and more inputs, the output voltage drops. What is worse, the current through the base–emitter junction of each input (what we have shown as diodes) drops substantially. As the number of inputs connected to the single output is increased, there will come a time when the base–emitter current at each input will not be sufficient to saturate the input transistor. When a transistor is no longer saturated, the voltage at the collector of that transistor will begin to rise. Thus, instead of having a collector voltage of 0.1 volt as in Figure 2.7, we might get a voltage of 1 or 2 volts. This would not appear as a low output, but would instead look like a high output. In effect, the circuit no longer functions properly. The number of inputs that a single output can *successfully* drive is called the fan-out of the output. This is an important consideration in designing logic gates, since it is useful to have an output drive as many inputs as possible.

2.3 COMMON LOGIC FAMILIES

We saw in Section 2.2 that both diodes and transistors can form the basis of digital logic gates. During the past several years, many electronics manufacturers have devised a variety of ways of constructing digital logic circuits using the basic characteristics of diodes and transistors. Discrete diodes and transistors are no longer used, however. Instead, integrated-circuit technology allows dozens or even hundreds of diodes, transistors, resistors, and even capacitors to be formed on a substrate of properly prepared silicon. Thus, circuits that implement everything from simple gates to highly complex logical functions have been designed and manufactured.

As a matter of commercial practicality, a number of families of digital circuits have evolved. Circuits within each family share similar part numbers and electrical characteristics. Some of the early families have fallen into disuse as better designs have been developed. At the present time, there are still several families available. Which family a designer chooses depends on a number of factors, such as cost, availability, and detailed electrical characteristics. The two most widely used families at the present time are the TTL family and the CMOS family. Therefore, we will concentrate our attention on these. Within these families there are subfamilies that we will examine. Before discussing TTL and CMOS logic, however, we need to consider in a general way what characteristics are important in a digital integrated circuit.

The reason there are several families of digital integrated circuits is because these circuits possess many, sometimes competing, characteristics. Some of the more important such characteristics are cost, power consumption, speed, ease of handling, voltage requirements, drive capabilities, and ease of interfacing with other electrical circuitry. Some of these characteristics tend to be mutually incompatible. For example, high speed usually means higher power consumption. Very high speed may require special voltages. Low power often means low drive capability. The TTL and CMOS families seem to possess the most attractive combinations of characteristics at the moment since these families currently dominate the commercial market. Let us now discuss these families in some detail.

TTL Logic

If you refer back to Figure 2.6, you will recall that the output of the logic circuit shown there consisted of a transistor in series with a resistor. The input caused the transistor to be either cut off or saturated. However, the resistor always remained fixed at 2 kilohms. This caused a problem when the output was supposed to be high, since any output current had to flow through the 2-kilohm resistor. TTL logic solves this problem by using a second transistor to replace the 2-kilohm resistor. This configuration is known as a totem-pole output. A second innovation of TTL logic is the use of the multiple emitter. Figure 2.10 illustrates a TTL NAND gate that shows the two features just mentioned. Let us consider Figure 2.10 in some detail.

First, let us assume that either input A or input B (or both) is made 0 volts. This will cause current to flow across the base–emitter junction of transistor Q_1, and thus this transistor will saturate. The voltage at the collector of Q_1 will thus fall to only a few tenths of a volt. This will be insufficient to allow current to flow across the base–emitter junction of transistor Q_2, and thus this transistor will be cut off. With Q_2 cut off, current will flow through the 1.6-kilohm resistor and

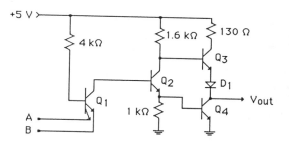

FIGURE 2.10 The equivalent circuit for a TTL NAND gate. Note the multiple-emitter input and the totem-pole output.

through the base–emitter junction of transistor Q_3, and thus this transistor will saturate. At the same time, with Q_2 cut off, no current can flow through it to the base of transistor Q_4. Thus Q_4 will be cut off. Thus at the output we have Q_3 saturated and Q_4 cut off. The output is said to be *actively pulled up* toward the 5-volt supply. Actually, the output will not reach 5 volts because the saturation voltage across Q_3 is a few tenths of a volt, and there is a 0.6-volt drop across the diode. Thus the output will be more like 3.5 to 4 volts.

Let us now consider what happens if *both* inputs A and B are at 5 volts (or both left unconnected, for that matter). In this case, no current will flow across the base–emitter junction of Q_1, so this transistor will remain cut off. This will tend to raise the voltage at the collector of Q_1, since the base–collector junction acts now like a forward-biased diode (base the anode, collector the cathode; see Figure 2.4). The higher voltage at the collector of Q_1 allows current to flow across the base–emitter junction of Q_2, which saturates this transistor. This does two things. First, it drops the voltage at the base of Q_3 so that this transistor is cut off. Second, it provides a path for current to flow across the base–emitter junction of Q_4, which causes this transistor to become saturated. Thus we have Q_3 cut off and Q_4 saturated. The output drops to whatever the saturation voltage of Q_4 is, usually a few tenths of a volt, and is said to be *actively pulled down toward ground*. If we take the high-voltage state to be logical 1 and the low-voltage state to be logical 0, we see that the circuit in Figure 2.10 does indeed perform the NAND function. That is, the output is low only when both inputs are high (or unconnected, since this will also leave Q_1 cut off). It might occur to you to wonder about the 130-ohm (Ω) resistor connected to the collector of Q_3. As the input conditions change, it may occur that the cutoff output transistor goes into saturation *before* the saturated output transistor cuts off. If this happens, without the 130-ohm resistor, there would be an effective short between the 5-volt supply and ground. This would not be good for the output transistors.

Although the circuit in Figure 2.10 is a NAND gate, the multiple-emitter input and the totem-pole output are characteristic of TTL logic in general. Thus we can use this circuit to analyze the input and output limitations of TTL logic. First, let us consider the input side of the circuit. It is clear that it is not difficult for another circuit to provide a high input since in this case there is very little current requirement. A circuit providing a high input will actually need to supply a few microamperes to the TTL input because of leakage current across the reverse-biased emitter–base junction. When a circuit must provide current, it is said to *source* current. Thus a circuit providing a high input to a TTL circuit must source a few microamperes.

However, when an input is low, current will flow from the 5-volt supply through the 4-kilohm resistor, through the base–emitter junction, and into the circuit providing the low input. The circuit providing the low input must be capable of accepting, or *sinking*, this current. We can get a good idea of the current involved by noting that, when a low is input to a TTL circuit such as that in Figure 2.10, the base of Q_1 will be 0.6 volt higher that the emitter. Assuming that the emitter is at a low voltage (a few tenths of a volt), this will leave something like 4 volts across the 4-kilohm resistor. This means that a current of about 1 milliampere will flow through the base of the transistor, through the emitter, and *into* the circuit providing the low input. The circuit providing the low input must therefore be capable of sinking about 1 milliampere of current. It is the input of the TTL circuit that is the source of this 1 milliampere of current. The input is said to place a 1-milliampere *load* on the circuit providing the low input. The actual industry standard for a TTL circuit requires that a standard TTL input place no more than 1.6-milliampere load on the circuit providing the low input, and that the input recognize as a low voltage anything below 0.8 volt. Likewise, the TTL requirement for a high input is that the input require no more than 40 microamperes of source current from the circuit providing the high input, and that the input recognize as a high voltage anything above 2.0 volts.

Now let us look at the output side of the circuit. First, consider the case of a low output. For the output to remain low, Q_4 must remain saturated. Recall that a transistor comes out of saturation when the collector current reaches a certain multiple of the base current (this multiple is called the *beta* of the transistor and depends on the detailed construction of the device). The base current of Q_4 is fixed and is determined by the 1.6-kilohm resistor, the saturation voltage of Q_2, and the 0.6-volt base–emitter drop at Q_4. Thus the amount of current that Q_4 can sink and still remain saturated is determined by its beta value. TTL standards guarantee that the beta value will be high enough so that the output can sink 16 milliamperes of current at a voltage of no more than 0.4 volt. For a high output, TTL standards guarantee that the output can source 400 microamperes of current at a voltage no less than 2.4 volts. The complete, standard TTL input–output characteristics are shown in the top line (labeled 74XX) of the table shown in Figure 2.11. This information shows that a standard TTL high output is capable of sourcing enough current to drive 10 standard TTL high inputs, and likewise that a standard TTL low output is capable of sinking enough current to drive 10 standard TTL low inputs. Thus we say that the fan-out of standard TTL logic driving standard TTL logic is 10.

Also shown in Figure 2.11 are two other numbers of significance to the designer, the *propagation delay* of a typical gate and the *power dissipation* per gate.

	Input Load		Output Drive		Propagation Delay	Power per Gate
	High	Low	High	Low		
	max. sink curr. (μA)	max. source curr. (mA)	max. sink curr. (μA)	max. source curr. (mA)	(ns)	(mW)
74XX	40	−1.6	−400	16	10	10
74LXX	10	−0.18	−200	3.6	33	1
74HXX	50	−2.0	−500	20	6	22
74SXX	50	−2.0	−1000	20	3	19
74LSXX	20	−0.4	−400	8	9	2
74ALSXX	20	−0.1	−400	8	5	1
74FXX	20	−0.6	−1000	20	3.7	.5.5

FIGURE 2.11 Some basic properties of TTL logic families.

The propagation delay of a gate is the time between a change in input and the resulting change in output.

The power dissipation per gate is the typical power dissipated by each gate on an integrated circuit when the gate is not changing state.

For standard TTL logic, the typical propagation delay is about 10 nanoseconds (ns), and the typical power dissipation per gate is about 10 milliwatts (mw).

Figure 2.12 summarizes the voltage levels for TTL logic. The figure shows that TTL logic requires a supply voltage of 5 volts ± 5%. We also see that a TTL circuit will accept anything up to 0.8 volt as a low input, whereas TTL logic is guaranteed to keep a low output under 0.4 volt. This provides a 0.4-volt margin of error or *noise margin* between the low generated at one TTL output and what will be accepted as a low at another TTL input.

Noise is unwanted voltage fluctuations that occur at circuit inputs or outputs.

Noise margin is the acceptable amount of noise that can be tolerated by a circuit without causing malfunction.

FIGURE 2.12 Basic TTL voltage requirements. Notice that there is a 0.4-volt noise margin at both the high and low input states.

IC Vcc voltage requirement: 4.75 – 5.25 volts

Max. low input	0.8 V
Max. low output	0.4 V
Min. high input	2.0 V
Min. high output	2.4 V

> 0.4-V noise margin (low)

> 0.4-V noise margin (high)

Figure 2.12 also shows that a TTL input will accept a voltage as low as 2.0 volts as a high input, whereas TTL logic is guaranteed to produce a high output of at least 2.4 volts. Thus, again, there is a 0.4-volt noise margin between the minimum acceptable high input and what a standard TTL high output is guaranteed to produce.

Other TTL Series

Since the original development of standard TTL logic, a number of more advanced TTL families have been developed. In general, these families either provide for faster operation, lower power dissipation per gate, or both (although usually there is a trade-off between reduced power and increased speed).

The 74LXX series is low-power logic, while 74HXX is high-speed logic. These two series have largely been supplanted by the other families in Figure 2.11. Three of these families are based on the Schottky barrier diode.

A Schottky barrier diode is formed by the junction of aluminum and N-type silicon. This diode has a forward-biased diode drop of only 0.3 volt.

To understand the importance of Schottky diodes, it is necessary to recall the structure of the typical totem-pole TTL output shown in Figure 2.10. In our discussion of this type of output, it was pointed out that one of the two output transistors is always saturated while the other is cut off. When a transistor is saturated, a significant amount of charge accumulates within the base region of the transistor. To turn the transistor off, this charge must be drained away, and this takes time. For a typical saturated transistor, the collector will be only about 0.2 volt above the emitter, while the base will be about 0.6 volt above the emitter. Thus the collector is actually at a lower voltage than the base and is therefore slightly forward biased. If a Schottky diode is placed between the collector and base of a transistor, oriented with the anode at the base and cathode at the collector, then the collector will not be able to get more than 0.3 volt more negative than the base. The transistor will thus be prevented from completely saturating, and this avoids the charge buildup in the base region. The transistor can therefore be turned off more quickly. The various Schottky TTL families incorporate Schottky diodes between the base and collector of the circuit transistors.

The three Schottky families are the 74SXX (Schottky), 74LSXX (low-power Schottky), and 74ALSXX (advanced low-power Schottky). These three families have become the most popular TTL families, with 74LSXX being the cheapest and most widely used and 74ALSXX being the most expensive. The 74FXX is the Fairchild fast TTL logic. It provides a slight increase in operating speed over the 74ALSXX at a significantly larger power dissipation and price.

TTL ICs from different series can be directly connected together, since the input and output voltages conform to the same set of standards (with some very minor variations). However, the input and output current specifications of the different families are different. Thus care must be taken when using different TTL families in the same circuit. Consider the following examples.

■ EXAMPLE 2.3–1

We wish to connect the output of a standard 74XX TTL IC to several 74LSXX inputs. How many inputs can we safely connect to? First, we examine the conditions when the 74XX output is high. According to Figure 2.11, a 74XX high output can provide 400 microamperes of current, while a 74LSXX input requires only 20 microamperes of current when the input is high. Thus one 74XX high output can provide sufficient current for 20 74LSXX inputs.

We must also examine the conditions when the 74XX output is low. According to Figure 2.11, a 74XX low output can sink 16 milliamperes of current, while a 74LSXX input will source only 0.4 milliampere of current when the input is low. We see that one 74XX low output can sink sufficient current to handle 40 74LSXX inputs.

The fan-out is determined by the lower of the two drive capabilities. Thus we conclude that the fan-out for 74XX logic driving 74LSXX logic is 20. ■

■ EXAMPLE 2.3–2

Suppose we wish to connect one 74LSXX output to a number of 74XX inputs. How many 74XX inputs can one 74LSXX output drive?

First, we consider the conditions when the 74LSXX output is high. Figure 2.11 shows that a high 74LSXX output can source 400 microamperes of current, while a 74XX high input requires 40 microamperes of current. Thus 74LSXX high output can drive 10 74XX inputs.

We must also examine the conditions when the 74LSXX output is low. In this case, Figure 2.11 shows that a low 74LSXX output can sink 8 milliamperes of current, while a low 74XX input will source 1.6 milliamperes of current. Thus a low 74LSXX output can only drive 5 74XX inputs.

The fan-out is determined by the lower of the two drive capabilities. Thus we conclude that the fan-out for 74LSXX logic driving 74XX logic is 5. ■

If it is necessary to connect the output of a TTL IC to more inputs than the known fan-out capability dictates, it is a simple matter to use intermediate buffers to increase fan-out. Figure 2.13 illustrates two ways of doing this. In Figure 2.13(a), inverters have been used to double the fan-out. If inversion is a problem, then noninverting buffers can be used as shown in Figure 2.13(b). The noninverting

FIGURE 2.13 Two simple ways to increase the fan-out of a TTL output. The number of outputs tied to each inverter (a) or non-inverting buffer (b) may be as large as 10.

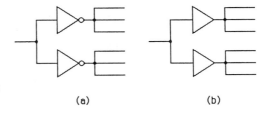

(a)   (b)

buffers will be a little slower, but the logic will not be inverted. Sometimes pairs of inverters in series are used to produce a noninverting buffer.

Open-collector Outputs

The output stage of most TTL logic circuits is the totem-pole output shown in Figure 2.10. The totem pole is an active output in the sense that one transistor is always saturated, and the output is then more or less directly connected either to ground or to 5 volts. Because of this situation, it is not possible to connect several totem-pole outputs to each other. If this were attempted, imagine what would happen if one output attempted to go high while another tried to go low. There would be a nearly direct short between the 5-volt supply voltage and ground. The resulting large current would probably destroy one or both outputs.

Because there are situations where it is desirable to tie outputs directly together, and because it is sometimes necessary to change voltage levels, the open-collector output is available on many TTL ICs. The open-collector output is shown in Figure 2.14, which is similar to Figure 2.10 except for the output stage. The open-collector output will indeed be pulled actively to ground whenever the output goes low. However, the high output simply amounts to the output transistor being cut off. Thus a number of outputs can be connected together in the manner shown in Figure 2.15. The resister R is called a *pull-up* resistor. The configuration shown in Figure 2.15 is called the *wired-OR* connection because the output will be pulled to ground if *any* of the individual outputs goes low. That is, the output is low if

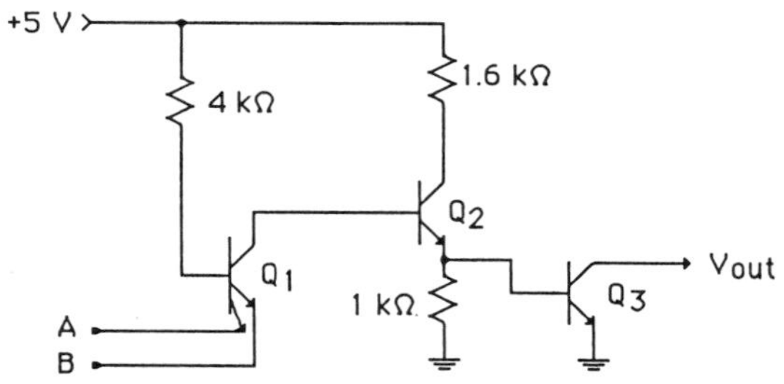

FIGURE 2.14 A TTL NAND gate with an open-collector output.

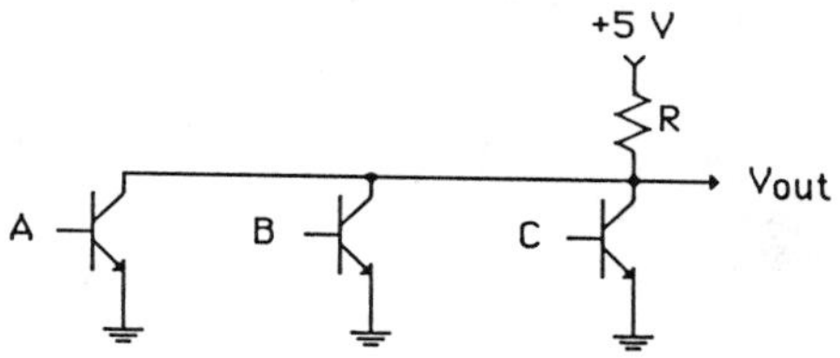

FIGURE 2.15 The wired-OR connection using open-collector outputs. If any individual output goes low, then the circuit output goes low.

output A OR B OR C is low, and is high otherwise. Another thing to notice about Figure 2.15 is that the high output voltage is determined by the 5-volt supply. If a different supply voltage were used, then this new supply voltage would become the high output voltage. Thus the open-collector output can be used to change the level of the high output from 5 volts to whatever is needed.

Tristate Outputs

Normal TTL logic has two output states, low and high. For some applications, there are times when it would be convenient to have several devices connected to the same input line, but have only one device at a time active. The solution to this problem is the tristate output.

The tristate output has three output states as the name implies. Two of these states are the normal high and low states. The third state is the *high-impedance* state. This state amounts to electrically disconnecting the output from the rest of the circuit. Thus, when a device is in the high-impedance output state, it acts as if it does not exist as far as any other circuitry connected to the output is concerned. Figure 2.16 shows two common ways that you are likely to encounter tristate circuitry. Figure 2.16(a) shows a tristate buffer. When the enable line is active (high in the example shown), the output of the buffer equals the input. When the enable line is inactive (low in this case), the output is electrically disconnected from the input. It is as if a switch had been opened between the input and output. Figure 2.16(b) shows an inverting tristate buffer. Again, when the enable line is active, the circuit behaves as an ordinary inverting gate. When the enable line is inactive, the output is disconnected from the input.

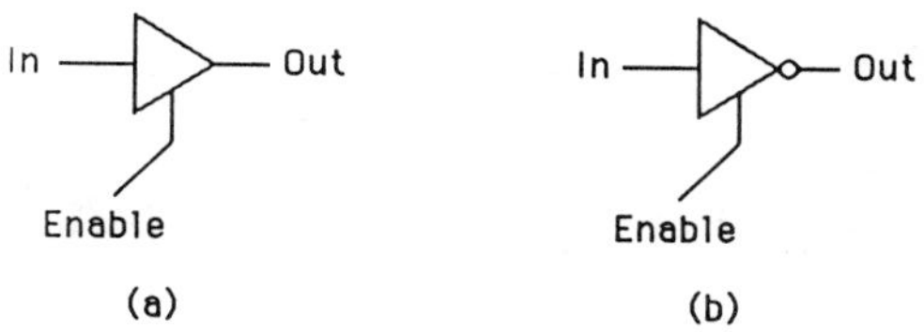

FIGURE 2.16 (a) Noninverting tristate buffer with active-high enable. (b) Inverting tristate buffer with active-high enable.

CMOS Logic

In the early 1970s, RCA introduced a new type of digital IC family called the CD4000 series. This IC family is based on *complementary metal oxide semiconductor* (CMOS) technology. CMOS technology is based on the idea of the *enhancement-mode, MOS, field-effect transistor* (MOSFET), and the fact that, like *NPN* and *PNP* transistors, there are two varieties of enhancement-mode MOSFETs. Figure 2.17 shows an enhancement-mode *N*-channel MOSFET. In a very rough way, the three terminals of this device are analogous to the base, emitter, and collector of the junction transistor, where the gate acts like the base and the drain acts like the collector. There are very striking differences, however. We will give a simplified version of the operation of the enhancement-mode *N*-type MOSFET.

First, we must point out that the source and drain are essentially *N*-type silicon embedded in a *P*-type material. This means that, as we proceed from the source to the drain, we must cross two *N–P* junctions. No matter which way voltage is applied between source and drain, at least one of these *N–P* junctions will behave as a reverse-biased diode. Thus current will not normally flow from the source to the drain because it must first cross a diode barrier. There is no direct good conduction path between the source and the drain.

Next we note that the gate is electrically insulated from the main body of the device by a thin metal oxide insulating layer. Hence there is no conducting path at all between the gate and the rest of the device. If no voltage is applied to the gate, then the source and the drain remain electrically isolated from each other, and no current flows between them. For this reason, the enhancement-mode MOS-FET is called a *normally off* device. However, if a sufficiently large positive voltage is applied to the gate, this will produce a fairly large electric field in the region just under the oxide layer. This field will pull electrons from within the body of the substrate material, and these electrons will form a conducting path between the source and the drain. The conducting path is said to be enhanced (thus the name enhancement mode). A sufficiently large applied voltage causes the connection between the source and drain to become very good, only a few hundred ohms. This is similar to the saturation of a junction transistor. Thus the MOSFET can be used as a two-state device. Summarizing this effect, then, the *N*-type MOS-

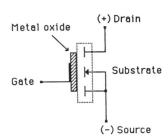

FIGURE 2.17 The basic construction of an enhancement-mode *N*-type metal oxide field-effect transistor (MOSFET).

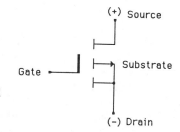

FIGURE 2.18 The basic circuit symbol for the enhancement-mode *P*-type MOSFET.

FET conducts only when the gate is made sufficiently positive with respect to the source.

A *P*-type MOSFET can be made as well. Such a device is shown in Figure 2.18. The operation is essentially complementary to the *N*-type device. Thus conduction occurs only when the gate is made sufficiently negative with respect to the source. Both types of MOSFETs have some very useful properties. First, since the gate is electrically insulated from the rest of the device, virtually no gate current flows. In addition, in the nonconducting state, virtually no current flows between the source and drain. Finally, MOSFETs can be fabricated quite easily within integrated circuits.

These properties of MOSFETs make them well suited for the construction of logic gates. Figure 2.19 shows a CMOS inverter made from an *N*-type and a *P*-type MOSFET. If the input is low, then the gate of the lower (*N*-type) MOSFET will be at the same voltage as the source, and this MOSFET will be in the nonconducting state. However, this low input voltage will be quite negative compared to the voltage at the source of the upper (*P*-type) MOSFET. Thus this upper MOSFET will be in the conducting state. The output will thus be isolated from ground and connected to $V+$ through the small resistance of the upper MOSFET. The value of $V+$ can be anywhere from 3 volts to about 15 volts. This is one advantage of CMOS logic: its supply voltage does not have to be 5 volts and thus the power supply need not be as precisely regulated (batteries, whose output slowly drops with time, work nicely, for example). When the input goes high, the lower MOSFET is turned on and the upper MOSFET is turned off. The output then goes

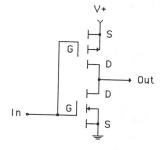

FIGURE 2.19 A *P*-type MOSFET (upper transistor) and *N*-type MOSFET (lower transistor) used as a complementary pair (thus the name CMOS) to produce an inverter.

essentially to ground. Because one MOSFET or the other is always off, virtually no current flows in either state of the inverter. Only during an actual change of state is there a short current pulse. Thus CMOS logic consumes very little power unless the gates are constantly changing state at a high frequency.

CMOS logic has been somewhat slower than TTL logic (CD4XXX propagation delays are roughly 100 nanoseconds). However, newer versions of CMOS are closing the gap. The 74CXX family is pin for pin compatible with the corresponding TTL numbered parts. Thus a 74C00 is pin for pin identical to a 7400 (or 74LS00, and so on) and can be used to replace it in an existing circuit. The 74C4XXX family is pin for pin equivalent to the CD4XXX series. The newest and fastest CMOS family is the 74HCXX series with a typical propagation delay of 8 nanoseconds. This series operates with a supply voltage that must be between 2 and 6 volts, unlike the other CMOS families, which can operate on any supply voltage between 3 and 15 volts.

CMOS ICs require careful handling and are always delivered in conducting foam to short the leads together or are delivered inside antistatic packages. The problem is that even a small static charge deposited on the gate of a MOSFET will produce a large voltage across the oxide layer that lies below the gate. This voltage may punch a conduction hole through the oxide layer to the substrate below. If this happens, the MOSFET becomes useless. Thus CMOS ICs are never left lying around outside of their conducting foam or antistatic packages.

Recall that for TTL logic, an unconnected input acts like a high input. For CMOS logic, all unused inputs should be connected either to $V+$ or to ground. An unconnected input can accumulate static charge that will cause it to behave unpredictably. For this reason, CMOS inputs that leave the circuit board should always have a resistor to ground to prevent static-charge buildup.

CMOS logic ICs do not possess as much drive capability as TTL logic. That is, they cannot source or sink as much current. This is no problem if CMOS circuits are connected to each other since relatively little drive capability is required. However, even assuming that a 5-volt supply is being used, one CMOS output can only drive one low-power Schottky (74LSXX) load. Thus some sort of buffering is usually required to connect a CMOS output to several TTL inputs. Going from TTL to CMOS can also be a problem since the guaranteed 2.4 volt-high TTL output may not be high enough to be recognized as a legitimate high by a CMOS IC. In this case, a pull-up resistor should be used on the TTL output to make sure that a high output is close to 5 volts.

2.4 MORE ON INTERFACING

Since interfacing ICs from one family to ICs from another family occurs quite frequently, we will consider it in more detail. When interfacing an IC from one TTL subfamily to an IC of another subfamily, there is no problem with voltage-level compatability since all TTL subfamilies use the same high- and low-voltage states. Thus the only important considerations are current drive and delay time. We have already discussed current-drive considerations in the previous section. It is simply a question of making sure that a given output has the current capability to drive all the inputs that it is connected to. If the output lacks sufficient current drive, then one or more buffers must be used in the manner shown in Figure 2.13.

Delay time is another matter. It has already been pointed out that some TTL subfamilies are faster than others. For example, the 74SXX TTL subfamily has typical gate delays of about 3 nanoseconds, while the 74LXX subfamily has gate delays of about 33 nanoseconds. These figures are only *typical* values. You should always consult the specific circuit specification sheet (the "spec sheet") provided by the manufacturer to get the actual values. The large range over which delay times vary from one subfamily to another does not pose a direct interfacing problem. However, it can indirectly be a concern. In a typical situation, there may be many signal pathways through a complex circuit. At a given point within the circuit, two signals may arrive at the two inputs to a given logic gate by very different routes. The circuit will function properly only if the two signals take on their desired values at the same time. However, if these two signals have followed very different paths through gates with considerably different gate delays, then there could be a significant length of time where one of the two signals has changed to its desired state while the other has not. This may not be a problem, but it may also lead to an unwanted output for some period of time. Thus the designer or service technician must not automatically assume that ICs from different TTL subfamilies can be used interchangeably even if current drive is adequate.

TTL to CMOS Interfacing

The majority of interfacing problems arise when it becomes necessary to interface ICs from one family to those of an entirely different family. In this case, there will be both voltage-level problems and current-drive problems. With this in mind, let us look at some examples of how TTL ICs can be interfaced to CMOS ICs.

When a TTL output must drive a CMOS input, the problem becomes one of producing the correct voltage levels. Current drive is not a concern, since CMOS inputs require very little input current. CMOS inputs are *high-impedance* inputs.

A high-impedance input is an input that behaves as if it were a very large resistance. It thus draws very little current from any circuit connected to it.

Let us consider a specific example. Suppose that we have a digital circuit that is to be composed of a TTL section that must then be interfaced to a CMOS section. Suppose that the CMOS section has a 12-volt supply voltage. The TTL section will produce a high-voltage state that is nowhere near 12 volts. There are a number of ways that a voltage conversion can be made. One straightforward approach is to

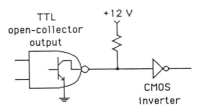

FIGURE 2.20 Using an open-collector TTL output to interface TTL circuitry to CMOS circuitry.

use a TTL open-collector output at the point of interface. This is illustrated in Figure 2.20, where an open-collector NAND gate (with the output transistor shown for emphasis) is connected to a CMOS inverter. The circuit works as follows. When the NAND output goes low, the transistor will be saturated. This will pull the output voltage to nearly 0 volts. This 0 volts will be recognized as a low by the CMOS inverter. When the NAND output goes high, this amounts to the transistor being cut off. Thus the output will be pulled to 12 volts by the 5-kilohm pull-up resistor. The 12 volts will be recognized as a high by the CMOS inverter. The interface is thus complete.

A second popular way of changing voltage levels is to use an optoisolator. This consists of a light-emitting diode (LED) and a phototransistor inside the same closed package. Figure 2.21 illustrates this approach. The LED will emit light whenever current flows through it. This requires an input voltage of about 2 volts or more. The phototransistor, on the other hand, normally has a very high resistance when no light strikes it. Thus, when the LED is off, the phototransistor in series with the 5-kilohm resistor produces an output of essentially 12 volts. If a significant amount of light strikes the phototransistor, then the phototransistor will essentially saturate at a very low output voltage. Circuit operation then proceeds as follows. When the TTL inverter output is high, this produces sufficient current to light the LED and thus saturate the transistor, which produces a low output. When the inverter output is low, the LED produces no light, and thus the transistor remains cut off. In this case, the input to the CMOS inverter will be 12 volts. The interface is thus complete. One advantage of using the optoisolator is the almost complete electrical isolation that it provides. There is virtually no electrical connection between the TTL part of the circuit and the CMOS part of the circuit. The connection is through the flow of light (rather than electricity) inside the optoisolator.

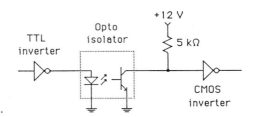

FIGURE 2.21 Using an optoisolator to interface a TTL output to a CMOS input.

CMOS to TTL Interfacing

When it is necessary to interface a CMOS output to a TTL input, both voltage conversion and current drive are a problem. If the CMOS device is operated at 5 volts, then a CMOS output will drive one 74LSXX input with no problem. However, even if the CMOS circuit used a 5-volt supply, the CMOS output would not provide sufficient output current to drive a standard TTL input. The main problem here is the CMOS low-output state.

Let us refer back to Figure 2.19, which shows a CMOS inverter. The operation of this circuit is such that one of the two MOSFETs will always be on and the other will be off. The MOSFET that is on will have a typical resistance of about 400 ohms (which is somewhat dependent on supply voltage). Let us suppose that the inverter is producing a low output so that the lower MOSFET is the one that is on. If this output is connected directly to a *standard* TTL input, then the lower MOSFET must sink 1.6 milliamperes of current to hold the TTL input low. This 1.6 milliamperes of current flowing through a 400-ohm resistor produces a voltage drop of at least 0.64 volt. This gets a little too close to the 0.8-volt TTL limit for comfort.

One answer to CMOS to TTL interfacing is to use a specific interface IC designed for the task, such as the CMOS CD4049 (which is an inverter) or the CD4050 (which is noninverting). In this case, the CD4049 or CD4050 is operated at the supply voltage of the TTL circuit (which *must* be at a lower voltage than the driving CMOS circuit). A second approach is to use the method of Figure 2.21. In this case, the input inverter would be CMOS, the 12-volt supply would be replaced by 5 volts, and the second inverter would be TTL.

Standard IC Packages

Before leaving the subject of interfacing, it would be useful to describe the way that digital ICs are typically packaged. Standard TTL and CMOS ICs are usually provided in *dual in-line packages*, or DIPs. Figure 2.22 shows a typical DIP. One end of the package will have a notch in it or perhaps a small circle set off to one side. Both of these features are shown in the figure. This is done to identify the numbering of the leads, or *pins*, connected to the package. Figure 2.22(b) shows the numbering convention, which is used regardless of how many pins the IC actually has. IC packages have anywhere from 8 pins to as many as 64. Most *small-scale integrated* (SSI) digital circuits have 14 or 16 pins. Small-scale integration refers to ICs containing a small number of relatively simple logic devices. *Medium-scale integrated* (MSI) circuits may have 18 to 24 pins, or even more. Medium-scale integration refers to ICs containing several logic gates configured to implement a more complex logic function. Technology has now advanced to *large-scale integration* (LSI) involving circuits with thousands of logic gates and *very large-scale integration* (VLSI) involving circuits with tens of thousands of logic gates.

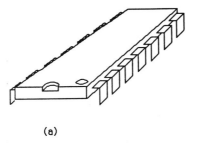
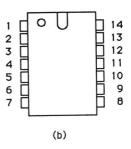

(a)   (b)

FIGURE 2.22 Standard DIP pin configuration. Pins are always numbered counterclockwise starting at left side of the notch end.

The pins of the standard DIP package are connected at standard separations both along the package and across the package. Thus the ICs can be easily inserted into standard printed circuit boards (PC boards) or into solderless breadboards for circuit fabrication and testing.

Recently, new packaging techniques have been developed for digital ICs. One increasingly popular package uses *surface-mount technology* (SMT). This type of package has no pins extending from the package. Instead, small metal pads are placed on one side of the IC package. The package is soldered directly to the circuit board by means of these small pads. This type of package is well suited to automated assembly of circuit boards, but is a nightmare for the circuit trouble-shooter who has to desolder such a package.

A second type of package continues to use metal pins extending from the package, but has the pins extending directly from one side of the package in a square pattern, rather than from the sides of the packages as with the DIP. This allows many more pins to extend from a given-sized package. This type of packaging is being used for some of the new, very large, microprocessor ICs, which have a large number of pins.

PROBLEMS

1. Consider the circuit shown in Figure 2.23. Suppose that inputs A and B can be either 0 or 5 V. Construct a truth table showing the output voltage for each possible input combination. What logic function does this circuit perform?

2. Repeat Problem 1 for the circuit shown in Figure 2.24.

3. Consider the circuit shown in Figure 2.25. Assume that inputs A and B can be either 0 or 10 V. Notice that the upper two MOSFETs are connected in

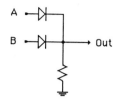

FIGURE 2.23 Use with Problem 1.

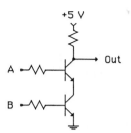

FIGURE 2.24 Use with Problem 2.

parallel, one to A and one to B. The lower two MOSFETs are connected in series, one to A and one to B.

a. When A is 10 V, what are the states of Q_2 and Q_3 (on or off)?

b. When B is 0 V, what are the states of Q_1 and Q_4?

c. Notice that, for the output to be pulled to ground, both Q_3 and Q_4 must be on. What does this imply about inputs A and B?

d. Construct a complete truth table for the circuit and identify the logical function being performed (assuming positive logic).

4. Assume that the transistor in the circuit shown in Figure 2.26 has a beta value of 20. That is, when the transistor operates in its linear region, the collector current will be 20 times the base current.

a. Assuming a 0.6-V drop between base and emitter, what is the voltage across the 50-kΩ resistor? How much base current does this allow to flow?

b. If the collector current is 20 times the base current, then what collector current flows?

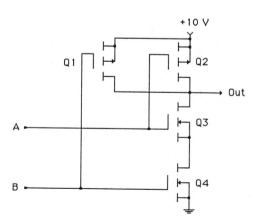

FIGURE 2.25 Use with Problem 3.

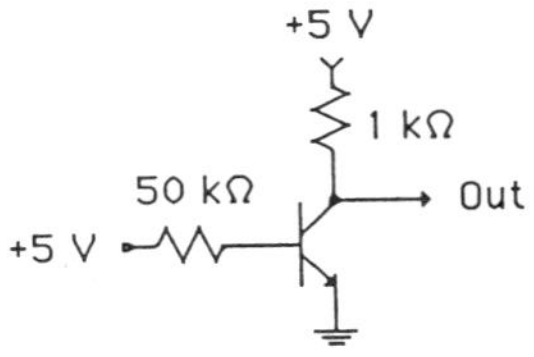

FIGURE 2.26 Use with Problem 4.

c. How large of a voltage drop does the collector current in part b produce across the 1-kΩ collector resistor? If this voltage drop is less than the available supply voltage (5 V), then the transistor is in its linear region. If the calculated voltage drop exceeds the supply voltage, then, in reality, less collector current will actually flow and the transistor will be saturated. Is the transistor in this case saturated?

d. Repeat parts a to c for a beta of 100.

5. What is the fan-out of a 74LXX TTL output driving 74XX TTL inputs? Which type of output (high or low) proves to be the limiting case?

6. A certain non-TTL circuit is to be used to directly drive a TTL input. This circuit typically provides a low output of 0.6 V or a high output of 3.4 V with plenty of current capability.

   a. Will these voltages successfully drive a TTL input?

   b. How much noise can be tolerated on these voltages and still have the output work correctly?

7. It is desired that a 74LSXX TTL output be connected to 20 74HXX inputs. Since a 74LSXX output does not possess sufficient current output to directly drive 20 74HXX inputs, design a workable interface using an appropriate buffer arrangement.

8. An LED (light-emitting diode) is a type of diode that emits light when sufficient current flows in the forward direction (typically a few milliamperes). Suppose that we have four inputs, A, B, C, and D. It is desired that an LED light whenever either A and B are both high or C and D are both high. Show how two open-collector NAND gates can be wired to accomplish this.

9. Suppose the following situation arises. There are two TTL input lines, A and B. Sometimes it is necessary to have input line A connected to a particular input pin on a TTL IC. Other times it is necessary to have input line B connected instead of line A. We want to be able to choose which input line, A or B, is connected by means of the state of a third control line, C. When C is high, then input A is to be connected. Otherwise, input B is to be connected. Show how to accomplish this with an inverter and two tristate buffers of the type shown in Figure 2.16(a).

10. Refer to Figure 2.5. Assume that the transistor has a beta of 50. Suppose that we want V_{out} to be 0.4 V.

 a. What would the voltage drop across the 2-kΩ collector resistor then be?

 b. How much collector current would be required to achieve the voltage drop in part a?

 c. How much base current does this imply must be flowing?

 d. How large a voltage drop does this base current produce across the 2-kΩ base resistor?

 e. Assuming a 0.6-V drop across the base–emitter junction, what must the battery voltage V be?

 f. If the battery voltage were smaller than the value you computed in part e, would V_{out} be larger or smaller? Justify your answer.

11. Refer to Figure 2.10. Assume that both inputs A and B are 4.5 V so that transistor Q_1 is cut off and the base–*collector* junction acts like a *forward-biased diode*. Observe that the base–emitter junctions of transistors Q_2 and Q_4 act like forward-biased diodes in this situation. Also assume that the collector–emitter voltage of any saturated transistor is 0.2 V.

 a. Determine the voltages at the base of transistors Q_1, Q_2, and Q_4.

 b. Which transistors do you expect to be saturated?

 c. Based on your answer to part b, determine the voltage at the collector of transistor Q_2. Will this voltage be sufficient to turn on transistor Q_3?

 d. What do you expect the output voltage of the circuit to be?

12. What is meant by saying that an input *sinks* current? What is meant by saying that an input *sources* current?

13. When a standard TTL input is low, does it sink or source current? How much current does it sink or source?

14. When a standard TTL output is high, does it sink or source current? How much current does it sink or source?

15. The output of a TTL gate is 0.1 V, which will be read as a logical 0 output. How much noise voltage could this output tolerate before it would be *certain* to indicate a logical 1 output?

16. A concept sometimes introduced to evaluate the performance of various TTL families is the *figure of merit*. The figure of merit is defined as:

Figure of merit = (propagation delay) × (power dissipation)

The propagation delay is measured in nanoseconds (ns), and the power dissipation (per gate) is measured in milliwatts (mW). The resulting unit for the figure of merit is picojoules (pJ) and is a measure of the amount of energy dissipated by one gate during a single change of state. Using the numbers in Figure 2.11, compute the figure of merit for each of the TTL families listed.

17. A given digital circuit consists of standard TTL ICs. It is desired to replace the standard TTL ICs with ICs from a different TTL family in order to reduce the power consumption of the board by at least 50%. Which other TTL families could be used to achieve this reduction?

18. In many products using TTL logic, the standard TTL family has been replaced by the LS TTL family. What is the chief benefit derived from this change?

19. A TTL 2-input NAND gate has one input at 0.3 V and the other input at 0.5 V. What is the *logical* state of the output?

20. A TTL 2-input NAND gate has one input at 0.3 V and the other input at 2.9 V. What is the *logical* state of the output?

21. A TTL 2-input NAND gate has one input at 0.2 V and the other input at 1.5 V. What is the *logical* state of the output?

22. Three standard TTL inverters are connected in series. That is, the output of the first is connected to the input of the second, and so on. The input of the first inverter changes state. How much time passes before the output of the third inverter changes state?

23. A digital signal must pass through a series of five gates (such as five inverters in series). For the circuit to operate properly, a delay of no more than 35 ns must occur between the change of state of the input to the first gate and the resulting change of state of the output of the final gate in the series. What are the possible choices for a TTL family that will achieve this result?

Chapter 3

BOOLEAN ALGEBRA

Learning Objectives

After completing this chapter you should know:

- The basic theorems and identities of Boolean algebra.

- How to manipulate Boolean functions using the rules of Boolean algebra.

- How to simplify or reduce a Boolean function by applying the rules of Boolean algebra.

- How to convert a Boolean expression into a form that uses only NAND gates or only NOR gates.

■ How to easily convert a circuit consisting of simple logic gates into a form using only NAND gates or only NOR gates without using Boolean algebra.

■ The two standard forms for expressing a Boolean function, the sum-of-products form and the product-of-sums form.

■ How to convert a logical truth table into a standard Boolean form.

3.1 INTRODUCTION

In the two previous chapters, you were introduced to the basic logic gates used in digital electronics and to some of the ways these gates are actually constructed. You also saw how to manipulate, in a limited way, the symbols used to represent digital functions. This ability to manipulate logical symbols is very important to the circuit designer or technician. In many cases, the logical task to be performed is first determined in terms of a complex Boolean expression, which, if implemented directly, would require many logic gates. One of the powers of Boolean algebra lies in the fact that it can be used to reduce complex logical expressions to much simpler form. The simpler form can then be implemented at much less expense.

A second task often faced in digital electronics is that of implementing a digital circuit by using, for example, only NAND gates or only NOR gates. To do this kind of thing, it is necessary to manipulate a Boolean expression to change it into a form that includes only NAND functions (and inversions) or only NOR functions (and inversions). To accomplish either of these two tasks, it is necessary to become familiar with the basic theorems of Boolean algebra. The central theme of this chapter is the development of these theorems and their application to the two tasks outlined.

3.2 THE BASIC THEOREMS OF BOOLEAN ALGEBRA

Boolean algebra is a formal branch of mathematics that has proved itself very useful in the field of digital logic design. Generally, we say that Boolean algebra concerns itself with a set of two elements, which we take to be 0 and 1. These two elements can be represented for convenience by symbols such as $A, B, C, X, Y,$ or Z. Each symbol can represent either 0 or 1, but only one of these at any given time.

Two Boolean or binary operations are defined for combining Boolean symbols or elements. These are the AND operation and the OR operation. The AND operation is represented by a dot between symbols or by two symbols written together with no intervening symbol (for example, AB for A and B). The AND operation is defined by the truth table given in Figure 1.4. The OR operation is represented by the $+$ symbol and is defined by the truth table given in Figure 1.7.

There is also a unary operation, the complement, represented by a bar over a symbol or a prime on the symbol (for example, $\overline{A}$). This operation is defined such that $\overline{0} = 1$ and $\overline{1} = 0$. This is obviously the NOT operation defined by the truth table in Figure 1.2.

With the definition of the basic Boolean operations, it becomes possible to introduce a number of theorems and identities. To make it easier to refer to these theorems and identities later in the chapter, they will be numbered according to the following conventions. Theorems will be numbered (T–1), (T–2), and so on, and identities will be numbered (I–1), (I–2), and so on.

It can be shown that Boolean algebra is *associative*. That is,

$$A + (B + C) = (A + B) + C \tag{T-1}$$

Thus we simply write $A + B + C$ instead of including the parentheses. And, likewise,

$$AB(C) = (AB)C \tag{T-2}$$

Thus we usually write simply ABC instead of including the parentheses. Boolean algebra is also *commutative*. Thus

$$A + B = B + A \tag{T-3}$$

and

$$AB = BA \tag{T-4}$$

And Boolean algebra is *distributive*. Thus

$$\begin{aligned} A(B + C) &= (AB) + (AC) \\ &= AB + AC \end{aligned} \tag{T-5}$$

Notice that in the second version of Theorem (T–5) the parentheses have been removed on the right. This follows the normal convention of giving the AND operation precedence over the OR operation. Thus it is understood that the AB term and the AC term are to be evaluated first and the results ORed together. These basic theorems probably seem obvious since they are also obeyed by ordinary algebra. However, many mathematical systems exist in which the basic operations do not obey these theorems.

There are also a number of one-variable identities that can easily be verified by truth table. They are

$$\overline{(\overline{A})} = A \quad \text{(I–1)} \qquad A + 0 = A \quad \text{(I–6)}$$
$$(A)(0) = 0 \quad \text{(I–2)} \qquad A + 1 = 1 \quad \text{(I–7)}$$
$$(A)(1) = A \quad \text{(I–3)} \qquad A + A = A \quad \text{(I–8)}$$
$$(A)(A) = A \quad \text{(I–4)} \qquad A + \overline{A} = 1 \quad \text{(I–9)}$$
$$(A)(\overline{A}) = 0 \quad \text{(I–5)}$$

Parentheses have been included here for ease of reading and are not really necessary.

Two useful theorems, known as the absorption theorems, are

$$A + AB = A \tag{T-6}$$

and

$$A(A + B) = A \tag{T-7}$$

The proof of these theorems by truth tables is left as an exercise. Another very useful theorem is

$$A + \overline{A}B = A + B \tag{T-8}$$

This theorem is used very frequently in Boolean reductions. However, its form is often disguised and seems to be difficult for many students to recognize. All the following equations are forms of Theorem (T–8).

$$\overline{A} + AB = \overline{A} + B \qquad C + \overline{C}\overline{A} \qquad = C + \overline{A}$$

$$\overline{A} + A\overline{B} = \overline{A} + \overline{B} \qquad CD + (\overline{CD})B = CD + B$$

In each of these examples, the first Boolean term is ORed with the product (AND) of its inverse and some other term. The result is always the logical sum (OR) of the first term and the other term.

Two exceedingly useful theorems are the two forms of *De Morgan's theorem*. These are

$$\overline{AB} = \overline{A} + \overline{B} \tag{T-9}$$

or

$$\overline{\overline{A}\,\overline{B}} = A + B$$

and

$$\overline{A + B} = \overline{A}\,\overline{B} \tag{T-10}$$

or

$$(\overline{\overline{A} + \overline{B}}) = AB$$

We have already seen these identities in a practical way when we observed that a NAND gate was equivalent to a low-level input OR gate (T–9), and a NOR gate was equivalent to a low-level input AND gate (T–10). The second form of the two preceding theorems is included for ease in identifying De Morgan's theorems when they are encountered. These theorems may be generalized to several variables according to

$$\overline{ABC\ldots} = \overline{A} + \overline{B} + \overline{C} + \cdots \tag{T-11}$$

and

$$\overline{A + B + C + \cdots} = \overline{A}\overline{B}\overline{C} \ldots \qquad \text{(T–12)}$$

The various identities and theorems that have been introduced can be used to reduce or manipulate Boolean expressions. In the next section we will see several examples of Boolean reduction.

3.3   EXAMPLES OF BOOLEAN REDUCTION

The general design problem involves many steps. Among these are (1) developing the appropriate Boolean expression, (2) reducing the expression, and (3) perhaps manipulating the expression. In this section we will concentrate on the second of these three steps. Our approach is to simply list several examples of Boolean reduction, which are illustrated step by step. Let us begin.

■ **EXAMPLE 3.3–1**

$$
\begin{aligned}
F &= ABC + \overline{A}BC + A\overline{B}C \\
  &= (A + \overline{A})BC + A\overline{B}C && \text{by (T–5)} \\
  &= BC + A\overline{B}C && \text{by (I–9)} \\
  &= (B + A\overline{B})C && \text{by (T–5)} \\
  &= (B + A)C && \text{by (T–8)}
\end{aligned}
$$
■

■ **EXAMPLE 3.3–2**

$$
\begin{aligned}
F &= \overline{A}\,(\overline{B}(B + A) + C) \\
  &= \overline{A}(\overline{B}B + \overline{B}A + C) && \text{by (T–5)} \\
  &= \overline{A}(\overline{B}A + C) && \text{by (I–5)} \\
  &= \overline{A}\overline{B}A + \overline{A}C && \text{by (T–5)} \\
  &= (\overline{A}A)\overline{B} + \overline{A}C && \text{by (T–4)} \\
  &= \overline{A}C && \text{by (I–5)}
\end{aligned}
$$
■

■ EXAMPLE 3.3–3

$$
\begin{aligned}
F &= \overline{(\overline{A}B)} + \overline{A\overline{B}} \\
&= (\overline{\overline{A}B})(\overline{A\overline{B}}) && \text{by (T–10)} \\
&= (A + \overline{B})(\overline{A} + B) && \text{by (T–9)} \\
&= A\overline{A} + AB + \overline{B}\,\overline{A} + \overline{B}B && \text{by (T–5)} \\
&= AB + \overline{A}\,\overline{B} && \text{by (I–5) and (T–4)}
\end{aligned}
$$ ■

■ EXAMPLE 3.3–4

$$
\begin{aligned}
F &= \overline{A}\,\overline{B}\,\overline{C} + \overline{A}\,\overline{B}C + A\overline{B}\,\overline{C} + A\overline{B}C + ABC \\
&= \overline{A}\,\overline{B}(\overline{C} + C) + A\overline{B}(\overline{C} + C) + ABC && \text{by (T–5)} \\
&= \overline{A}\,\overline{B} + A\overline{B} + ABC && \text{by (I–9)} \\
&= (\overline{A} + A)\overline{B} + ABC && \text{by (T–5)} \\
&= \overline{B} + B(AC) && \text{by (I–9) and (T–4)} \\
&= \overline{B} + AC && \text{by (T–8)}
\end{aligned}
$$ ■

■ EXAMPLE 3.3–5

$$
\begin{aligned}
F &= \overline{(\overline{A\overline{B}C}) + \overline{B}\overline{C} + A} \\
&= (\overline{\overline{\overline{A\overline{B}C}}})\,(\overline{\overline{\overline{B}\overline{C}}})\,\overline{A} && \text{by (T–10)} \\
&= (A\overline{B}C)(BC)(\overline{A}) && \text{by (I–1)} \\
&= (A\overline{A})(\overline{B}B)(CC) && \text{by (T–4)} \\
&= 0 && \text{by (I–5)}
\end{aligned}
$$ ■

■ EXAMPLE 3.3–6

Recall that in Chapter 1 it was asserted that four NAND gates could be connected in such a way that the resulting circuit was in fact an XOR gate. The circuit is shown in Figure 1.15, and the direct Boolean expression for this circuit is

$$F = \overline{(\overline{AB})A} \; \overline{(\overline{AB})B}$$

$$= (\overline{AB})A \; + \; (\overline{AB})B \qquad \text{by (T–9)}$$

$$= (\overline{AB})(A \; + \; B) \qquad \text{by (T–5)}$$

$$= (\overline{A} \; + \; \overline{B}) \; (A \; + \; B) \qquad \text{by (T–9)}$$

$$= \overline{A}A \; + \; \overline{A}B \; + \; \overline{B}A \; + \; \overline{B}B \qquad \text{by (T–5)}$$

$$= \overline{A}B \; + \; \overline{B}A \qquad \text{by (T–5)}$$

This final result is just the XOR function.   ■

3.4   NAND AND NOR GATE CIRCUIT IMPLEMENTATIONS

The examples in Section 3.5 were mainly concerned with reducing a Boolean expression to a simpler form. Sometimes it is necessary to change the form of an already reduced expression into something more useful. For example, TTL NAND gates are essentially the fastest and cheapest form of SSI TTL gate. Thus it is often useful to implement a simple circuit completely in the form of NAND gates. On other occasions, perhaps when using CMOS logic, it might be more appropriate to use only NOR gates. The following examples illustrate this type of problem.

■ EXAMPLE 3.4–1

Let us try to implement the following Boolean expression using only NAND gates and inverters (which can easily be made from a NAND gate).

$$F = AB + \overline{A}C + BC$$

The basic line of attack is to eliminate the OR symbols from the equation. What is left must then be either AND or NAND functions. An AND function can be turned into a NAND function by using an inverter and thus poses no problem. De Morgan's theorem is what we need.

We see that our expression resembles the right side of the second version of (T–9). Thus

$$F = \overline{AB + \overline{A}C + BC}$$

$$= (\overline{AB})(\overline{\overline{A}C})(\overline{BC})$$

This expression is entirely in NAND form except for an inversion that is necessary on A. The direct circuit implementation does, however, require a 3-input NAND gate. ■

■ **EXAMPLE 3.4–2**

Let us see how a combination of 2-input NAND gates can be used to implement the expression found in Example 3.4–1. To do this, it is necessary to isolate the first two terms in the expression from the third term. We could proceed as follows where brackets ([ ) have been introduced for visual clarity.

$$F = \overline{(\overline{AB})(\overline{\overline{A}C})(\overline{BC})}$$

$$= \overline{[(\overline{AB})(\overline{\overline{A}C})](\overline{BC})} \qquad \text{by (T–2)}$$

$$= \overline{[\overline{\overline{(\overline{AB})(\overline{\overline{A}C})}}](\overline{BC})} \qquad \text{by (I–1)}$$

This expression, as messy as it might seem, can be directly implemented with 2-input NAND gates and two inverters. Note that in the last step use was made of double inversion to get the two terms inside the [ ] to be NANDed together. This is a common technique to convert an AND to a NAND followed by inversion or an OR to a NOR followed by inversion. We can implement the final result using the approach outlined in Section 1.4. The circuit is shown in Figure 3.1, where the low-level input OR symbol for the NAND gate has been used in two cases. ■

Looking at Figure 3.1, we are immediately struck by the upper combination of two NANDs feeding a third NAND, shown as a low-level input OR. The implied inverters at the NAND outputs are essentially canceled by the implied inverters at the low-level OR input. This suggests another way of implementing a circuit using all NAND gates and inverters. Let us go back to the original Boolean expression that we started with, which was $F = AB + \overline{A}C + BC$. If we implement this directly as it is read, confining ourselves to 2-input gates, we get the circuit shown in Figure 3.2. The task is to get rid of the OR gates and turn them into NAND gates. This can be done by putting inverters on the inputs of the OR gates. Of course, introducing a single inverter at any point would change the logic. But a *pair* of inverters inserted at any point leaves the logic unchanged. Thus, for each inverter introduced at the input of an OR gate, there must be another inverter introduced somewhere before that point in the same logic line. A second task is to convert all AND gates to NAND gates. This can be done by placing an inverter

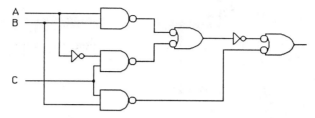

FIGURE 3.1   Implementation of the expression $F = AB + \overline{A}C + BC$ using only NAND gates and inverters.

at the output of all AND gates. Again, this inverter must be one of a pair of inverters introduced on the same logic line.

■ EXAMPLE 3.4–3

Let us see how to use the approach just described with the circuit in Figure 3.2. We first attempt to simply introduce an inverter at the output of each AND gate and at the input of each OR gate. We then check to see that each inverter thus introduced has been one of a pair of inverters. Our check shows us that the inverter on the output of AND gate 1 forms a pair with the inverter on the input of OR gate 4. Likewise, the inverter on the output of AND gate 2 forms a pair with the inverter on the input of OR gate 4. And the inverter on the output of AND gate 3 forms a pair with the inverter on the input to OR gate 5. The only problem is the upper input to OR gate 5. An inverter on this input is not matched by an inverter on the output of OR gate 4. Thus we must introduce an actual inverter into the logic line that feeds the upper input to OR gate 5. A comparison of the circuits in Figures 3.1 and 3.2 illustrates the procedure we have just outlined.   ■

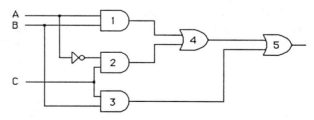

FIGURE 3.2   The direct implementation of the Boolean expression $F = AB + \overline{A}C + BC$ using 2-input gates.

■ EXAMPLE 3.4–4

Suppose that we wished to implement the following function using only 2-input NOR gates.

$$F = AB + \overline{A}\,\overline{B}$$

In this case, we need to get rid of any AND functions. Again, De Morgan's theorem is useful. We use both versions of (T–10). Thus

$$
\begin{aligned}
F &= AB + \overline{A}\,\overline{B} \\
&= \overline{\overline{A} + \overline{B}} + \overline{(A + B)} \qquad \text{by (T–10)} \\
&= \overline{\overline{(\overline{A} + \overline{B})} + \overline{(A + B)}} \qquad \text{by (I–1)}
\end{aligned}
$$

This expression can be implemented directly and is shown in Figure 3.3. Again notice that in two cases the low-level input AND symbol has been used for a NOR gate.   ■

   If we implement the original Boolean expression from Example 3.4–4 directly, as shown in Figure 3.4, we see a close correspondence to the NOR version shown in Figure 3.3. Again, the method of conversion suggests itself. In Figure 3.4, to eliminate OR gates, simply put an inverter on the output, making sure that it is one of a pair of inverters introduced. To eliminate AND gates, simply put inverters on the inputs, again making sure that each inverter introduced is one of a pair.

■ EXAMPLE 3.4–5

Let us apply this procedure to the circuit in Figure 3.4. We first consider AND gate 1. Here it is sufficient to simply merge the real inverters at the inputs to AND gate 1 with the AND gate, thus producing an OR gate (in low-level input form). For AND gate 2, it is necessary to place inverters on the inputs (in the form of small circles). But these inverters are not matched by a second inverter preceding the input. Thus we must add real inverters to both input lines. Finally, we need

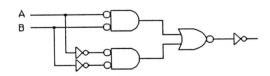

FIGURE 3.3   Implementation of the Boolean function $F = AB + \overline{A}\,\overline{B}$ using 2-input NOR gates and inverters.

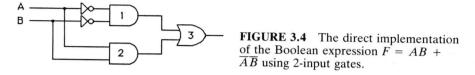

FIGURE 3.4 The direct implementation of the Boolean expression $F = AB + \overline{A}\overline{B}$ using 2-input gates.

an inverter on the output of OR gate 3. This must be one of a pair. Thus we get a small circle as the first inverter and an actual inverter as the second. A comparison of Figures 3.4 and 3.3 illustrates the method. ∎

■ EXAMPLE 3.4—6

Let us try to implement the following Boolean expression in terms of 2-input NAND gates and inverters.

$$F = \overline{A + \overline{B}C}$$

This time, we will first implement the expression directly as a circuit. This is shown in Figure 3.5. Now let us use the approach developed in Example 3.4–4 to eliminate the AND gate and NOR gate. To change the AND gate into a NAND gate, we put an inverter on its output. To change the NOR gate into a NAND gate, we must put inverters on each input and *get rid of the inverter on the output*. Putting an inverter on the lower input of the NOR gate completes the inverter pair that began with the inverter at the AND gate output. Putting an inverter on the upper input of the NOR gate requires that an actual inverter be added also to complete the pair. To eliminate the inverter on the output, simply break it free and consider it an actual inverter. Or add a pair of inverters and use the first to cancel the circle at the NOR output. Either way, the circuit in Figure 3.6 results.

Now let us reconsider our Boolean function and use De Morgan's theorem to convert it to NAND functions and inversions.

$$F = \overline{A + \overline{B}C}$$
$$= \overline{A}(\overline{\overline{B}C}) \qquad \text{by (T–10)}$$

FIGURE 3.5 The direct circuit implementation of the Boolean expression $F = \overline{(A + \overline{B}C)}$ using 2-input gates.

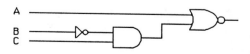

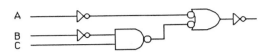

FIGURE
3.6 Implementation of the expression $F = \overline{(\overline{A} + \overline{BC})}$ using 2-input NAND gates and inverters.

$$= \overline{\overline{(\overline{A})}(\overline{\overline{BC}})} \qquad \text{by (I–1)}$$

This final form is directly implemented in Figure 3.6. ∎

■ **EXAMPLE 3.4–7**

As a final example, let us implement the following expression in terms of 2-input NOR gates.

$$F = (A + B)(\overline{A} + C)$$

Of course, this expression can be somewhat simplified, but let us work with it just as it is. First, let us implement the circuit directly. This is shown in Figure 3.7. Now let us apply the method developed in Example 3.4–5 to change the OR gates to NOR gates and the AND gates to NOR gates.

We place inverters on the outputs of the OR gates and on the inputs of the AND gate. These inverters form pairs with each other, so no other inverters need be added. The circuit becomes as shown in Figure 3.8. We can arrive at this same circuit by applying De Morgan's theorem to the original Boolean function. Thus

$$F = (A + B)(\overline{A} + C)$$

$$= \overline{\overline{(A + B)} + \overline{(\overline{A} + C)}} \qquad \text{by (T–10), second version}$$

This final expression is directly implemented in Figure 3.8. ∎

FIGURE 3.7 The direct circuit implementation of the Boolean expression $F = (A + B)(\overline{A} + C)$.

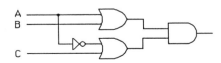

FIGURE 3.8 Implementation of the Boolean expression $F = (A + B)(\overline{A} + C)$ using 2-input NOR gates and inverters.

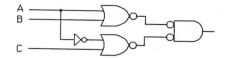

3.5 DEVELOPING THE BOOLEAN EXPRESSION

So far we have seen how to reduce complex Boolean expressions and to manipulate expressions once they have been reduced. Before either of these two steps can be taken, however, it is first necessary to have a Boolean expression to work with. In this section, we will investigate two methods for doing this, the sum-of-products method and the product-of-sums method.

The Sum-of-Products Method

It turns out that any Boolean function can always be written in either of two standard forms. These forms are the *sum-of-products form* and the *product-of-sums form*. We will first concentrate on the sum-of-products form. In this form, the Boolean function of N variables is manipulated so that it consists of a series of products. Each product must contain all N variables, either directly or as an inverse. Let us proceed by example.

■ EXAMPLE 3.5–1

Consider the truth table shown in Figure 3.9. We assume that this truth table has been arrived at by a designer and that it accomplishes some task. Now that we have the truth table, we can use it to obtain the Boolean function needed to implement the desired task. The truth table indicates that there are three input variables. Thus the sum-of-products form will consist of products of three terms each. For three input variables, there are eight unique products of the form ABC, $\overline{A}BC$, $AB\overline{C}$ and so on. Which of these terms are to be included in the sum-of-products form?

We examine the truth table and identify all the entries under F that are 1's. Each such entry must be represented in the sum-of-products form. That is, F is

A	B	C	F	
0	0	0	0	
0	0	1	1	← $\overline{A}\overline{B}C$
0	1	0	1	← $\overline{A}B\overline{C}$
0	1	1	0	
1	0	0	0	
1	0	1	0	
1	1	0	1	← $AB\overline{C}$
1	1	1	0	

FIGURE 3.9 Truth table used to develop the sum-of-products Boolean expression $F = \overline{A}\overline{B}C + \overline{A}B\overline{C} + AB\overline{C}$.

clearly 1 when $A = 0$, $B = 0$, and $C = 1$ (that is, when $\overline{A}\overline{B}C$ is 1). F is also 1 when $\overline{A}B\overline{C}$ is 1, and when $AB\overline{C}$ is 1. Thus F can be written as

$$F = \overline{A}\overline{B}C + \overline{A}B\overline{C} + AB\overline{C}$$

This is the sum-of-products form for the Boolean expression that represents the truth table in Figure 3.9. It has been very easy to arrive at, but it is not usually in a minimum form. In this case the expression could be reduced slightly to

$$F = \overline{A}\overline{B}C + \overline{A}B\overline{C} + AB\overline{C}$$
$$= \overline{A}\overline{B}C + (\overline{A} + A)B\overline{C} = \overline{A}\overline{B}C + B\overline{C} \qquad \blacksquare$$

■ EXAMPLE 3.5–2

Consider the truth table shown in Figure 3.10. To obtain the sum-of-products form, we simply identify all the entries under F that are 1 and then include the corresponding combination of A, B, and C for each entry. Thus

$$F = \overline{A}\overline{B}\overline{C} + \overline{A}\overline{B}C + \overline{A}B\overline{C} + \overline{A}BC + AB\overline{C} + ABC$$

A	B	C	F	
0	0	0	1	← $\overline{A}\overline{B}\overline{C}$
0	0	1	1	← $\overline{A}\overline{B}C$
0	1	0	1	← $\overline{A}B\overline{C}$
0	1	1	1	← $\overline{A}BC$
1	0	0	0	
1	0	1	0	
1	1	0	1	← $AB\overline{C}$
1	1	1	1	← ABC

FIGURE 3.10 Truth table used to develop the sum-of-products Boolean expression $F = \overline{A}\overline{B}\overline{C} + \overline{A}\overline{B}C + \overline{A}B\overline{C} + \overline{A}BC + AB\overline{C} + ABC$.

This can be greatly reduced by the techniques of Section 3.3. Thus

$$F = \overline{A}\,\overline{B}(\overline{C} + C) + \overline{A}B(\overline{C} + C) + AB(\overline{C} + C)$$
$$= \overline{A}\,\overline{B} + \overline{A}B + AB$$
$$= \overline{A}(\overline{B} + B) + AB$$
$$= \overline{A} + AB = \overline{A} + B$$

Obviously, the sum-of-products form is by no means minimum in this case. ■

The Product-of-Sums Form

The product-of-sums form amounts to writing the Boolean expression as the product of a series of sums. If the Boolean function contains N variables, then each sum must contain all N variables either directly or inverted. For a three-variable Boolean function, the product-of-sums form would consist of the product of a series of three-term sums such as $(A + \overline{B} + C)(\overline{A} + B + \overline{C})$. . . . The easiest way to arrive at the product-of-sums form from a truth table is to look for entries where the output function is 0.

■ EXAMPLE 3.5–3

If we look again at Figure 3.10, we see that there are two entries where F is 0. These are $A = 1$, $B = 0$, $C = 0$ and $A = 1$, $B = 0$, $C = 1$. That is, *for F to be low (0)*, we must have either $A\overline{B}\,\overline{C} = 1$ or $A\overline{B}C = 1$. This can be written as

$$\overline{F} = A\overline{B}\,\overline{C} + A\overline{B}C$$

Notice the inversion on F. Strictly speaking, this means that NOT F is high when $A\overline{B}\,\overline{C}$ is high OR $A\overline{B}C$ is high. But NOT F high implies that F is low. Thus the preceding expression says that F is low when either $A\overline{B}\,\overline{C}$ is high OR $A\overline{B}C$ is high. Let us invert our expression and then apply De Morgan's theorem.

$$\overline{F} = A\overline{B}\,\overline{C} + A\overline{B}C$$
$$F = \overline{(A\overline{B}\,\overline{C} + A\overline{B}C)}$$
$$F = (\overline{A\overline{B}\,\overline{C}})(\overline{A\overline{B}C})$$
$$F = (\overline{A} + B + C)(\overline{A} + B + \overline{C})$$

The last result is in the product-of-sums form. Examining this result more closely suggests a quicker way to arrive at the product-of-sums form. The first term in the product of sums, $(\overline{A} + B + C)$, came from the first 0 term in the truth table. This term was $A\overline{B}\overline{C}$. Notice that the state of each variable in $(\overline{A} + B + C)$ is *precisely the inverse* of the each variable in $A\overline{B}\overline{C}$. Thus, to create a product-of-sums form, we first identify the entries where F is 0. We then examine the states of A, B, and C for each of these entries and invert these states. We then form the product-of-sums expression. Another example will be useful. ∎

■ EXAMPLE 3.5–4

Let us examine the truth table shown in Figure 3.11 and use it to arrive at a product-of-sums expression. First, we identify the entries where F is 0. These are $\overline{A}BC$, $A\overline{B}\overline{C}$, and $AB\overline{C}$. Then we invert the states of A, B, and C for each entry and form a sum. Thus

$$\overline{A}BC \rightarrow A + \overline{B} + \overline{C}$$

$$A\overline{B}\overline{C} \rightarrow \overline{A} + B + C$$

$$AB\overline{C} \rightarrow \overline{A} + \overline{B} + C$$

The product-of-sums expression is just the product of these three terms. Thus

$$F = (A + \overline{B} + \overline{C})(\overline{A} + B + C)(\overline{A} + \overline{B} + C)$$

This expression is entirely equivalent to the sum-of-products form obtained from the same truth table. The sum-of-products form in this case would be

$$F = \overline{A}\overline{B}\overline{C} + \overline{A}\overline{B}C + \overline{A}B\overline{C} + A\overline{B}C + ABC$$

FIGURE 3.11 Truth table used to develop the product-of-sums Boolean expression $F = (A + \overline{B} + \overline{C})(\overline{A} + B + C)(\overline{A} + \overline{B} + C)$.

A	B	C	F		
0	0	0	1		
0	0	1	1		
0	1	0	1		
0	1	1	0	$\leftarrow \overline{A}BC \rightarrow$	$(A + \overline{B} + \overline{C})$
1	0	0	0	$\leftarrow A\overline{B}\overline{C} \rightarrow$	$(\overline{A} + B + C)$
1	0	1	1		
1	1	0	0	$\leftarrow AB\overline{C} \rightarrow$	$(\overline{A} + \overline{B} + C)$
1	1	1	1		

Neither of the two preceding expressions is in minimum form. Both could be reduced by the techniques developed in Section 3.3. Such a reduction would yield

$$F = \overline{A}\overline{B} + \overline{A}\,\overline{C} + AC$$

∎

Some Final Comments on Standard Forms

We have seen how a truth table can be used to generate a Boolean expression in either a sum-of-products or product-of-sums form. Since either form can be used, which is the method of choice? The answer depends on which approach yields results more quickly.

Whichever approach is used, the expression obtained is unlikely to be in minimum form. Therefore, it will need to be reduced. If the number of inputs is large, reducing a product-of-sums form can be quite cumbersome. The successive sums must be "multiplied" out to begin the reduction. This can be a messy process unless there are a small number of sums in the original product. Thus it is usually simpler to use the sum-of-products form unless the truth table contains only a small number of zeros in the function column. Of the three truth tables shown in Figures 3.9, 3.10, and 3.11, only the one in Figure 3.10 would be a certain candidate for the product-of-sums form. To illustrate the work involved in the two approaches, let us return to the expressions we developed previously for the truth table in Figure 3.11. We begin with the product-of-sums form.

∎ **EXAMPLE 3.5–5**

$$F = (A + \overline{B} + \overline{C})(\overline{A} + B + C)(\overline{A} + \overline{B} + C)$$
$$= (A + \overline{B} + \overline{C})(\overline{A}\overline{A} + \overline{A}B + \overline{A}C + B\overline{A} + B\overline{B}$$
$$+ BC + C\overline{A} + C\overline{B} + CC)$$
$$= (A + \overline{B} + \overline{C})(\overline{A} + C)$$

$$F = A\overline{A} + AC + \overline{B}\overline{A} + \overline{B}C + \overline{C}\overline{A} + \overline{C}C$$
$$= AC + \overline{B}\overline{A} + \overline{B}C + \overline{C}\overline{A}$$

It is difficult to be certain if this last result is in minimum form or not. Actually it isn't. We can write

$$F = AC + \overline{B}\overline{A} + \overline{B}C + \overline{A}\overline{C}$$
$$= AC + \overline{B}(\overline{A} + C) + \overline{A}\overline{C}$$
$$= AC + \overline{B}(\overline{A} + AC) + \overline{A}\overline{C}$$
$$= AC + \overline{B}\overline{A} + \overline{B}AC + \overline{A}\overline{C}$$
$$= (1 + \overline{B})AC + \overline{B}\overline{A} + \overline{A}\overline{C}$$
$$= AC + \overline{B}\overline{A} + \overline{A}\overline{C}$$

This result is now in minimum form. Let us do this example again, beginning with the sum-of-products form. ■

■ **EXAMPLE 3.5–6**

$$F = \overline{A}\,\overline{B}\,\overline{C} + \overline{A}BC + \overline{A}B\overline{C} + A\overline{B}C + ABC$$
$$= \overline{A}B(\overline{C} + C) + \overline{A}B\overline{C} + AC(\overline{B} + B)$$
$$= \overline{A}B + \overline{A}B\overline{C} + AC$$
$$= A(B + B\overline{C}) + AC$$
$$= \overline{A}(\overline{B} + \overline{C}) + AC$$
$$= \overline{A}B + \overline{A}\overline{C} + AC$$

This is the same result as we obtained previously. It is a matter of personal choice as to which of the two reductions seems simpler. ■

PROBLEMS

1. Verify the distributive theorem (T–5) for Boolean algebra by using a truth table.
2. Verify the absorption theorems (T–6) and (T–7) by truth tables.
3. Verify theorem (T–8) by truth table.
4. Reduce the following expressions:
   a. $A((\overline{\overline{A} + B})C) + BC$
   b. $ABC + A + \overline{A}C + AB + AC$
   c. $(\overline{A} + \overline{B})A\overline{B}$

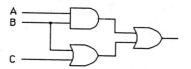

FIGURE 3.12   Use with Problems 7 and 8.

d. $\overline{\overline{AB} + \overline{\overline{C}}}$

e. $A\overline{B}C + \overline{A}B + \overline{B}C + \overline{B}\overline{C}$

f. $\overline{A\overline{B}C} + \overline{A} + B\overline{C}$

5. Using Boolean theorems and identities, change the following expressions into a form that uses only the 2-input NAND function and inversion. Then draw the appropriate circuit for each of your results.

   a. $A + BC$

   b. $A + B + C$

   c. $A\overline{B} + B\overline{C}$

   d. $\overline{A + \overline{\overline{B}} + AC}$

6. Repeat Problem 5 using only the 2-input NOR function and inversion.

7. By introducing pairs of inverters in appropriate locations, redraw the circuit in Figure 3.12 using only 2-input NAND gates and inverters.

8. By introducing pairs of inverters in appropriate locations, redraw the circuit in Figure 3.12 using only 2-input NOR gates and inverters.

9. Repeat Problem 7 using the circuit in Figure 3.13.

10. Repeat Problem 8 using the circuit in Figure 3.13.

11. For the truth table shown in Figure 3.14, derive the sum-of-products and the product-of-sums Boolean expressions. Show that both expressions reduce to

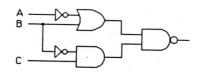

FIGURE 3.13   Use with Problems 9 and 10.

A	B	C	F
0	0	0	1
0	0	1	1
0	1	0	0
0	1	1	1
1	0	0	1
1	0	1	1
1	1	0	0
1	1	1	1

FIGURE 3.14 Use with Problem 11.

the same simplified form. Implement the reduced expression in terms of 2-input NAND gates and inverters and then in terms of 2-input NOR gates and inverters.

12. Imagine a small town council consisting of three council members. Each member of this council votes on various issues by toggling a small switch at their seat to the yes or no position. Assume the yes position produces a logical 1, and the no position produces a logical 0. A circuit must be designed that produces a logical 1 output any time two or more of the council members vote yes.

   a. Using A, B, and C as the states of each switch, complete a truth table that shows all the possible switch settings and the desired output for each setting.

   b. From the truth table in part a, determine the sum-of-products Boolean expression for the "majority-detector" function.

   c. Reduce the function using Boolean algebra, and then implement the function directly, using gates with whatever number of inputs is convenient.

13. An odd-parity detector is a circuit that examines the state of all the inputs to the circuit and produces a logical 1 output if the number of logical 1 inputs is odd and a logical 0 output otherwise. Consider a 3-input odd-parity detector with inputs labeled A, B, and C.

   a. Develop the truth table for the odd-parity detector.

   b. From the truth table, develop the sum-of-products Boolean expression for the 3-input odd-parity detector.

   c. You will find that the Boolean expression from part b cannot be reduced any further. Implement this expression directly using gates with whatever number of inputs is convenient. Later you will see that this function can be implemented much more directly with XOR gates.

14. An equality detector is a circuit that compares two multidigit binary numbers and produces a high output only if the two numbers are equal. For example, consider the two-digit number AB and the second two-digit number CD (where A, B, C, and D are 0's and 1's).

a. Develop the truth table for the equality detector that compares the number *AB* with the number *CD*.

b. From the truth table, develop the sum-of-products Boolean expression for the equality detector.

c. You will find that the Boolean expression from part b cannot be reduced any further. Implement this expression directly using gates with whatever number of inputs is convenient. Later you will see that this function can be implemented much more directly using XNOR gates.

Chapter 4

COMBINATIONAL LOGIC CIRCUITS

Learning Objectives

After completing this chapter you should know:

- What a decoder is and how it functions.

- How to construct a decoder from simple logic gates.

- Some examples of MSI decoders and how they can be used.

- How to combine MSI decoders to fabricate larger-scale decoders.

- How to use decoders to implement arbitrary Boolean functions.

- What an encoder is and how it functions.

- What is meant by code conversion and how it can be accomplished by using simple logic gates.

- How decoders can be used to achieve code conversion.

- Some examples of MSI decoder circuits.

- What a 7-segment LED is and how it can be controlled using a MSI decoder circuit.

- How to construct a display from a series of LEDs.

- What a multiplexer is and how it functions.

- What a demultiplexer is and how it functions.

- What parity is and how XOR gates can be used to generate and detect parity.

- How XOR gates can be used to construct adder circuits and how these circuits function.

- How XNOR gates can be used to construct magnitude-comparer circuits.

- How XOR or XNOR gates can be used to construct code-comparer circuits.

4.1 INTRODUCTION

In the first three chapters, we developed the tools necessary to understand most digital circuitry. We will now put those tools to work and apply them to several common types of digital circuits.

There are basically two general classes of digital logic circuits. One is called *combinational* logic and the other is called *sequential* logic. In a combinational logic circuit, there may be several inputs and perhaps several outputs. In any case, the state of the output(s) is uniquely defined by the state of the inputs. That is, the prior state of the inputs plays no role at all in determining the present output. Typical combinational logic circuits include decoders, encoders, multiplexers, de-multiplexers, adders, parity detectors, and equality detectors, to name just a few.

In a sequential logic circuit, the output generally cycles through a series of states as a series of inputs is received. The state of the output at any given time will depend on the input history. That is, the present output depends not only on the present input, but also on inputs received in the past. Flip-flops form the basis of sequential circuits and may be combined to form counters of various kinds, shift registers, and parallel-to-serial or serial-to-parallel converters.

In this chapter, we will focus our attention on combinational logic circuits. We will consider a series of circuit types, such as decoders, encoders and multiplexers. Although all these kinds of circuits can be constructed from SSI gates, our main concern will be with MSI single-chip implementations. As we proceed, it will become apparent that modern MSI circuits offer greatly enhanced circuit flexibility with far fewer parts.

4.2 DECODERS

A decoder is difficult to define in a precise way since, in a sense, most combinational circuits are decoders. Usually, when we speak of a decoder, we mean a circuit with several inputs. The circuit then "examines" these inputs and responds uniquely to one particular input combination. That is, the circuit looks for a particular *code* and responds to it.

Most decoders actually have several outputs rather than just one. Each output responds to a unique input code. We can illustrate these ideas best by example.

■ EXAMPLE 4.2—1

Figure 4.1 shows a 4-input AND gate used as a decoder. The circuit will respond with a high output only when the input is $AB\overline{C}D$. That is, it decodes for the state $AB\overline{C}D$. In this case, there are four inputs and only one output. We could develop a decoder with several outputs. ■

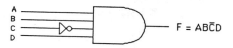

FIGURE 4.1 A 4-input AND gate used as a decoder for the state $AB\bar{C}D$.

$F = AB\bar{C}D$

■ **EXAMPLE 4.2–2**

Figure 4.2 shows four 2-input AND gates used as a *2-line to 1-of-4 decoder*, or simply a *1-of-4 decoder*. There are two inputs to the circuit, labeled *A* and *B*. With two inputs, there are four possible input combinations. Notice that the circuit is arranged so that each AND gate will respond with a high output to a different input combination. If we think of the inputs as binary *numbers* (with *A* the most significant bit and *B* the least significant bit), then the four possible inputs become 00, 01, 10, and 11. These correspond to the decimal numbers 0, 1, 2, and 3. Thus each output can be thought of as decoding for a unique numerical input or code. The type of decoder shown in Figure 4.2 is called a binary decoder because it has a unique response to every possible binary input combination. ■

MSI Binary Decoders

It takes little imagination to visualize how the principles used in constructing the 1-of-4 decoder in Figure 4.2 can be expanded to a 1-of-8 decoder. We simply have three inputs and eight 3-input AND gates. We could also construct a 1-of-16 decoder using sixteen 4-input AND gates. Obviously, both of these decoders require an increasing number of SSI gates. Fortunately, many decoders are available as MSI ICs. We will review two of these in some detail.

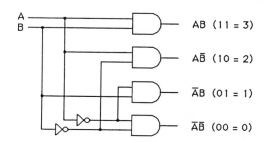

FIGURE 4.2 An SSI implementation of a 1-of-4 decoder circuit.

AB (11 = 3)

$A\bar{B}$ (10 = 2)

$\bar{A}B$ (01 = 1)

$\bar{A}\bar{B}$ (00 = 0)

The 74155 Dual 1-of-4 Decoder

The 74155 TTL IC contains two 1-of-4 decoders in a single IC package. The IC itself is in a 16-pin, dual in-line package (DIP), such as that shown in Figure 2.22. Manufacturers of the *chip* (as these packages are referred to) provide detailed specifications on all aspects of the circuit's function and electrical characteristics. For most purposes, however, the electrical characteristics are just those of the TTL family of which the chip is a member (that is, 74XX, 74LSXX, and so on). It then remains to understand the actual function of the circuit.

To understand how a particular IC is to be used, we need to know its logical function and the exact pin locations of the various circuit inputs and outputs. Figure 4.3 illustrates both of these features for the 74155 (or 74LS155, 74S155, and so on). Figure 4.3(a) shows the basic logic diagram for the circuit. Notice that there are two input lines, *A* and *B*, that carry the code to be decoded. Both of these lines feed each decoder, so the two decoders are not really completely independent. Each decoder, *a* and *b*, consists of four 3-input NAND gates, one for each possible input combination on the two code lines. The third input on each NAND gate comes from an *enable* AND gate. There is a separate enable AND gate for each decoder. Each enable AND gate has two inputs (one labeled *strobe* and the other labeled *data*), *both of which must be active* if the AND gate's output is to go high. Until the output of the enable AND gate goes high, all outputs will be held in the high state.

If we examine the enable AND gate for the *a* decoder, we find that it has an active-high input and an active-low input. Both of these inputs must be in their active state in order for decoder *a* to be enabled. Once decoder *a* is enabled, one of its four outputs will go low, depending on the code present on the two input lines *A* and *B*. Notice that decoder *b* is enabled by an entirely different AND gate, which has two active-low inputs. We can enable either decoder *a* or *b*, or both at one time.

Since the AND gate that enables each decoder has two inputs, this gives the circuit designer the option of requiring two conditions to be met in order for a particular decoder to be enabled. Often it is convenient to have a particular decoder permanently enabled. In such a case, it is customary to simply connect the enable inputs to +5 volts or ground (0 volts), as appropriate.

If you examine Figure 4.3(a) closely, you will notice that the two code inputs are each connected to two inverters in series. That is, neither code input is directly connected to any of the 3-input NAND gates. This is done for buffering purposes. In effect, only one TTL load is placed on each input line. This must be done since each input line to a TTL IC should constitute only one TTL load.

Although Figure 4.3(a) illustrates the function of the 74155, it does not identify the actual pins on the IC package. These numbers could have been added to the figure, perhaps in parentheses next to each input and output. It is more usual, however, to show the *pin-outs* in the manner of Figure 4.3(b). This figure has two sets of labels on it. Inside the package are functional labels. Outside the package,

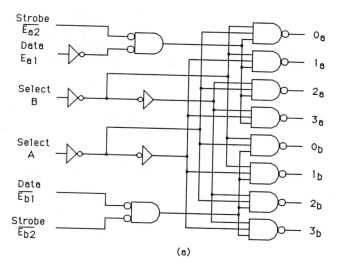

FIGURE 4.3 (a) Logic diagram, (b) connection diagram, and (c) truth table for the 74155 dual 1-of-4 decoder.

ADDRESS		ENABLE a		OUTPUT a				ENABLE b		OUTPUT b			
A	B	E_{a1}	$\overline{E_{a2}}$	$\bar{0}$	$\bar{1}$	$\bar{2}$	$\bar{3}$	$\overline{E_{b1}}$	$\overline{E_{b2}}$	$\bar{0}$	$\bar{1}$	$\bar{2}$	$\bar{3}$
X	X	L	X	H	H	H	H	H	X	H	H	H	H
X	X	X	H	H	H	H	H	X	H	H	H	H	H
L	L	H	L	L	H	H	H	L	L	L	H	H	H
H	L	H	L	H	L	H	H	L	L	H	L	H	H
L	H	H	L	H	H	L	H	L	L	H	H	L	H
H	H	H	L	H	H	H	L	L	L	H	H	H	L

next to the lines leading in or out of the package, are the pin numbers associated with each function. Some of the input lines and all the output lines have small circles where they enter or leave the package. These circles indicate that these lines are active low. Absence of such a circle indicates that a line is active high.

Although Figure 4.3(a) clearly identifies the function of the 74155, another more common way of doing this is by means of a truth table. Such a table is shown in Figure 4.3(c). This table shows all the possible input combinations, including the enable inputs, and the corresponding outputs that result. In the table the address lines are shown only once since they are common to both decoders. The enable inputs and decoder outputs are shown separately for each decoder. Notice that, in the table headings, active-low inputs such as $\overline{E}_{a2}$ and active-low outputs (in this case all the outputs) are shown with a bar over their label. This indicates that these inputs and outputs are indeed active low.

Reading the truth table in Figure 4.3(c) is relatively straightforward. The top two lines of the table show that if the enable inputs are not in their active state then all the outputs will be high. The next four lines of the truth table show which outputs are active (in this case low) for each input address or code.

An IC such as the 74155 can be used for a number of things. The fact that the two input lines, A and B, are referred to as address lines gives us a clue to one common use. In microcomputer interfacing, address lines are used to select different memory locations to be accessed or different input/output ports to be activated. An IC like the 74155 could be connected to the lowest two address lines of a microcomputer to select between one of four I/O ports to be activated. That is, the state of the low two address lines (00, 01, 10, or 11) would determine which output of the 74155 was active, and thus which I/O port was turned on.

■ EXAMPLE 4.2–3

A 74155 dual 1-of-4 decoder has the following inputs: A, B, E_{a1}, and $\overline{E}_{a2}$, all high, and $\overline{E}_{b1}$ and $\overline{E}_{b2}$ low. What are the states of the outputs?

We can attack this problem either by examining the logic diagram for the IC [Figure 4.3(a)] or the truth table [Figure 4.3(c)]. Perhaps the easiest approach is to use the truth table. First, we look at the state of the enable inputs for decoder a. We see that they are both high. The truth table indicates that when $\overline{E}_{a2}$ is high then all a outputs are high, *regardless of the state of* $\overline{E}_{a1}$.

Next we look at the state of the enable inputs for the b decoder. We see that they are both low. The truth table indicates that when both $\overline{E}_{b1}$ and $\overline{E}_{b2}$ are low the b decoder is enabled, and the output of the b decoder is determined by the select inputs A and B. In particular, with A and B both high, the truth table indicates that output 3_b is low and all other b outputs are high. ■

■ EXAMPLE 4.2–4

The two enable lines for each decoder on the 74155 IC are called *strobe* and *data*, respectively. To see what this might mean, consider decoder b. Imagine that its strobe input, $\overline{E}_{b2}$, is controlled by some other circuit that wishes to send data (either a logic 0 or a logic 1) to one of four possible locations. Let us suppose that these four locations are connected to the four b outputs of the 74155. Let us further imagine that the same circuit that controls the $\overline{E}_{b2}$ strobe input also controls the $\overline{E}_{b1}$ data input. How might this circuit arrange to send a logic 0 from output 2_b?

To send a logic 0 from output 2_b, decoder b must be enabled and output 2 must be selected. To select output 2_b, the select lines must be made $A = 0$ and B

= 1. To enable decoder b, and therefore allow output 2_b to go low, the strobe input $\overline{E}_{b2}$ must be made low, and so must the data input $\overline{E}_{b1}$.

What would be different about the preceding procedure if the datum to be sent from output 2_b was a logic 1 instead of a logic 0? Notice that if the data input $\overline{E}_{b1}$ were made high (logic 1) then decoder b would be disabled and output 2_b would be high (logic 1). Thus we see that, if the circuit controlling the 74155 selects output 2 by setting $A = 0$ and $B = 1$ and activates the strobe input $\overline{E}_{b2}$, then, whatever data are placed on the $\overline{E}_{b1}$ input, these data will be output from 2_b. ∎

The 74154 1-of-16 Decoder

The 74154 is a single 1-of-16 decoder in one IC package. Its function is similar to each of the 1-of-4 decoders on the 74155 chip. The logic diagram for the 74154 is shown in Figure 4.4. The connection diagram of the IC and the truth table for the circuit are shown in Figure 4.5.

Like each 1-of-4 decoder on the 74155, the 74154 consists of a series of NAND gates (in this case 16). Each NAND gate has four inputs from the four address or code lines and one input from an enable AND gate, for a total of five inputs. The enable gate has two active-low inputs. All the outputs are active low. That is, when a particular address or code is on the input lines, the appropriate output goes low, assuming the enable lines are active.

Figure 4.5(a) shows the pin-outs for the 74154. Notice that the pins on the figure are not shown in their actual physical arrangement, but rather in a more easily understood logical grouping. This is typical of many pin-out diagrams. The two enable lines are clearly identified as active low by the circles on the enable functions $\overline{1G}$ and $\overline{2G}$. All 16 outputs are also identified as active low. The supply voltage pin and the ground pin are identified separately.

Figure 4.5(b) is the truth table for the 74154. It clearly shows, by the bar over the symbols, that the enable inputs and all outputs are active low, while the address inputs are active high. Figure 4.5 contains all that is really needed to understand the function and use of the 74154. The information in Figure 4.4 is interesting, but not essential. Many "cookbooks" exist that list information of the kind shown in Figure 4.5 for a whole series of common 74XX circuits. Such a book is a must for anyone who wishes to design or troubleshoot digital circuits. More extensive books of detailed circuit specifications are available from circuit manufacturers. Companies such as Texas Instruments, RCA, National Semiconductor, and Motorola publish several useful compilations of the integrated circuits they produce. For critical design, these more extensive publications are useful because they contain very detailed electrical and timing information. Often, however, this more extensive information is not necessary for most practical work.

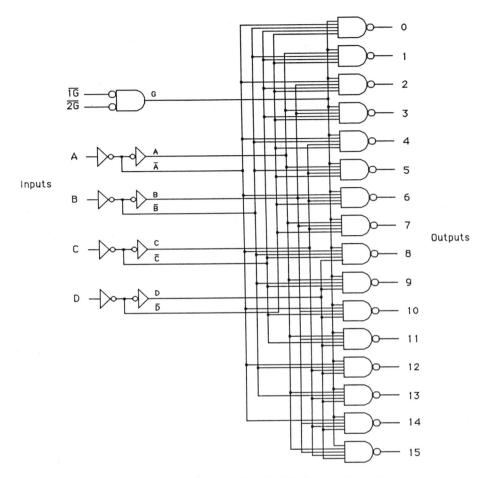

FIGURE 4.4 Logic diagram of the 74154 decoder/demultiplexer.

Other Decoders

Several other decoders are available as 74XX ICs. To mention a few, the 74139 is a dual 1-of-4 decoder that has a single enable input rather than two. The 74138 is a 1-of-8 decoder with two active-low enables and an active-high enable. And it is, of course, possible to use decoders in groups to achieve a 1-of-32 decoder or whatever is necessary.

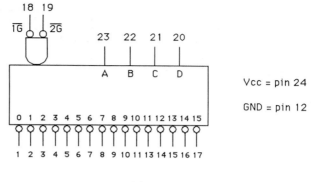

Vcc = pin 24

GND = pin 12

(a)

INPUTS						OUTPUTS															
$\overline{1G}$	$\overline{2G}$	D	C	B	A	$\overline{0}$	$\overline{1}$	$\overline{2}$	$\overline{3}$	$\overline{4}$	$\overline{5}$	$\overline{6}$	$\overline{7}$	$\overline{8}$	$\overline{9}$	$\overline{10}$	$\overline{11}$	$\overline{12}$	$\overline{13}$	$\overline{14}$	$\overline{15}$
L	H	X	X	X	X	H	H	H	H	H	H	H	H	H	H	H	H	H	H	H	H
H	L	X	X	X	X	H	H	H	H	H	H	H	H	H	H	H	H	H	H	H	H
H	H	X	X	X	X	H	H	H	H	H	H	H	H	H	H	H	H	H	H	H	H
L	L	L	L	L	L	L	H	H	H	H	H	H	H	H	H	H	H	H	H	H	H
L	L	L	L	L	H	H	L	H	H	H	H	H	H	H	H	H	H	H	H	H	H
L	L	L	L	H	L	H	H	L	H	H	H	H	H	H	H	H	H	H	H	H	H
L	L	L	L	H	H	H	H	H	L	H	H	H	H	H	H	H	H	H	H	H	H
L	L	L	H	L	L	H	H	H	H	L	H	H	H	H	H	H	H	H	H	H	H
L	L	L	H	L	H	H	H	H	H	H	L	H	H	H	H	H	H	H	H	H	H
L	L	L	H	H	L	H	H	H	H	H	H	L	H	H	H	H	H	H	H	H	H
L	L	L	H	H	H	H	H	H	H	H	H	H	L	H	H	H	H	H	H	H	H
L	L	H	L	L	L	H	H	H	H	H	H	H	H	L	H	H	H	H	H	H	H
L	L	H	L	L	H	H	H	H	H	H	H	H	H	H	L	H	H	H	H	H	H
L	L	H	L	H	L	H	H	H	H	H	H	H	H	H	H	L	H	H	H	H	H
L	L	H	L	H	H	H	H	H	H	H	H	H	H	H	H	H	L	H	H	H	H
L	L	H	H	L	L	H	H	H	H	H	H	H	H	H	H	H	H	L	H	H	H
L	L	H	H	L	H	H	H	H	H	H	H	H	H	H	H	H	H	H	L	H	H
L	L	H	H	H	L	H	H	H	H	H	H	H	H	H	H	H	H	H	H	L	H
L	L	H	H	H	H	H	H	H	H	H	H	H	H	H	H	H	H	H	H	H	L

(b)

FIGURE 4.5 (a) Connection diagram, and (b) truth table for the 74154 1-of-16 decoder/demultiplexer.

■ EXAMPLE 4.2–5

Design a 1-of-32 decoder using a 1-of-4 decoder and four 1-of-8 decoders. A 1-of-32 decoder would require five input address lines since $2^5 = 32$. If we think of the input address as $EDCBA$, from most significant to least significant bits, then lines E and D would be the select lines to the 1-of-4 decoder. These two lines would activate *one* of the four outputs from the 1-of-4 decoder. Each output from the 1-or-4 decoder would be used as an enable line for a *different* 1-of-8 decoder. The low three address lines, *CBA*, would be connected to *each* 1-of-8 decoder. However, only the *active* 1-of-8 decoder will produce an output. Figure 4.6 shows the circuit implementation using a 74139 1-of-4 decoder and four 74138 1-of-8 decoders. ∎

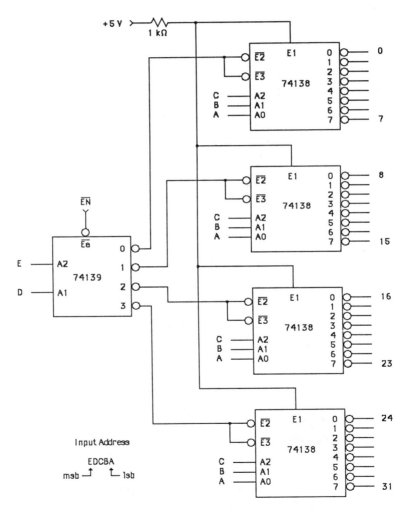

FIGURE 4.6   A 1-of-32 decoder implemented with a 1-of-4 decoder and four 1-of-8 decoders.

A 1-of-32 decoder could also be constructed by using a single input line (and its inverse) to select one of two 74154 1-of-16 decoders. The implementation of this circuit is left as an exercise.

■ **EXAMPLE 4.2–6**

Show how a 1-of-8 decoder and a series of NAND gates can be used to implement the four Boolean functions represented in Figure 4.7(a). Each of the four outputs F_1 to F_4 in the truth table represents a separate Boolean function. Using the methods of Chapter 3, we can quickly determine each Boolean function.

$$F_1 = \overline{A}B + A\overline{B}\overline{C} + ABC$$

$$F_2 = \overline{A}\overline{B} + \overline{B}\overline{C} + ABC$$

$$F_3 = AB + \overline{A}\overline{C}$$

$$F_4 = \overline{A}\overline{C} + \overline{B}\overline{C} + ABC$$

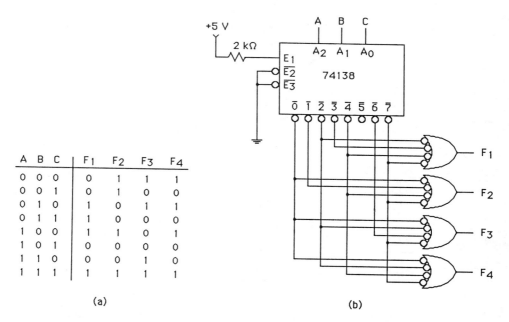

A	B	C	F₁	F₂	F₃	F₄
0	0	0	0	1	1	1
0	0	1	0	1	0	0
0	1	0	1	0	1	1
0	1	1	1	0	0	0
1	0	0	1	1	0	1
1	0	1	0	0	0	0
1	1	0	0	0	1	0
1	1	1	1	1	1	1

(a)

(b)

FIGURE 4.7 (a) Truth table, and (b) decoder implementation.

If we implement each of these functions with standard gates, we see that there are actually seven distinct AND terms (two occur twice and one three times), and each expression also requires an OR gate. Thus we would need 11 separate gates to implement the four functions. On the other hand, if we use a 1-of-8 decoder, we need at most four other gates (one for each function) to complete the circuit. The circuit is shown in Figure 4.7(b). The implementation is quite simple. Notice from the table that each function, F_1 to F_4, is high for certain combinations of A, B, and C. But the inputs ABC select active-*low* outputs. Thus we need simply use a low-level input OR gate (that is, a NAND gate) to OR the different active-low outputs for which a given function is high. Examine the circuit in Figure 4.7(b) closely to be sure that you understand how it works. ■

■ EXAMPLE 4.2–7

Show how a 1-of-8 decoder and three 4-input NAND gates can be used to convert the standard binary code 000 to 111 into any other code sequence. Let us look at the truth table in Figure 4.7(a) again, and imagine that the F_4 column is not there. In this case, the first three output columns can be thought of as representing a 3-bit code, which is simply a rearrangement of the input code ABC. The three outputs F_1 to F_3 constitute a code converter. That is, if we consider the input code ABC and the output code F_1, F_2, F_3, then when the input code is 000 the output code is 011, and so on.

It is clear from this line of thought that a 1-of-16 decoder could be used in this way to convert one 4-bit code into any other 4-bit code. We will have more to say about this later in a special section on codes and code conversion. ■

4.3 ENCODERS

In Section 4.4, we saw that the purpose of an decoder is to examine an input code or address and respond by activating one particular output. An encoder essentially accomplishes the reverse of this process. For example, an encoder might have eight inputs numbered 0 to 7 and three outputs labeled ABC. Assuming only one input can be active at any given time, the basic function of the encoder is to detect the active input and then generate a binary number as output that gives the number (0 to 7) of the active input. Thus, if input 5 is active, then the output would be 101 (the binary code for 5).

As a practical matter, the circuit just described would not be very useful. For one thing, it would be hard to guarantee that only one input was active at a time. In addition, a way is needed to signal that no inputs at all are active. These problems are overcome by the 74148 8-input priority encoder. Figure 4.8(a) shows the equivalent circuit for this unit. There are 8 inputs labeled $\overline{0}$ to $\overline{7}$ that are all active-low.

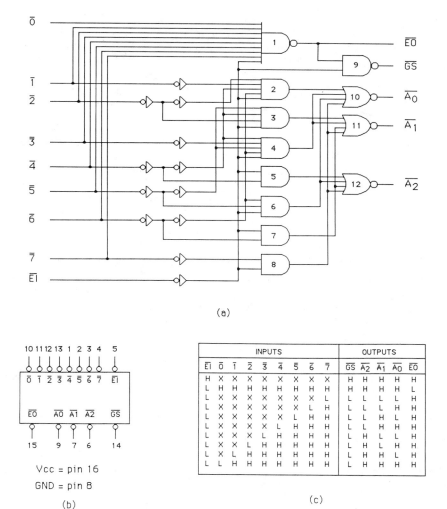

FIGURE 4.8 (a) Logic diagram, (b) connection diagram, and (c) truth table for the 74148 8-input priority encoder.

There is also an active-low enable input labeled $\overline{EI}$. There are five outputs. Three of these outputs $\overline{A}_0$ to $\overline{A}_2$, are used to output the code of the *highest* active input. This code is in negative logic form. That is, 0's are to be read as 1's, and vice versa. The other two outputs are $\overline{EO}$ and $\overline{GS}$. These outputs are used to signal various possible input conditions.

Let us examine the function of the 74148 in some detail. The truth table in Figure 4.8(c) should be consulted along with the equivalent circuit in Figure 4.8(a).

There are essentially three possible kinds of input conditions. The first is that the chip is not enabled (that is, $\overline{EI}$ is high). This condition causes both $\overline{EO}$ and $\overline{GS}$ to be high. We can see this either from the truth table or by looking at where $\overline{EI}$ is directed on the equivalent circuit.

The second type of input condition is for the chip to be enabled, but with no input, $\overline{0}$ to $\overline{7}$, active. This condition, and only this condition, causes $\overline{GS}$ to be high and $\overline{EO}$ to be low.

The third type of input condition is for the chip to be enabled and to have at least one input active. Under these conditions, $\overline{GS}$ will be low and $\overline{EO}$ will be high. In addition, the code of the *highest-numbered active input* will be output in negative logic form on $\overline{A}_0$ to $\overline{A}_2$. This last aspect of the circuit operation is what gives it the name priority encoder. Input $\overline{7}$ has the highest priority, and input $\overline{0}$ has the lowest priority. It is obvious from the truth table that this is how the circuit functions. It is not so obvious from the circuit diagram. You may wish to test your skill by verifying the truth table from the circuit diagram.

The main purpose of a priority encoder is to identify devices that need service. With the 74148, there can be up to eight devices connected to inputs $\overline{0}$ to $\overline{7}$. These devices might be thermostats, pressure switches, printers, or almost anything. When a device needs attention, it signals by producing a low output that becomes a low input to the 74148. The encoder then produces two important signals, assuming it is enabled. It takes $\overline{GS}$ low and $\overline{EO}$ high to indicate that at least one device requires service. It also outputs the negative logic binary code of the highest-priority device requiring attention. These outputs are most frequently sent to a computer, which first detects that some device needs service and then reads the code that identifies the device. The computer can then take appropriate action.

4.4 CODE CONVERSION

Because digital circuits have only two states, we have seen that standard binary arithmetic is a natural way to represent numbers in digital circuits. On the other hand, our society has developed a dependence on decimal arithmetic. In response to this situation, a number of binary codes have been developed to explicitly represent decimal digits. At the end of Chapter 1, three codes of this type were briefly introduced without much elaboration. Figure 4.9 lists four codes of this kind, which we will now examine in more detail.

The most common way of representing decimal digits is the binary-coded decimal (BCD) code. This code is simply the standard binary code for digits 0 to 9. To represent a decimal number using this code, *each digit* is converted to its binary-coded equivalent. Thus the number 9472 becomes

$$9 \to 1001 \qquad 4 \to 0100 \qquad 7 \to 0111 \qquad 2 \to 0010$$

$$9472 \to 1001\ 0100\ 0111\ 0010 \qquad \text{in BCD}$$

Decimal	Binary Coded Dec. 8421 BCD	2421	Excess 3 code (XS3)	XS3 Gray Code
0	0000	0000	0011	0010
1	0001	0001	0100	0110
2	0010	0010	0101	0111
3	0011	0011	0110	0101
4	0100	0100	0111	0100
5	0101	1011	1000	1100
6	0110	1100	1001	1101
7	0111	1101	1010	1111
8	1000	1110	1011	1110
9	1001	1111	1100	1010

FIGURE 4.9   Some common binary codes.

The same number in straight binary would be

$$9472 \rightarrow 10010100000000 \qquad \text{in straight binary}$$

Notice that the BCD method of representing 9472 requires 16 digits, while the straight binary representation requires only 14 digits. This highlights one weakness of any code that uses four binary digits to represent each decimal digit. The code is inefficient. On the other hand, it is much easier to convert a decimal number into a code such as BCD than it is to generate the equivalent straight binary code. This is a major advantage.

■ EXAMPLE 4.4–1

Convert the decimal numbers 529 and 1108 into BCD code. First, we convert each digit into its 4-bit binary equivalent. Thus

$$5 \rightarrow 0101 \qquad 2 \rightarrow 0010 \qquad 9 \rightarrow 1001$$

Therefore,

$$529 \rightarrow 0101\ 0010\ 1001 \qquad \text{in BCD}$$

$$1 \rightarrow 0001 \qquad 1 \rightarrow 0001 \qquad 0 \rightarrow 0000 \qquad 8 \rightarrow 1000$$

Therefore,

$$1108 \rightarrow 0001\ 0001\ 0000\ 1000 \qquad \text{in BCD} \qquad ■$$

Both straight binary and BCD are weighted codes. That is, each bit position within the code is worth a certain value or weight. Thus, given the binary code, it is only necessary to add the weights of the 1's in the code to determine the decimal digit being coded. The 2421 code shown in Figure 4.9 is also a weighted code, with weights 2, 4, 2, 1, respectivley. The excess 3 code and the XS3 Gray code in Figure 4.9 are not weighted codes.

■ EXAMPLE 4.4–2

Since the weights of the 2421 code are 2, 4, 2, 1, respectively, what is the decimal equivalent of the following 2421 binary numbers; 1111, 1011, and 0100?

We simply add the weights of each position where a 1 occurs. Thus

$$1111 = 2 + 4 + 2 + 1 = 9$$
$$1011 = 2 + 0 + 2 + 1 = 5$$
$$0100 = 0 + 4 + 0 + 0 = 4$$ ■

The 2421 code and the excess 3 code are self-complementing codes. The idea of a self-complementing code involves the concept of the *nine's complement* of a number. The nine's complement of a digit is just 9 minus the digit. Thus the nine's complement of 7 is 2, for example. Returning to the idea of self-complementing codes, in these codes the nine's complement of a digit is literally the complement (1's become 0's, 0's becomes 1's) of the code for the digit itself. Thus, in 2421 code, 7 is 1101, while 2 is 0010. Likewise, in XS3 code, 7 is 1010, while 2 is 0101. Self-complementing codes were developed for use in older computers where many decimal arithmetic operations were necessary. Such codes find different uses today. Notice that, while both the 2421 code and the XS3 code are self-complementing, the XS3 code is not a weighted code. The XS3 code is arrived at by adding 3 to the BCD code.

■ EXAMPLE 4.4–3

Find the nine's complement of the following digits: 4, 9, and 3. To find the nine's complement, we simply subtract the digit from 9. Thus

$$\text{Nine's complement of } 4 = 9 - 4 = 5$$
$$\text{Nine's complement of } 9 = 9 - 9 = 0$$
$$\text{Nine's complement of } 3 = 9 - 3 = 6$$ ■

■ EXAMPLE 4.4—4

For the number 6, first determine the nine's complement. Then compare the code for 6 with its nine's complement in both the 2421 code and the XS3 code to verify the self-complementing nature of the 2421 and XS3 codes.

$$\text{Nine's complement of } 6 = 9 - 6 = 3$$

$$\text{In 2421 code, } 6 = 1100 \text{ while } 3 = 0011$$

$$\text{In XS3 code, } 6 = 1001 \text{ while } 3 = 0110$$

The XS3 Gray code is neither weighted nor self-complementing, but it has another useful property. In the XS3 Gray code, as we go from one digit to the next, only 1 bit changes. This is unlike all the other codes in Figure 4.9, where at times all 4 bits change in going from one digit to the next. The problem here comes from various counting circuits that produce a binary output that is supposed to represent the number of counts received. In such a circuit, all the components that generate the count output are sometimes not perfectly synchronized. To see how this might lead to a problem, consider the following situation.

Suppose a counter simply counts in BCD. If the count is at 7, then the output will be 0111. The next count will be 8, or 1000. However, suppose the circuit that produces the most significant bit changes its state first. Then the count will, for a very short period of time, be 1111 until the last 3 bits change to 000. This is not even a valid BCD code. This kind of short intermediate state is not usually a problem, since the circuit reading the code is generally designed to take a reading only at a time when the code is known to be stable. However, there are situations where the circuit reading the code is in a constant reading mode and is fast enough to react to invalid intermediate codes. A code like the XS3 Gray code removes the difficulty. Adjacent numbers are only different by 1 bit. Thus any reading must read one of two adjacent numbers, depending on whether the count has changed or not. No invalid intermediate states are possible.   ■

Code-conversion and Decoder Circuits

Before discussing code-conversion circuits, we will return briefly to the subject of decoders. We saw in Section 4.2 that a number of 74XX ICs decode 2-, 3-, or 4-bit straight binary codes and activate the appropriate output. Thus we had 1-of-4, 1-of-8, and 1-of-16 decoders. Similar circuits exist to decode different 4-bit codes representing decimal digits.

One commonly used IC is the 7442, which accepts a 4-bit BCD input and activates (active low) 1 of 10 outputs accordingly. There are, of course, only 10 valid BCD inputs out of a possible 16 binary inputs. The six non-BCD inputs are called invalid input states. If any of these states are input into the 7442, all 10 outputs are held high. Figure 4.10(a) shows the logic diagram of the 7442. It essentially consists of ten 4-input NAND gates connected in such a way that one NAND gate becomes active for each valid BCD code. Figure 4.10(b) shows the connection diagram for the 7442, and Figure 4.10(c) is the truth table for this IC.

The 7443 is essentially the same as the 7442, except that it accepts XS3 code instead of BCD code. The logic diagram is very similar to that for the 7442, except

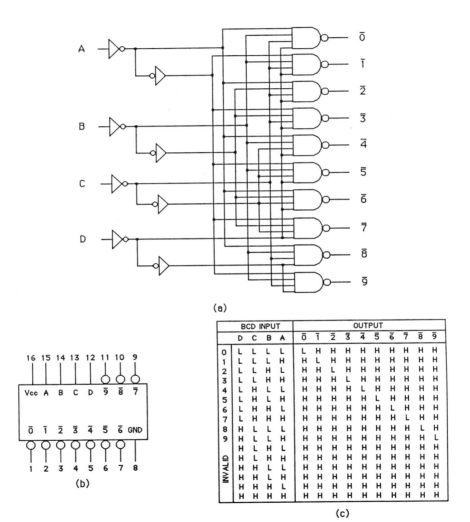

(a)

(b)

	BCD INPUT				OUTPUT									
	D	C	B	A	$\bar{0}$	$\bar{1}$	$\bar{2}$	$\bar{3}$	$\bar{4}$	$\bar{5}$	$\bar{6}$	$\bar{7}$	$\bar{8}$	$\bar{9}$
0	L	L	L	L	L	H	H	H	H	H	H	H	H	H
1	L	L	L	H	H	L	H	H	H	H	H	H	H	H
2	L	L	H	L	H	H	L	H	H	H	H	H	H	H
3	L	L	H	H	H	H	H	L	H	H	H	H	H	H
4	L	H	L	L	H	H	H	H	L	H	H	H	H	H
5	L	H	L	H	H	H	H	H	H	L	H	H	H	H
6	L	H	H	L	H	H	H	H	H	H	L	H	H	H
7	L	H	H	H	H	H	H	H	H	H	H	L	H	H
8	H	L	L	L	H	H	H	H	H	H	H	H	L	H
9	H	L	L	H	H	H	H	H	H	H	H	H	H	L
INVALID	H	L	H	L	H	H	H	H	H	H	H	H	H	H
	H	L	H	H	H	H	H	H	H	H	H	H	H	H
	H	H	L	L	H	H	H	H	H	H	H	H	H	H
	H	H	L	H	H	H	H	H	H	H	H	H	H	H
	H	H	H	L	H	H	H	H	H	H	H	H	H	H
	H	H	H	H	H	H	H	H	H	H	H	H	H	H

(c)

FIGURE 4.10   (a) Logic diagram for the 7442 BCD to decimal decoder. (b) Connection diagram for the 7442, 7443, and 7444 decoders. (c) Truth table for the 7442.

that the inputs to the 10 NAND gates are set to recognize XS3 code instead of BCD code. The connection diagram is identical to that of the 7442. The truth table has the same output states, but uses the valid XS3 states listed in Figure 4.9.

Finally, the 7444 is designed to accept XS3 Gray code. It also has a logic diagram similar to the 7442, with inputs set to recognize valid XS3 Gray code. The connection diagram is identical to the 7442, and the truth table contains the same output states as the 7442, but has valid inputs given by the XS3 Gray states listed in Figure 4.9. The 7442, 7443, and 7444 function is essentially the same way.

Code Converters

Since there are so many different codes, it is useful to be able to convert one code to another. Later we will develop a method that will permit the efficient design of code converters using SSI ICs. For now, we will examine two 74XX code converter chips.

Figure 4.11(a) gives the connection diagram for the 74184 BCD to binary converter. Figure 4.11(b) shows how this IC can be connected to convert a two-digit BCD-coded number into straight binary code. Notice that the most significant BCD digit can only contain 2 bits and is thus restricted to the numbers 0 to 3. Also notice that the least significant *bit* of the least significant BCD *digit* in Figure 4.11(b) is common to both input and output and is in fact not even connected to the 74184. The circuit in Figure 4.11(b) can convert any BCD number in the range 0 to 39 into straight binary code. Figure 4.11(c) gives a partial truth table for the 74184. We see from the table that the chip has an active-low enable. If this enable is high, all outputs will be high. Also, any set of inputs not listed on the truth table will cause all outputs to go high.

The 74185 binary to BCD converter has the same connection diagram as the 74184. However, the circuit functions in the reverse way, Figure 4.12(a) shows the truth table for the 74185. It is hard to understand this truth table until you see how the circuit is actually to be used. Figure 4.12(b) illustrates how to connect the 74185 to convert straight binary to BCD. Again, the least significant bit is common to both input and output and is not actually connected to the 74185. Once this is realized, the truth table makes sense. Outputs $Y3$ to $Y1$ actually form the three most significant bits of the least significant BCD output digit (the fourth bit of this output is coming straight from the input). Outputs $Y6$ to $Y4$ form the least significant 3 bits of the most significant BCD output digit. Figure 4.12(b) illustrates how a 6-bit binary input can be converted directly to a two-bit BCD output. Since there are six binary inputs, the numerical input range is 0 to 63. The truth table in Figure 4.12(a) shows that the outputs $Y6$ to $Y4$ form BCD numbers in the range 0 to 6 (assuming that the most significant bit of this BCD digit is 0). The 74185 also has an active-low enable input. If this enable is high, all outputs are forced high.

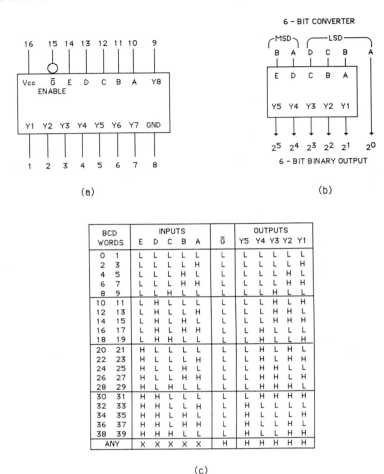

BCD WORDS		INPUTS E	D	C	B	A	$\overline{G}$	OUTPUTS Y5	Y4	Y3	Y2	Y1
0	1	L	L	L	L	L	L	L	L	L	L	L
2	3	L	L	L	L	H	L	L	L	L	L	H
4	5	L	L	L	H	L	L	L	L	L	H	L
6	7	L	L	L	H	H	L	L	L	L	H	H
8	9	L	L	H	L	L	L	L	L	H	L	L
10	11	L	H	L	L	L	L	L	L	H	L	H
12	13	L	H	L	L	H	L	L	L	H	H	L
14	15	L	H	L	H	L	L	L	L	H	H	H
16	17	L	H	L	H	H	L	L	H	L	L	L
18	19	L	H	H	L	L	L	L	H	L	L	H
20	21	H	L	L	L	L	L	L	H	L	H	L
22	23	H	L	L	L	H	L	L	H	L	H	H
24	25	H	L	L	H	L	L	L	H	H	L	L
26	27	H	L	L	H	H	L	L	H	H	L	H
28	29	H	L	H	L	L	L	L	H	H	H	L
30	31	H	H	L	L	L	L	L	H	H	H	H
32	33	H	H	L	L	H	L	H	L	L	L	L
34	35	H	H	L	H	L	L	H	L	L	L	H
36	37	H	H	L	H	H	L	H	L	L	H	L
38	39	H	H	H	L	L	L	H	L	L	H	H
ANY		X	X	X	X	X	H	H	H	H	H	H

(c)

FIGURE 4.11 (a) Connection diagram for the 74184 BCD to binary converter and 74185 binary to BCD converter. (b) Circuit using the 74184 for BCD to binary conversion. (c) Partial truth table for 74184.

Code Conversion Using Decoders

We saw in Section 4.2 that a decoder could be used along with a number of NAND gates to generate an arbitrary Boolean function. In fact, we saw an example of converting one 3-bit code into another 3-bit code. The principle applied at that time can be used to convert any 4-bit code into any other 4-bit code. This requires a 4-input decoder, such as a 1-of-16 decoder or a BCD to decimal decoder.

BINARY WORDS		INPUTS						OUTPUTS							
		E	D	C	B	A	Ḡ	Y8	Y7	Y6	Y5	Y4	Y3	Y2	Y1
0	1	L	L	L	L	L	L	H	H	L	L	L	L	L	L
2	3	L	L	L	L	H	L	H	H	L	L	L	L	L	H
4	5	L	L	L	H	L	L	H	H	L	L	L	L	H	L
6	7	L	L	L	H	H	L	H	H	L	L	L	L	H	H
8	9	L	L	H	L	L	L	H	H	L	L	L	H	L	L
10	11	L	L	H	L	H	L	H	H	L	L	H	L	L	L
12	13	L	L	H	H	L	L	H	H	L	L	H	L	L	H
14	15	L	L	H	H	H	L	H	H	L	L	H	L	H	L
16	17	L	H	L	L	L	L	H	H	L	L	H	L	H	H
18	19	L	H	L	L	H	L	H	H	L	L	H	H	L	L
20	21	L	H	L	H	L	L	H	H	L	H	L	L	L	L
22	23	L	H	L	H	H	L	H	H	L	H	L	L	L	H
24	25	L	H	H	L	L	L	H	H	L	H	L	L	H	L
26	27	L	H	H	L	H	L	H	H	L	H	L	L	H	H
28	29	L	H	H	H	L	L	H	H	L	H	L	H	L	L
30	31	L	H	H	H	H	L	H	H	L	H	H	L	L	L
32	33	H	L	L	L	L	L	H	H	L	H	H	L	L	H
34	35	H	L	L	L	H	L	H	H	L	H	H	L	H	L
36	37	H	L	L	H	L	L	H	H	L	H	H	L	H	H
38	39	H	L	L	H	H	L	H	H	L	H	H	H	L	L
40	41	H	L	H	L	L	L	H	H	H	L	L	L	L	L
42	43	H	L	H	L	H	L	H	H	H	L	L	L	L	H
44	45	H	L	H	H	L	L	H	H	H	L	L	L	H	L
46	47	H	L	H	H	H	L	H	H	H	L	L	L	H	H
48	49	H	H	L	L	L	L	H	H	H	L	L	H	L	L
50	51	H	H	L	L	H	L	H	H	H	L	H	L	L	L
52	53	H	H	L	H	L	L	H	H	H	L	H	L	L	H
54	55	H	H	L	H	H	L	H	H	H	L	H	L	H	L
56	57	H	H	H	L	L	L	H	H	H	L	H	L	H	H
58	59	H	H	H	L	H	L	H	H	H	L	H	H	L	L
60	61	H	H	H	H	L	L	H	H	H	H	L	L	L	L
62	63	H	H	H	H	H	L	H	H	H	H	L	L	L	H
ALL		X	X	X	X	X	H	H	H	H	H	H	H	H	H

(a)

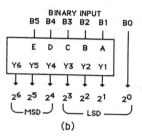

(b)

FIGURE 4.12   (a) Truth table for the 74185 binary to BCD converter. (b) Circuit to convert 6-bit binary code to 2-digit BCD code.

■ EXAMPLE 4.4–5

As a specific example, let us design a circuit that converts BCD code into XS3 code. Figure 4.13(a) shows the two codes. We will make use of the 7442 BCD to decimal decoder. Recall that this circuit (shown in Figure 4.10) has four inputs and 10 active-low outputs. When a valid BCD code is input to the circuit, the output numbered according to the BCD input goes low. The circuit in Figure 4.13(b) makes use of this fact. The four NAND gates (shown as low-level input OR gates) respond to any low input by going high. You should be able to satisfy yourself that the circuit does indeed convert BCD code to XS3 code. Invalid input combinations result in all the outputs of the 7442 going high. This causes all the NAND gate outputs to go low. Thus an invalid BCD code produces an output of 0000. It should be pointed out that the circuit in Figure 4.13(b) is not the most efficient way to produce BCD to XS3 code conversion. For example, output $F4$ could be more easily generated as simply $\overline{A}$. The point is, decoders can be used for code conversion in a fairly straightforward manner. ■

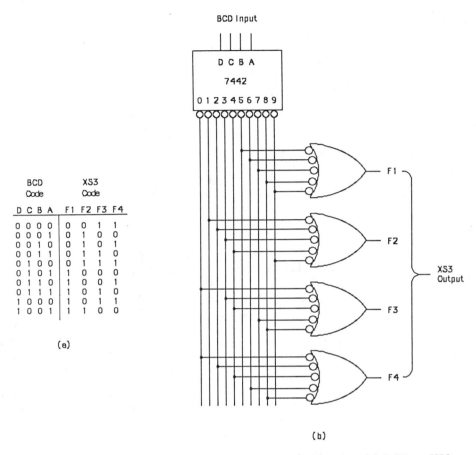

BCD Code				XS3 Code			
D	C	B	A	F1	F2	F3	F4
0	0	0	0	0	0	1	1
0	0	0	1	0	1	0	0
0	0	1	0	0	1	0	1
0	0	1	1	0	1	1	0
0	1	0	0	0	1	1	1
0	1	0	1	1	0	0	0
0	1	1	0	1	0	0	1
0	1	1	1	1	0	1	0
1	0	0	0	1	0	1	1
1	0	0	1	1	1	0	0

(a)

(b)

FIGURE 4.13 (a) Truth table for BCD and XS3 codes. (b) BCD to XS3 code converter circuit.

Seven-segment LEDs

For many digital circuits, it is ultimately necessary for a human operator to read some output. This output can take many forms, but one of the most common is decimal numerical output. A very useful device for outputing decimal digits is the 7-segment light-emitting diode (7-segment LED). The seven segments of this device are arranged as shown in Figure 4.14(a). Sometimes an eighth segment is added as a decimal point.

The seven segments illustrated in Figure 4.14(a) are in fact seven separate LEDs that can be individually controlled. Light-emitting diodes are just a special type of diode that have the property of emitting light when current flows in the forward direction. We encountered this device in Chapter 2 when we considered the optoisolator. Each segment shown in Figure 4.14(a) can be completely illuminated by causing a current to flow through the appropriate LED. From the pattern in the figure, it is clear that the digits 0 to 9 can be produced in a very readable fashion by lighting the appropriate LEDs.

There are two common ways of arranging the seven LEDs in a single IC package, the *common-anode* method and the *common-cathode* method. In the common-anode form, the anodes of all the LEDs are tied together in common, and a single supply voltage is applied to this connection. Figure 4.14(b) shows the pin connections for an IC of this type. Pin 3 is the common supply connection. Other pins control individual LEDs. Two pins are not connected to anything. On some other versions of this kind of IC, a decimal point LED may be provided. In this case, there will be eight LEDs on the chip and usually two separate voltage-supply pins, one each for four LEDs. Figure 4.14(c) shows how the individual LEDs are actually connected to each other at the supply pin. From this diagram it is clear that, to light up a given LED, the cathode of the LED must be grounded, usually through a small resistor that limits the current to a safe level.

The common-cathode version of the 7-segment LED has the cathodes of all the LEDs tied together and grounded. That is, the cathodes are all tied to a common pin that is grounded. To light a given LED on this type of circuit, a high voltage must be applied to the pin connected to the LED. Again this is done through a small resistor in order to limit the current to a safe level.

When using a 7-segment LED, the practical task is one of taking an input code, say a BCD digit, and causing the corresponding decimal digit to appear on

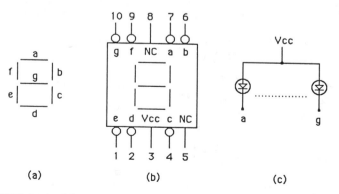

(a)   (b)   (c)

FIGURE 4.14   (a) Segment arrangement for a 7-segment LED. (b) Connection diagram and (c) internal connections for a 7-segment, common-anode LED.

the LED. This task can be thought of as either decoding or code conversion. A number of MSI circuits are available to accomplish this task. One of the most common is the 7447A BCD to 7-segment decoder driver. This circuit has open-collector, active-low outputs and is thus designed to drive a common-anode, 7-segment LED. The circuit shown in Figure 4.15 illustrates a minimum-configuration circuit for driving a common-anode, 7-segment LED with a 7447A. The voltage V may be up to 30 volts, and the size of the resistors is chosen to provide the correct LED current when the open-collector output for a given segment becomes active, effectively shorting the resistor to ground.

The 7447A is actually a fairly versatile circuit. The logic diagram for this circuit is shown in Figure 4.16. There are a number of important features on this diagram besides the BCD input $DCBA$. Let us review these one at a time. First, there is the *lamp test* input. This is an active-low input that causes all the segment outputs of the IC to go low. This should cause every segment on the common-anode, 7-segment LED to light up.

Next there is the $\overline{RBI}$ input (RBI stands for *ripple blanking input*). This is an active-low input that works in the following way. Suppose that the 7-segment LED is one of a series of such ICs. The usual practice is to display decimal numbers with leading zeros suppressed. That is, a number like 0084 would be displayed as simply 84. This is done by tying the $\overline{RBI}$ input of the 7447A driving the first LED to ground. Then, *if* the BCD input is zero, the 7447A outputs will go high, blanking the 7-segment LED. In addition, the $\overline{BI/RBO}$ output will go low (BI/RBO stands for blanking input/ripple blanking output). This output will normally be connected to the $\overline{RBI}$ input of the next 7447A. Thus, if the first digit is zero, it will be

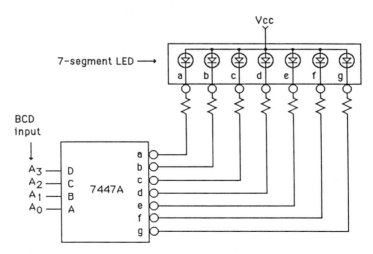

FIGURE 4.15 Minimum configuration circuit necessary to drive a 7-segment LED with a 7447A decoder/driver.

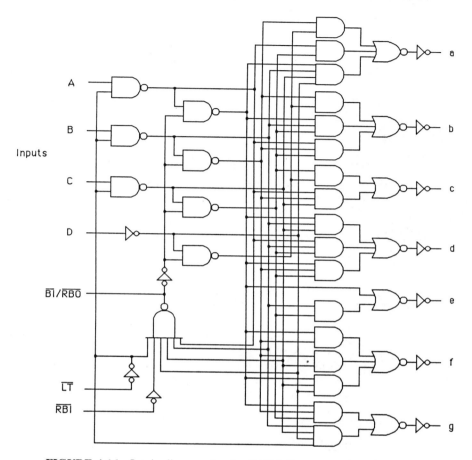

FIGURE 4.16 Logic diagram for the 7447A 7-segment decoder/driver.

suppressed and will signal the next 7447A with a low on its $\overline{\text{RBI}}$ input so that, if the second digit is zero, it too will be suppressed, and so on.

Finally, let us consider the $\overline{\text{BI/RBO}}$ connection in more detail. This connection is both an input and an output. We have already considered its output function. It is also called the *blanking input*, because if it is pulled low as an input, it causes all segment outputs to go high. This would turn off all LED segments. Thus this input can be used as an enable input. In fact, one way of controlling the brightness of the LED segments is to provide a signal on the $\overline{\text{BI}}$ input that varies between high and low very rapidly. By controlling the relative amount of time this signal is high versus low, the average brightness of the display can be controlled.

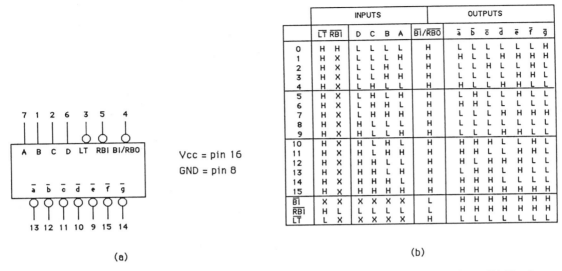

	INPUTS						OUTPUTS						
	L̄T̄ RB̄I	D	C	B	A	B̄I/RBO	ā	b̄	c̄	d̄	ē	f̄	ḡ
0	H H	L	L	L	L	H	L	L	L	L	L	L	H
1	H X	L	L	L	H	H	H	L	L	H	H	H	H
2	H X	L	L	H	L	H	L	L	H	L	L	H	L
3	H X	L	L	H	H	H	L	L	L	L	H	H	L
4	H X	L	H	L	L	H	H	L	L	H	H	L	L
5	H X	L	H	L	H	H	L	H	L	L	H	L	L
6	H X	L	H	H	L	H	H	H	L	L	L	L	L
7	H X	L	H	H	H	H	L	L	L	H	H	H	H
8	H X	H	L	L	L	H	L	L	L	L	L	L	L
9	H X	H	L	L	H	H	L	L	L	L	H	L	L
10	H X	H	L	H	L	H	H	H	H	L	L	H	L
11	H X	H	L	H	H	H	H	H	L	L	H	H	L
12	H X	H	H	L	L	H	H	L	L	H	H	L	L
13	H X	H	H	L	H	H	L	H	H	L	H	L	L
14	H X	H	H	H	L	H	H	H	H	L	L	L	L
15	H X	H	H	H	H	H	H	H	H	H	H	H	H
B̄I	X X	X	X	X	X	L	H	H	H	H	H	H	H
RB̄I	H L	L	L	L	L	L	H	H	H	H	H	H	H
L̄T̄	L X	X	X	X	X	H	L	L	L	L	L	L	L

(a)

(b)

Vcc = pin 16

GND = pin 8

FIGURE 4.17 (a) Connection diagram for the 7447A 7-segment decoder/driver. (b) Truth table for the 7447A.

If you examine Figure 4.16 very carefully, you should be able to verify that it functions as described. Figure 4.17(a) shows the pin connections on the 7447A, and Figure 4.17(b) is the truth table for the circuit.

Practical Display Circuits

In Figure 4.15, we saw a simple circuit that could be used to drive a 7-segment LED display. This circuit by itself would be fine to drive a single 7-segment LED. However, most digital displays consist of several digits. In some cases the BCD code for each digit comes from a separate BCD counter. If this is the case, the output from each counter can be directly input to a separate 7447A, which in turn drives a single-digit display. On the other hand, many times the decimal digits to be displayed are being computed within a microprocessor or some other arithmetic circuit. This poses a different problem.

A microprocessor is usually connected to the outside world through an input/ output port, which is simply a set of parallel wires (frequently eight) that are used to carry digital information into and out of the microprocessor. This means that to send a series of BCD digits to a series of 7-segment LEDs the information must travel over the same set of wires. There is not a separate set of wires running from the microprocessor to each 7-segment LED driver. How do we handle a situation like this? There are many ways to proceed, two of which we will now examine.

■ EXAMPLE 4.4–6

Let us suppose that we wished to send a series of four BCD digits to a display consisting of four 7-segment LEDs. One way to do this is shown in Figure 4.18. In this figure a new type of digital IC has been introduced, the 74100 dual 4-bit latch. Each 4-bit latch on the 74100 consists of four inputs and four outputs. Each

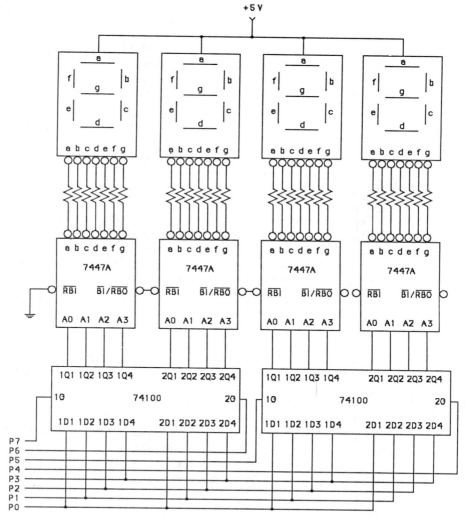

FIGURE 4.18 A system for displaying four decimal digits using a single 8-bit I/O port.

latch also has an active-high enable. The latch works in the following way. When the enable is high, the outputs follow (are equal to) the inputs. When the enable line goes low, the outputs are fixed or latched at the value they had when the enable line went low (at the *falling edge* of the enable input).

In Figure 4.18, there are two 74100s, and thus four 4-bit latches, one for each 7447A. The circuit functions as follows. To display a digit on one of the 7-segment LEDs, the microcomputer places the BCD digit on the low 4 bits of the I/O port. At the same time, the microcomputer sets *one* of the high 4 bits of the I/O port high while keeping the rest low. The one high line will enable one of the four latches to receive data. This high line is then taken low to latch the BCD digit into the latch. The 74100 output continues to send this digit to the appropriate 7447A until a new digit is sent to the latch. In this procedure, the same four I/O lines are used to send BCD digits to all four latches, but only one latch is enabled at a time. Once the digits have been sent to the latches, the microcomputer need not concern itself with the display until new digits are to be displayed. This is a major advantage since it frees the microcomputer to do other things. On the other hand, a fair amount of hardware is involved in the circuit, since each display has it own 7447A and half of a 74100. ■

■ EXAMPLE 4.4–7

The circuit in Figure 4.19 represents somewhat of a hardware reduction in the four-digit display (especially in terms of component cost). In this case, a single 7447A is used to control all four displays. However, *only one display can be on at a time.* Each display is controlled by an optocoupler. The optocoupler acts like a switch, allowing current to flow only when the input to the coupler is high. The circuit works like this. Suppose a digit is to be displayed on the first LED. Then the code for this digit is placed on the low 4 bits of the I/O port. At the same time, I/O line *P4* is taken high to activate the optocoupler controlling the first LED. This LED will now display the digit desired as long as the code for this digit remains on the low 4 bits of the I/O port and line *P4* remains high. After a very short time, perhaps a few milliseconds, the first LED is turned off and a new digit is displayed on the second LED. This process continues until all four digits have been displayed, one after the other. The process is then repeated over and over again, perhaps 60 times each second, for as long as a display is needed. To the observer, the display looks steady. In reality, only one digit is displayed at a time. To make the display look sufficiently bright, more current than normal is allowed to flow through each LED when it is on. This will cause no problem as long as each LED is only on for a short time. ■

There are a couple of points to be made about Figure 4.19. First, although the circuit requires less costly hardware than the circuit shown in Figure 4.18, the

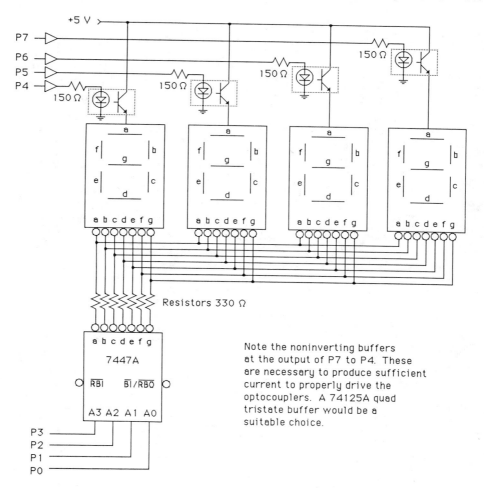

FIGURE 4.19 A different method of displaying four decimal digits using an 8-bit I/O port.

microprocessor will be more involved in maintaining the display. The program that controls the display will need to update the display every few milliseconds. This may pose no problem, but in some cases the time lost to updating the display may be intolerable. A second point involves how we might display more than four digits with an eight-line I/O port. The answer is to use the upper 4 bits of the I/O port more efficiently. Instead of connecting each of these four lines directly to an optocoupler, we could use the lines as an input to a 1-of-8 or even 1-of-16 decoder. This would allow the upper four I/O lines to carry a code that could enable any one of 16 optocouplers controlling 16 different LEDs. Thus a single eight-line I/O port could quite comfortably a maintain 16-digit display.

4.5 MULTIPLEXERS AND DEMULTIPLEXERS

The procedure used in Figures 4.18 and 4.19 is really a form of demultiplexing. The idea of demultiplexing is to use the same wire or group of wires to send data to several places. This is what is being done in both figures, where data coming from a group of four I/O lines is being routed to four different LEDs, one after the other. Multiplexing is the reverse of this process. In multiplexing, data coming from several different places are fed over a single line or group of lines.

Perhaps the easiest way to understand the function of a multiplexer is to examine a typical TTL version of this circuit. Figure 4.20 shows the logic diagram of the 74151 8-input multiplexer. As can be seen from the figure, the circuit has three data select inputs labeled CBA. There are also eight data inputs labeled $D0$ to $D7$. And, finally, there is an active-low enable input. The circuit functions in a very straightforward manner. If the enable is active (low), then the data-select code enables one of the eight AND gates. For example, if the data-select code is

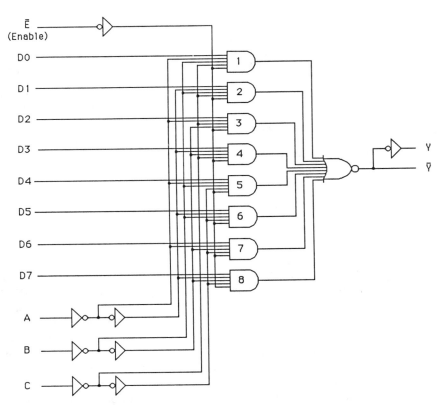

FIGURE 4.20 Logic diagram for the 74151 8-input multiplexer.

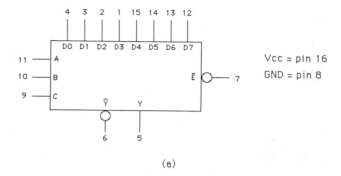

(a)

INPUTS												OUTPUTS	
$\bar{E}$	C	B	A	D0	D1	D2	D3	D4	D5	D6	D7	$\bar{Y}$	Y
H	X	X	X	X	X	X	X	X	X	X	X	H	L
L	L	L	L	L	X	X	X	X	X	X	X	H	L
L	L	L	L	H	X	X	X	X	X	X	X	L	H
L	L	L	H	X	L	X	X	X	X	X	X	H	L
L	L	L	H	X	H	X	X	X	X	X	X	L	H
L	L	H	L	X	X	L	X	X	X	X	X	H	L
L	L	H	L	X	X	H	X	X	X	X	X	L	H
L	L	H	H	X	X	X	L	X	X	X	X	H	L
L	L	H	H	X	X	X	H	X	X	X	X	L	H
L	H	L	L	X	X	X	X	L	X	X	X	H	L
L	H	L	L	X	X	X	X	H	X	X	X	L	H
L	H	L	H	X	X	X	X	X	L	X	X	H	L
L	H	L	H	X	X	X	X	X	H	X	X	L	H
L	H	H	L	X	X	X	X	X	X	L	X	H	L
L	H	H	L	X	X	X	X	X	X	H	X	L	H
L	H	H	H	X	X	X	X	X	X	X	L	H	L
L	H	H	H	X	X	X	X	X	X	X	H	L	H

(b)

FIGURE 4.21   (a) Connection diagram for the 74151 8-input multiplexer.
(b) Truth table for the 74151.

000, then AND gate 1 is enabled. The term enabled in this context means that all
the inputs to the AND gate coming from the data-select lines are high. The final
input to AND gate 1 comes from data input $D0$. If this input is high, then the
output of AND gate 1 will be high. This will cause the output of the 8-input NOR
gate to go low, and thus output Y will go high. The net result is that, when the
data-select code is 000, whatever datum is on input $D0$ will show up at output Y.
Output $\bar{Y}$ will be the inverse of this.

By changing the data-select code, data from any of the eight inputs can be
routed to output Y and sent to wherever this line is connected. Figure 4.21(a)
shows the connection diagram for the 74151, and Figure 4.21(b) shows the truth
table for the circuit. Notice that when the enable input is high the Y output is
forced low and, naturally, $\bar{Y}$ is forced high.

■ **EXAMPLE 4.5–1**

What should the state of the inputs to a 74151 multiplexer be if it is desired to send data from input $D3$ out through output Y? The select lines CBA must select input $D3$. From the truth table or logic diagram, it is clear that the CBA inputs must be 011. In addition, the chip enable, E, must be active, in this case low. ■

The 74151 is an 8-input multiplexer. Other types of multiplexers are available as TTL circuits as well. For example, the 74153 is a dual 4-input multiplexer. Figure 4.22(a) shows the truth table for this IC, and Figure 4.22(b) shows the connection diagram. The 74150 is a 16-input multiplexer with an active-low output. That is, it passes the inverse of the selected data through to its single output. Unlike the 74151, the 74150 has only one output, not a complementary pair. Figure 4.23(a) shows the connection diagram for the 74150. Notice that there is a single active-low enable. When this enable is high, the single output $\overline{Y}$ is forced high. When

INPUTS (a or b)							OUTPUT
$\overline{E}$	B	A	D0	D1	D2	D3	Y
H	X	X	X	X	X	X	L
L	L	L	L	X	X	X	L
L	L	L	H	X	X	X	H
L	L	H	X	L	X	X	L
L	L	H	X	H	X	X	H
L	H	L	X	X	L	X	L
L	H	L	X	X	H	X	H
L	H	H	X	X	X	L	L
L	H	H	X	X	X	H	H

(a)

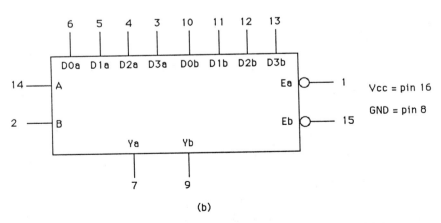

(b)

FIGURE 4.22 (a) Truth table for the 74153 dual 4-input multiplexer. (b) Connection diagram for the 74153.

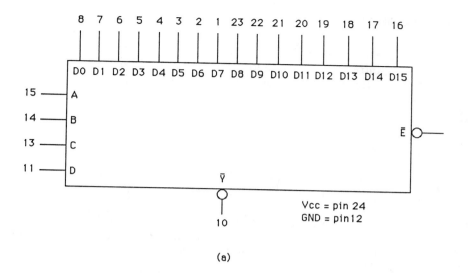

(a)

$\bar{E}$	D	C	B	A	D0	D1	D2	D3	D4	D5	D6	D7	D8	D9	D10	D11	D12	D13	D14	D15	$\bar{Y}$
										INPUTS											OUTPUT
H	X	X	X	X	X	X	X	X	X	X	X	X	X	X	X	X	X	X	X	X	H
L	L	L	L	L	L	X	X	X	X	X	X	X	X	X	X	X	X	X	X	X	H
L	L	L	L	L	H	X	X	X	X	X	X	X	X	X	X	X	X	X	X	X	L
L	L	L	L	H	X	L	X	X	X	X	X	X	X	X	X	X	X	X	X	X	H
L	L	L	L	H	X	H	X	X	X	X	X	X	X	X	X	X	X	X	X	X	L
L	L	L	H	L	X	X	L	X	X	X	X	X	X	X	X	X	X	X	X	X	H
L	L	L	H	L	X	X	H	X	X	X	X	X	X	X	X	X	X	X	X	X	L
L	L	L	H	H	X	X	X	L	X	X	X	X	X	X	X	X	X	X	X	X	H
L	L	L	H	H	X	X	X	H	X	X	X	X	X	X	X	X	X	X	X	X	L
L	L	H	L	L	X	X	X	X	L	X	X	X	X	X	X	X	X	X	X	X	H
L	L	H	L	L	X	X	X	X	H	X	X	X	X	X	X	X	X	X	X	X	L
L	L	H	L	H	X	X	X	X	X	L	X	X	X	X	X	X	X	X	X	X	H
L	L	H	L	H	X	X	X	X	X	H	X	X	X	X	X	X	X	X	X	X	L
L	L	H	H	L	X	X	X	X	X	X	L	X	X	X	X	X	X	X	X	X	H
L	L	H	H	L	X	X	X	X	X	X	H	X	X	X	X	X	X	X	X	X	L
L	L	H	H	H	X	X	X	X	X	X	X	L	X	X	X	X	X	X	X	X	H
L	L	H	H	H	X	X	X	X	X	X	X	H	X	X	X	X	X	X	X	X	L
L	H	L	L	L	X	X	X	X	X	X	X	X	L	X	X	X	X	X	X	X	H
L	H	L	L	L	X	X	X	X	X	X	X	X	H	X	X	X	X	X	X	X	L
L	H	L	L	H	X	X	X	X	X	X	X	X	X	L	X	X	X	X	X	X	H
L	H	L	L	H	X	X	X	X	X	X	X	X	X	H	X	X	X	X	X	X	L
L	H	L	H	L	X	X	X	X	X	X	X	X	X	X	L	X	X	X	X	X	H
L	H	L	H	L	X	X	X	X	X	X	X	X	X	X	H	X	X	X	X	X	L
L	H	L	H	H	X	X	X	X	X	X	X	X	X	X	X	L	X	X	X	X	H
L	H	L	H	H	X	X	X	X	X	X	X	X	X	X	X	H	X	X	X	X	L
L	H	H	L	L	X	X	X	X	X	X	X	X	X	X	X	X	L	X	X	X	H
L	H	H	L	L	X	X	X	X	X	X	X	X	X	X	X	X	H	X	X	X	L
L	H	H	L	H	X	X	X	X	X	X	X	X	X	X	X	X	X	L	X	X	H
L	H	H	L	H	X	X	X	X	X	X	X	X	X	X	X	X	X	H	X	X	L
L	H	H	H	L	X	X	X	X	X	X	X	X	X	X	X	X	X	X	L	X	H
L	H	H	H	L	X	X	X	X	X	X	X	X	X	X	X	X	X	X	H	X	L
L	H	H	H	H	X	X	X	X	X	X	X	X	X	X	X	X	X	X	X	L	H
L	H	H	H	H	X	X	X	X	X	X	X	X	X	X	X	X	X	X	X	H	L

(b)

FIGURE 4.23 (a) Connection diagram for the 74150 16-input multiplexer. (b) Truth table for the 74150.

the enable is low, the four data-select lines choose which input will be passed through, in inverted form, to output $\overline{Y}$. Thus, if the data-select lines carry the code 1010, then the data on input line $D10$ will be inverted and output at $\overline{Y}$.

■ **EXAMPLE 4.5-2**

The $DCBA$ inputs of a 74150 are 1100. The odd data inputs are all low, while the even data inputs are all high. The chip enable is low. What is the $\overline{Y}$ output of the IC?

The $DCBA$ inputs have selected input $D12$, which is an even input and therefore high. Since the chip is enabled, the *inverse* of the data on input $D12$ will be output at $\overline{Y}$. Since $D12$ is high, $\overline{Y}$ will be low. ■

A primary function of a multiplexer is to make data transmission more efficient. Suppose, for example, that a circuit needs to monitor the condition of eight remote devices. Let us imagine that these devices can be either on or off at any given time, and that they will signal their condition by placing either 5 volts or 0 volts on a wire and sending it to a distant monitoring circuit. Eight devices would require eight wires going to the circuit and eight different circuit inputs. On the other hand, a multiplexer located near the devices could be used to send the status of the devices to the remote circuit over a single wire, one at a time. The circuit would need to send the data-select code to the remote multiplexer, of course, and this would require three wires for controlling an 8-input multiplexer. Nevertheless, the number of wires required is only half of the eight required without the multiplexer. The savings becomes even greater as the number of devices increases.

A multiplexer can also be used to generate an arbitrary Boolean function. Consider the truth table shown in Figure 4.24(a). The Boolean function F could of course be implemented by first arriving at the sum-of-products Boolean form and then reducing this expression by the theorems of Boolean algebra. However, a very simple approach is to take advantage of the multiplexer's ability to distribute data from any one of several inputs. In this case, an 8-input 74151 multiplexer serves the purpose. The circuit in Figure 4.24(b) will implement the Boolean function shown in the truth table. We simply tie each multiplexer input high or low according to whether the corresponding term in the truth table is 1 or 0.

Demultiplexers

Sometimes it is necessary to send data from many different devices a considerable distance and have the data arrive on separate lines. In this case, a multiplexer is useful in combination with a demultiplexer.

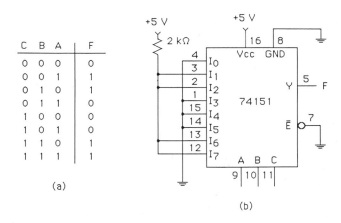

C	B	A	F
0	0	0	0
0	0	1	1
0	1	0	1
0	1	1	0
1	0	0	0
1	0	1	0
1	1	0	1
1	1	1	1

(a)

(b)

FIGURE 4.24 (a) Truth table implemented by the 74151 8-input multiplexer in (b).

A demultiplexer works in just the opposite fashion from a multiplexer. The circuit has a single input that can be routed to any one of several outputs. Again, a set of data-select inputs are used to determine which output is to receive the incoming data. Actually, we have already seen a demultiplexer in the 74154. This circuit was first discussed in Section 4.2, where it was called a 1-of-16 decoder. However, if you examine Figure 4.5 closely, you will see that either enable input could be used as a data input. To see how this works, imagine that enable $\overline{E_0}$ is tied low. Now think of enable $\overline{E_1}$ as the data input. Imagine that the data-select inputs are 0000. The truth table shows us that if $\overline{E_1}$ is high then output 0 will be high, but if $\overline{E_1}$ is low then output 0 will be low. Thus, whatever data is on $\overline{E_1}$, that is what will appear on output 0 when the data-select input is 0000. If the data-select input is changed to 0011, then whatever is on $\overline{E_1}$ will appear on output 3. Most decoders can be used as demultiplexers.

Figure 4.25 shows how a multiplexer can be combined with a demultiplexer to greatly reduce the number of lines needed to transmit data over a distance. In this case, only five lines are needed to do the work of sixteen. The data-select lines are connected to both the multiplexer and demultiplexer.

■ EXAMPLE 4.5–3

A logic 1 is present on input $D13$ of the 74150 in Figure 4.25. What should the status of the data-select lines be in order to have this datum show up at output 13 of the 74154? To select input $D13$, $DCBA$ must be the binary equivalent of 13, or 1101. If this is so, then $\overline{Y}$ will be the inverse of $D13$, or a logic 0. The inverter

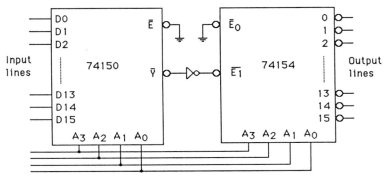

FIGURE 4.25   Data transmission using a multiplexer–demultiplexer combination.

following $\overline{Y}$ will invert the logic 0 to a logic 1. This logic 1 will be input to $\overline{E_1}$ of the 74154, which will disable the chip and cause a logic 1 to appear at output 13 of the 74154.   ■

Generally, the code on the data-select lines is cycled on a periodic basis through all 16 select codes so that all 16 inputs are sent through on a regular periodic basis. A variation of this circuit would use a sixth common line to enable both the 74150 at $\overline{E}$ and the 74154 at $\overline{E_0}$ simultaneously whenever data were to be transmitted. This approach would not leave the two circuits permanently enabled, as is shown in Figure 4.25.

4.6   XOR CIRCUITS

In Section 1.3 we were introduced to the XOR gate. Figure 4.26(a) shows the connection diagram for the 74136 quad XOR gate. It can be seen that the package holds four independent XOR gates. Figure 4.26(b) gives the truth table for each gate. XOR gates have a number of uses, such as parity generators or checkers, adders, code comparers, and magnitude comparers. We will investigate each of these functions in this section.

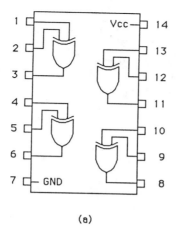

INPUTS		OUTPUT
A	B	Y
L	L	L
L	H	H
H	L	H
H	H	L

(a)   (b)

FIGURE 4.26 (a) Connection diagram for the 74136 quad XOR gate. (b) Truth table for each XOR gate.

Parity

Let us first consider the matter of parity. Imagine four input lines labeled *ABCD*. There are 16 possible input combinations that can exist on these four lines. Some input combinations contain an even number of 1's, and some combinations contain an odd number of 1's. The *parity* of a given input is said to be even if the number of 1's in the input code is even. Otherwise, the parity is said to be odd.

The concept of parity is used as a simple means of checking for errors in data transmission. Data are often sent in a serial fashion, one bit after another over a single line. We saw this concept implemented in the multiplexer–demultiplexer circuit of Figure 4.25. There is always a small chance that electrical noise on the transmission line might cause an occasional bit to be improperly received. That is, a transmitted 1 might be received as a 0, or vice versa. As long as this is a rare event, parity provides a good check for such an error. It works like this. Say that we wish to transmit a 4-bit code in a serial fashion. First, we input this code into a circuit that determines the parity of the code to be sent. We will then send an extra fifth bit over the transmission line. The idea of this fifth bit is to make the parity of *all 5 bits* always even or always odd, whatever has been agreed on ahead of time. Let us illustrate this with a specific example.

■ EXAMPLE 4.6–1

Suppose that the 4-bit code to be sent is 0110. Suppose further that we have agreed on even parity (that is, the parity of the *five* transmitted bits will always be even).

Since the code 0110 contains an even number of 1's, it already has even parity. Thus we add a fifth bit, which is 0, and send the code 00110. At the receiving circuit, the parity is checked to see if it is even. If no bits have been changed in transit, the parity should still be even and no error is signaled. However, if any *single* bit has changed the parity will have become odd. This causes an error to be signaled, and retransmission of the data should be requested. ∎

∎ EXAMPLE 4.6–2

As a second example, suppose that the code to be sent is 1110. Since this code contains an odd number of 1's, it has odd parity. Thus we add a fifth bit of 1 and send the code 11110. Again, the parity of the 5 bits should be even. If the receiving circuit does not detect even parity, it will request retransmission of the data. This system works well as long as errors only occur in 1 bit. If 2 bits are changed, their effect cancels and no error is detected. This is why a parity-detecting scheme is only effective if errors are relatively rare, and thus double errors are extremely rare. ∎

The XOR gate is ideally suited for parity detection. Figure 4.27(a) shows a 3-bit parity detector. The notation $F = A \oplus B \oplus C$ in the figure is just another way of saying F will be high if the parity is odd. That is, if a particular combination of ABC contains an odd number of 1's, then $A \oplus B \oplus C$ will be high. Figure 4.27(b) shows a 4-bit parity detector. In this case the output is $F = A \oplus B \oplus C \oplus D$, and again this expression means that F is high if the parity is odd. To convert either of the circuits in Figure 4.27 to an active-high, even-parity detector, simply invert the output.

There are MSI circuits that check parity. For example, the 74180 has eight inputs that can be checked for parity. It has two enable lines and two parity output lines. Depending on how the enable lines are connected, the parity outputs can be made active high or active low. For example, if the outputs are made active high,

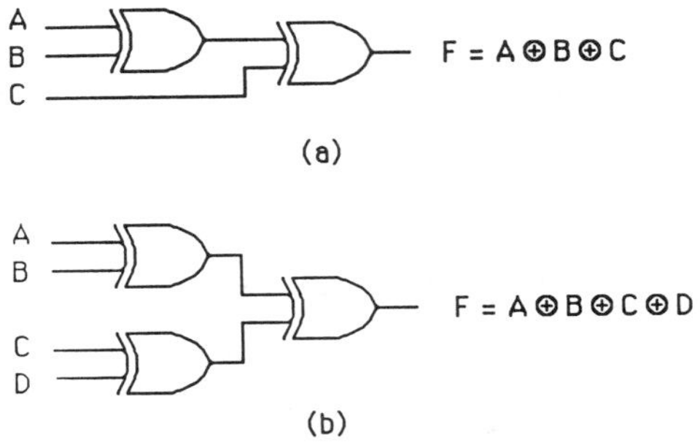

(a)

FIGURE 4.27 (a) A 3-bit parity detector, with F high for odd parity. (b) A 4-bit parity detector, with F high for odd parity.

(b)

then if the input code has even parity the even-parity output will go high while the odd-parity output will remain low.

Half-adders and Full Adders

XOR gates are ideally suited to implement binary addition. Consider the problem of adding two binary bits. The results are summarized in the table in Figure 4.28(a). The sum column is clearly just the XOR of A and B. There is also a carry that must be accounted for. The carry column is clearly just AB. The circuit that implements the truth table is shown in Figure 4.28(b). This circuit is called a half-adder.

  The half-adder circuit is suitable for adding two binary bits, but has no provision for adding in the carry bit from a previous addition step. That is, we are usually interested in adding two binary numbers that consist of several bits each. The two least significant bits can be added with a half-adder since there would be no carry from a previous addition step. However, the output of the half-adder may now contain a carry that must be added along with the 2 bits of the next addition step.

■ **EXAMPLE 4.6–3**

Consider adding the two binary numbers 10011 and 01001. The result would be

$$
\begin{array}{r}
\text{carries} \rightarrow \quad 0011 \\
10011 \\
+ \ 01001 \\
\hline
11100
\end{array}
$$

■ **EXAMPLE 4.6–4**   ■

Add the two numbers 10110 and 10011. The result is

FIGURE 4.28 (a) Truth table for 2-bit binary addition. (b) A half-adder circuit to accomplish the 2-bit addition.

A	B	Sum	Carry
0	0	0	0
0	1	1	0
1	0	1	0
1	1	0	1

(a)

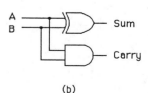

(b)

$$\begin{array}{r}
\text{carries} \rightarrow \quad 1011 \\
10110 \\
+ \ 10011 \\
\hline
101001
\end{array}$$

Notice that the final result has one more bit than either of the two original numbers. This is because the addition of the most significant bits has resulted in a carry, and the carry adds another bit to the number. ■

In essence, except for the addition of the first 2 bits, we need a circuit that will add 3 binary bits, producing both the sum and any carry. Such a circuit is called a full adder. The truth table that summarizes the function of the full adder is shown in Figure 4.29(a). The sum column in this table is essentially an odd-parity function. That is, the sum is 1 when the parity of the 3 bits is odd. Next we consider the carry. One way of approaching the carry is to realize that the carry will be 1 if A and B are both 1, or if the XOR of A and B is 1 AND the *carry-in* is also 1. In Boolean terms,

$$\text{Carry-out} = AB + (A \oplus B) \text{ (carry-in)} \tag{4.1}$$

The full-adder circuit is shown in Figure 4.29(b). This figure shows that the OR gate that generates the carry-out has two inputs, one being AB and the other being $(A \oplus B)$(carry-in). The carry-out OR gate thus directly implements Eq. (4.1). The sum is created by an odd-parity detector exactly like the one in Figure 4.27(a).

The addition of two multibit binary numbers can easily be accomplished by chaining together a half-adder and several full adders. A circuit to add two 3-bit

A	B	Carry-in	Sum	Carry-out
0	0	0	0	0
0	0	1	1	0
0	1	0	1	0
0	1	1	0	1
1	0	0	1	0
1	0	1	0	1
1	1	0	0	1
1	1	1	1	1

(a)   (b)

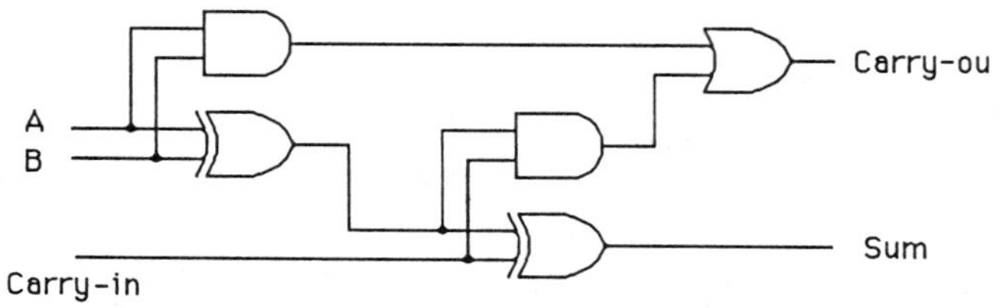

FIGURE 4.29 (a) Truth table for 3-bit binary addition. (b) Full-adder circuit to implement the 3-bit addition.

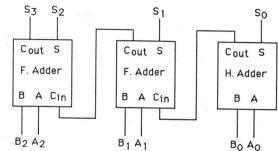

FIGURE 4.30 Two full adders and a half-adder used to add two 3-bit numbers.

binary numbers is shown in Figure 4.30. Let us examine the operation of this circuit through an example.

■ **EXAMPLE 4.6–5**

Determine the sum and carry outputs of the half-adder and the two full adders in Figure 4.30 when the two numbers to be added are $A = 011$ and $B = 101$.

We consider the least significant bits first. These are $A_0 = 1$ and $B_0 = 1$ and are added by the half-adder. This produces a sum of 0 and a carry of 1. Thus S_0 is 0, and the carry of 1 becomes the carry-in to the first full adder.

Now we consider the first full adder. This adder adds the bits $A_1 = 1$, $B_1 = 0$, and the carry-in of 1. The sum of these 3 bits is 0 with a carry of 1. Thus the first full adder produces a sum bit, S_1, of 0 and a carry-out of 1. This carry of 1 becomes the carry-in to the second full adder.

Finally, we consider the second full adder. This adder adds the bits $A_2 = 0$, $B_1 = 1$, and the carry-in of 1. The sum of these 3 bits is 0 with a carry of 1. Thus the final full adder produces a sum bit, S_2, of 0 and a carry of 1. Since this is the last adder in the chain, the carry from this adder is considered the fourth bit of the sum and is labeled S_3. ■

Although Figure 4.30 shows how to add two 3-bit numbers, it should be clear that it would be straightforward to expand the concept to the addition of two n-bit numbers. This would simply require the addition of more full adders at the left end of the chain.

It is not necessary to use individual full and half-adders to accomplish multibit addition. There are MSI adder circuits available. The 7483 is a 4-bit adder with a carry-in, a carry-out, two sets of four inputs for the numbers to be added, and a set of four *sum* outputs. The circuit directly adds two 4-bit numbers (and the carry-in), and a number of these circuits can be chained together to achieve the addition of 8, 12, 16, or more bit numbers. The first adder in the chain has its carry-in input tied to ground. The carry-out from each adder is then connected to the carry-in of the next adder. Two 16-bit numbers can be added in about 45 nanoseconds.

Magnitude Detectors

One important arithmetic function that must often be performed is the determination of which of two numbers is the larger or if the two numbers are equal. This function can be achieved with the help of the XNOR gate or equality detector. Figure 4.31(a) shows the truth table for the XNOR gate, and Figure 4.31(b) shows two standard circuit symbols used for this gate.

It is very easy to see if two multibit numbers are equal. We simply use a series of XNOR gates to check equivalent bits from each number for equality. The outputs of these XNOR gates are ANDed together. If all the XNOR gates detect equality, the AND gate will have a high output indicating that the two numbers are equal. If even one XNOR gate detects an inequality, the two numbers are unequal.

Additional circuitry is required to determine which of the two numbers is larger in the case of inequality. To see what additional circuitry is needed, let us consider two multibit binary numbers, A and B. Suppose that we want to design a circuit whose output goes high if $A > B$. The approach is straightforward. First, we check to see if the most significant bit (msb) of A is greater than the msb of B. If it is, then we know that $A > B$. If this test fails, *and if the msb's of A and B are equal*, then we check the next msb of A and B. If this test fails, and if the *two msb's of A and B are equal*, then we check the next msb of A and B. We continue this process until we get a positive result or run out of bits. The circuit in Figure 4.32 shows a 3-bit equality and relative-magnitude detector. Let us consider this circuit in some detail.

First, the output of each equality detector (XNOR gate) is input to AND gate 1. The output of this gate will go high only if $A = B$ (all bit pairs are equal). AND gate 2 compares A2 with $\overline{B2}$. If the output of this AND gate is high, then $A > B$ since the msb of A must be 1 and the msb of B must be 0. Since the output of AND gate 2 is input to the OR gate, a high from AND gate 2 will cause the output of the OR gate to go high, indicating that $A > B$. Now consider AND gate

A	B	$\overline{A \oplus B}$
0	0	1
0	1	0
1	0	0
1	1	1

(a)

FIGURE 4.31 (a) Truth table and (b) circuit symbols for the XNOR gate or equality detector.

A —⟫o— $\overline{A \oplus B}$ ⟶ A, B = $A = B$

(b)

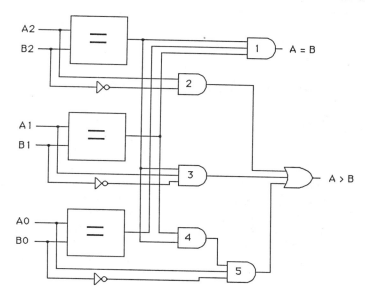

FIGURE 4.32 An equality and relative-magnitude detector.

3. This gate is enabled by the output from the msb equality detector. That is, AND gate 3 can only produce a high output if A2 = B2. If this condition is met, then AND gate 3 goes on to compare A1 with $\overline{B1}$. If these two are found equal (and the AND gate is enabled) then $A > B$. Finally, we come to AND gate 5. This gate is enabled by AND gate 4, which requires that A2 = B2 AND A1 = B1. If AND gate 5 is enabled, then it goes on to compare A0 with $\overline{B0}$. If these two are found equal, then $A > B$. If the outputs of AND gates 2, 3, and 5 are all low, then the output of the OR gate will also be low. This indicates that A is not greater than B. It is still possible that $A = B$. This will be indicated if the output of AND gate 1 is high. If this too is low, then the only remaining possibility is $B > A$.

Rather than build a magnitude comparator from individual gates, as in Figure 4.32, it is possible to use a MSI circuit. The 7485 is a 4-bit magnitude comparator. This circuit has two sets of four inputs for the two 4-bit numbers to be compared. There are separate outputs for $A > B$, $A = B$, and $A < B$, which are active high and signal the relative magnitude of the two 4-bit numbers being compared. Provision is also made to chain several 7485s together to compare two 8, 12, 16, or greater bit numbers. This is accomplished by providing $A > B$, $A = B$, and $A < B$ *inputs*, which are also active high. These inputs tell the circuit the relative magnitude of the preceding bits of the two being compared. That is, the $A > B$, $A = B$, and $A < B$ outputs from one 7485 are connected to the corresponding inputs of the next higher 7485. The 7485 handling the least significant 4 bits must have its $A = B$ input activated (tied high) to tell it that the preceding bits of the two numbers are equal.

Code Comparers

There are times when it would be useful to have a decoder that can be set by manual switches to decode any desired code. XOR gates can help accomplish this task. The circuit shown in Figure 4.33 decodes the inputs CBA and produces a high output only if the CBA inputs match the switch settings (where open = high and closed = low). The circuit works like this. When a switch is open, the resistor pulls the XOR input to 5 volts. When a switch is closed, the XOR input is grounded. To activate the NOR gate (shown as a low-level-input AND gate), the output of all three XOR gates must be low. This means that the CBA inputs must exactly match the switch inputs. Thus the circuit decodes for the code established by the switch settings. These can be changed according to the needs of the occasion.

PROBLEMS

1. Using 2-input NAND gates and inverters, design a 1-of-4 decoder with active-low outputs.

2. A 74155 dual 1-of-4 decoder has the following inputs: B, $E_{a1}, \overline{E}_{b1}$, all high, and A, $\overline{E}_{a2}$, $\overline{E}_{b2}$, all low. What are the states of all the outputs?

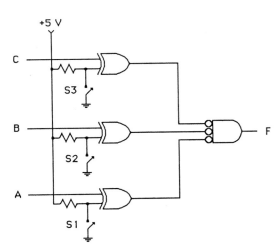

FIGURE 4.33 A 3-bit code comparer. F is high only when CBA matches the switch settings.

C	B	A	F1	F2
0	0	0	0	1
0	0	1	0	1
0	1	0	1	1
0	1	1	0	0
1	0	0	1	0
1	0	1	0	0
1	1	0	0	0
1	1	1	0	1

FIGURE 4.34 Use with Problems 5 and 20.

3. A 74154 1-of-16 decoder/demultiplexer is to be used to activate 16 different I/O ports connected to a microprocessor. The low four lines of the microprocessor address bus are connected to inputs $DCBA$ of the 74154. The fifth address line is connected to the $1G$ input of the 74154 (refer to Figure 4.4). The sixth address line is *inverted* and then connected to the $2G$ input of the 74154. What code must be on these low six address lines to activate output 7 on the 74154?

4. Design a 1-of-32 decoder using two 74154 ICs and five input lines. Remember, only one of the 74154s can be enabled at a time.

5. Use a 74138 1-of-8 decoder and two NAND gates with the appropriate number of inputs to implement the Boolean functions in Figure 4.34.

6. Eight devices, numbered 0 to 7, are connected to a 74148 priority encoder. The $\overline{EI}$ is low, and devices 5 and 3 are requesting service. What is the state of all the chip outputs?

7. Repeat Problem 6 assuming that $\overline{EI}$ is high and device 2 is requesting service.

8. Repeat Problem 6 assuming that $\overline{EI}$ is low and no device is requesting service.

9. Three switches are numbered 1 to 3. When a switch is closed, it produces a low on the line it is connected to. When the switch is open, it produces a high. Design a circuit that operates as follows. The circuit has two outputs. If no switch is closed, the circuit output is 00. If one or more switches are closed, the circuit produces an output that is the binary number of the highest-number switch that is closed. That is, if switches 1 and 2 are closed, the output should be 10 (binary 2).

10. Design a priority encoder that has three active-low inputs and three active-low outputs and operates as follows. The three inputs are numbered 1 to 3, and so are the three outputs. If no inputs are active, then no outputs are active. If one or more inputs are active, then only the output corresponding to the highest-numbered input is active. For example, inputs 1 and 3 are active, then only output 3 should be active.

11. Using a 7442 BCD to decimal decoder and four NAND gates, design a BCD to 2421 code converter.

12. List the nine's complement for the following numbers: 2, 5, and 8. List the 2421 and XS3 codes for these numbers and verify that the codes are self-complementing.

13. A 74184 BCD to binary converter has the following inputs: enable is low; EDCBA inputs are 01100. What are the outputs $Y1$ to $Y5$? Assume that the low input bit [pass-through input A in Figure 4.11(b)] is 1. What are the two BCD input digits? What is the binary output? Suppose inputs $EDCBA$ are 10110. Then what are the outputs $Y1$ to $Y5$?

14. Suppose you are using a common-anode, 7-segment LED of the type shown in Figure 4.14. What input code on the inputs $a \rightarrow g$ will result in only the center bar being lit? What code would be necessary to produce a capital H? What about a capital F?

15. In Figure 4.15, suppose V is 20 volts. Assume that when each LED is conducting it has a forward voltage drop of 2 volts. Assume also that when an output becomes active on the 7447A, the output drops to 0 volts. Finally, suppose that the maximum current that an LED can safely handle is 20 milliamperes. What should the size of the resistors be in the figure?

16. Six 7447A 7-segment decoder/drivers are connected to six 7-segment LEDs to form a six-digit display. The input to each 7447A comes from a separate counter. The $\overline{RBI}$ input of the first (left, most significant digit) 7447A is grounded. The $\overline{BI}/\overline{RBO}$ output of each 7447A is connected to the $\overline{RBI}$ input of the next 7447A, except for the last 7447A, which has its $\overline{RBI}$ input open. What will appear on the display for each of the following counter outputs: 800943, 040000, 000445, 000000?

17. A 7447A 7-segment decoder/driver is connected to a common-anode, 7-segment LED in the manner shown in Figure 4.15. What pattern will appear on the LED if the input code to the 7447A is the following: 1010, 1011, 1100, 1101, 1110, 1111?

18. Refer to Figure 4.19. The second LED from the right is displaying the number 7. What is the status of each of the eight lines from the I/O port? What is the status of the 7447A outputs?

19. Refer to Figure 4.20. The enable input is active (low), the $D0$ to $D7$ inputs are 01100111, and the data-select inputs (ABC) are 101. What are the states of the Y and $\overline{Y}$ outputs?

20. Use a 74151 8-input multiplexer to implement the F_1 Boolean function in Figure 4.34.

21. Show how to use a BCD to decimal decoder as a 1-to-8 demultiplexer.

22. Show how you could use a 1-of-8 decoder as a 1-of-4 decoder.

23. Consider the following design task. The 74153 is a dual 4-input multiplexer. Figure 4.35 shows the pin connections for one of the multiplexers on this IC. The task is to design a 32-input multiplexer using one 74153 and four 74151 ICs. There will be 5 data-select lines and 32 data-input lines. Complete the design for this circuit.

24. Design a 5-bit odd-parity detector. That is, the circuit should produce a high output only when the parity of the input is odd.

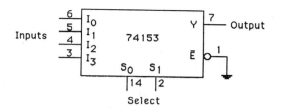

FIGURE 4.35 Use with
Problem 23.

25. A circuit has six input lines. The purpose of the circuit is to produce a 3-bit binary number equal to the number of high inputs. For example, if three of the input lines are high, then the circuit should produce 011 as an output. Design the circuit.

26. Add whatever circuitry is necessary to the circuit in Figure 4.32 to produce an output that goes high when $B > A$.

27. The circuit in Figure 4.33 produces an active-high output when the code on the input lines, CBA, matches the code on the switch settings. Replace the NOR gate on this circuit with a single gate of some other type so that the circuit has an active-low output instead of an active-high output.

28. Construct a circuit that adds two 6-bit binary numbers. You may show adders and half-adders as boxes. How many bits does the output have?

Chapter 5

FLIP-FLOPS

Learning Objectives

After completing this chapter you should know:

- What an *R–S* flip-flop is and how it functions.
- How an *R–S* flip-flop can be constructed from NAND gates or NOR gates and how these two constructions differ in performance.
- Some uses of the *R–S* flip-flop.
- What a clocked *R–S* flip-flop is.
- What a *D*-type flip-flop is and how it is constructed from an *R–S* flip-flop.

- The concept of a D-type flip-flop as a data latch.

- What an edge-triggered D-type flip-flop is and how it is used.

- What a $J-K$ master–slave flip-flop is and how it functions.

5.1 INTRODUCTION

Up to now we have considered only combinational logic circuits. The output of a combinational logic circuit is determined entirely by the present state of the circuit's inputs. A second important type of digital circuit, which we will now explore, is the sequential logic circuit. The output of this type of circuit depends not only on the present input, but also on the sequence of past inputs. A counter is typical of a sequential logic circuit. A binary counter is a circuit that produces a binary output that is numerically equal to the number of clock pulses that have arrived at the clock input. As a specific example, consider a 4-bit binary counter. This type of circuit has four outputs, one clock input, and perhaps a few other inputs as well (such as reset, preset, and four data inputs). At some point, suppose that the output of the counter is set to zero (all outputs are zero, 0000). As the clock input receives pulses, the counter output simply increments in standard binary code. Thus the output at any given instant depends on how many clock pulses have been received.

Counters of various kinds are just one kind of sequential logic circuit. Other sequential circuits include data latches, shift registers, and ring counters. Virtually all these circuits have one thing in common. They are constructed to some extent from some type of flip-flop circuit. To understand what is meant by a flip-flop, we will consider the simplest version of this type of circuit, the R–S flip-flop.

5.2 THE R–S FLIP-FLOP

NAND Gate Implementation

An R–S flip-flop (FF) can be constructed in many different ways using either discrete components (transistors or vacuum tubes) or digital logic gates. One of the simplest ways to construct an R–S FF is with 2-input NAND gates. Such a circuit is shown in Figure 5.1(a). The function of the circuit is summarized in the truth table shown in Figure 5.1(b).

Let us examine the behavior of the NAND gate R–S FF in some detail. First notice that the circuit has two outputs labeled Q and $\overline{Q}$. The state of the Q output generally defines the state of the FF. When Q is high (1), the flip-flop is said to be *set*, and when Q is low (0), the flip-flop is said to be *reset*. Normally, Q and $\overline{Q}$ should be in opposite states if the FF is functioning *as a flip-flop*. The S input of an R–S FF is called the *set* input, and the R input is called the *reset* input. In the NAND gate version of the R–S FF in Figure 5.1(a), both the R and S inputs are normally high and are active low. That is, an effect is produced by making R or S low. In Figure 5.1(a), the negative pulse shown on each input indicates that it is a negative pulse (active low) that sets or resets the FF. With the R–S inputs held

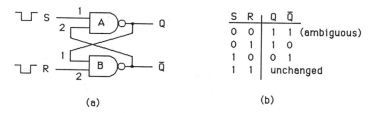

FIGURE 5.1   (a) NAND gate implementation of an R–S flip-flop. (b) Truth table for the circuit in (a).

high, the output of the FF will remain unchanged (either set or reset) indefinitely. This is indicated in the truth table of Figure 5.1(b) by the notation "unchanged."

To set or reset the FF, it is necessary to activate the S or R input, respectively. Both R and S are active low for the NAND gate FF. Thus, if S is made low, this sets the FF regardless of its previous state. Once S is made high again, the FF remains set. If R is made low, the FF is reset (Q becomes 0). When R is made high again, the FF remains reset. To make sure that the output state of the FF is unambiguous (Q and $\overline{Q}$ in opposite states), we must make sure that *R and S do not both go low at the same time*. Should this occur, both Q and $\overline{Q}$ would go high at the same time. Such a state is perfectly acceptable from an electrical viewpoint, but is ambiguous in the sense that a FF should maintain Q and $\overline{Q}$ in opposite states.

To satisfy ourselves that the NAND gate FF does indeed function as described, let us consider the response of the circuit to various kinds of inputs.

■ EXAMPLE 5.2–1

Let us suppose that R and S are both held high and that the FF is reset (Q = 0). In this situation, the low from Q is fed to input 1 of NAND B and holds its output high. NAND A has a high on input 1 from S and a high on input 2 from $\overline{Q}$. The two highs force the output of NAND A low, as assumed previously. The analysis shows that the assumed outputs are consistent with the assumed inputs. Thus we conclude that the FF is stable in this state.   ■

■ EXAMPLE 5.2–2

Suppose that, with the FF in the state defined in Example 5.2-1, input R is made low. Since input 1 of NAND B is *already low*, this action will have no effect. The FF will remain reset.   ■

■ EXAMPLE 5.2–3

Suppose that, with the FF in the state defined in Example 5.2-1, input S is made low. This will cause Q to go high. This will also remove the low from input 1 of NAND B, leaving NAND B with both inputs high. This will force $\overline{Q}$ low. Thus the FF will be set by making S low. The FF can be reset by taking R low. ■

In the normal functioning of a FF (that is, when Q and $\overline{Q}$ are opposite), we see that the low output on either Q or $\overline{Q}$ is what holds the other NAND gate in its high state. For example, when the FF is set, the low from $\overline{Q}$ is input to NAND A and holds the output of NAND A high. This action is sometimes referred to as "sealing the NAND gate." The seal is said to be "broken" when $\overline{Q}$ goes high.

NOR Gate Implementation

A FF can also be constructed from NOR gates. This type of circuit is shown in Figure 5.2(a) and the resulting truth table in Figure 5.2(b). There are some differences between the function of the circuit in Figure 5.2(a) and that of Figure 5.1(a). In the NOR gate version of the $R–S$ FF, the inputs are active high instead of active low. Again, this is indicated by the positive-going pulse shown at each input in Figure 5.2(a). Thus the inputs are normally kept low (inactive). Also, the R input is connected to the gate with the Q output rather than the $\overline{Q}$ output. In the case of the NOR gate version of the FF, it is the high output of either Q or $\overline{Q}$ that seals the output of the other gate to a low. Finally, the ambiguous input condition is both R and S high instead of low, resulting in Q and $\overline{Q}$ both low instead of both high.

On the other hand, the fundamental operation of both versions of the FF is the same. Both can be either set or reset, and both are stable in either state as long as the inputs remain inactive. By assuming that the FF is reset and that both R and S are low, you should satisfy yourself that this is a stable condition for the NOR gate FF and that the FF will function as summarized in the truth table.

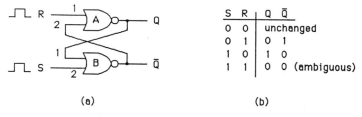

S	R	Q	Q̄
0	0	unchanged	
0	1	0	1
1	0	1	0
1	1	0	0 (ambiguous)

(a)   (b)

FIGURE 5.2 (a) NOR gate implementation of an $R–S$ flip-flop. (b) Truth table for the circuit in (a).

Uses of R–S Flip-flops

One use of R–S FFs is in constructing other more sophisticated flip-flops. We will take up this topic shortly. R–S FFs, or any other type of FF for that matter, can also be used to construct counters, registers, and sequencers. We will delay these applications until J–K FFs are introduced. For now, let us examine two simple applications of R–S FFs.

■ **EXAMPLE 5.2–4**

Consider the simple switch shown in Figure 5.3(a). The switch is a double-pole, double-throw (DPDT) switch. When the switch is at position A, V_{out} will be 5 volts. After the switch is thrown to position B, the output will be 0 volts. However, switch contact will not be made cleanly. Thus, when the switch is switched from one position to the other, the output voltage will fluctuate rapidly (bounce) as contact is made and broken at positions A and B. If the switch output is connected to a typical TTL input, the greatest difficulty occurs as the switch is in the process of making or breaking contact at position B. Whenever contact is made with position B, V_{out} will be zero volts. When contact breaks, V_{out} will *float*, but this will appear as a logical high to a TTL circuit. Thus, as contact is made and broken at position B, the TTL input will see a rapid series of high and low pulses. If the TTL circuit is some kind of counting circuit, the counting circuit will receive several pulses instead of just one. This phenomena is called *key bounce*.

The circuit shown in Figure 5.3(b) corrects the key-bounce problem. Let us suppose that the switch is at position A. The S input will thus be grounded, and Q will be high. Now suppose we throw the switch toward position B. As the switch breaks contact at position A, the S input bounces between 0 and 5 volts several times. However, the R input is still 5 volts because no contact has been made at position B yet. Since the R–S FF is already set (Q is high), the bounce on breaking contact at position A causes no change in output. Now the switch moves across

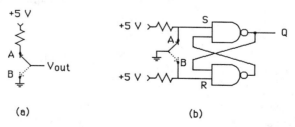

(a) (b)

FIGURE 5.3 (a) Switch connection that will exhibit key bounce. (b) A bounceless switch.

the open space between contacts (the switch *must* be a break-before-make type of switch). When the first contact is made at position B, the R input will be taken low momentarily. This will immediately reset the FF (Q will go low). As contact is made and broken several times, R will bounce between 0 and 5 volts. However, once the FF is reset, it will remain this way since the S input is still held high. When the switch is thrown back to position A, Q will go high when first contact is made at position A. The Q output will thus change cleanly from 0 to 5 volts (or vice versa) at each throw of the switch. No bounce will occur. ∎

■ **EXAMPLE 5.2–5**

A home consists of two heating zones, each with its own thermostat and electrical heating system. Because of billing rates, it is undesirable to have both zones receiving heat at the same time. On the other hand, neither zone has any priority for heat. Thus we need a circuit that will work as follows. If zone A requests heat and zone B does not, then zone A gets heat (and vice versa). On the other hand, if zone A requests heat but zone B is already receiving heat, then zone A must wait until zone B is finished (and vice versa). Let us see how a FF can solve this problem.

We will assume that a request for heat is a logical low. We also assume that a logical high will turn on a heating zone. Refer to Figure 5.2(a). Let us connect the thermostat output from zone A to the S input of the FF. The heating unit for zone A is connected to the $\overline{Q}$ output. Likewise, the thermostat for zone B is connected to the R input, and the heating unit for zone B is connected to the Q output. First, assume that neither zone requests heat. Then both the R and S inputs will be high, and both Q and $\overline{Q}$ will be low. Now suppose that zone A wants heat. It asserts a low at the S input. The truth table shows that with $R = 1$ and $S = 0$, $\overline{Q}$ will go high, sending heat to zone A. Now suppose that, while heat is being sent to zone A, zone B requests heat. It does this by asserting a low at input R. Now both R and S inputs are low. The truth table shows us that this leaves Q and $\overline{Q}$ unchanged. Thus zone B must wait. When zone A is finished, S will go high, leaving $R = 0$ and $S = 1$. The truth table shows us that this forces $\overline{Q}$ low and Q high, sending heat to zone B. Finally, when B is satisfied, it makes R high. With both R and S high, both Q and $\overline{Q}$ are forced low, turning off all the heat. ∎

The Clocked R–S Flip-flop

One problem with the R–S flip-flop is its susceptibility to voltage noise. For the NAND version, for example, any momentary negative voltage spike on the S input will set the FF. It is often useful to be able to disable the FF, and only enable it when the possibility exists that its state should be changed.

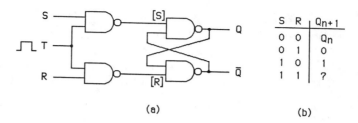

FIGURE 5.4   (a) Clocked *R–S* flip-flop. (b) Truth table for the circuit in (a).

The clocked *R–S* FF shown in Figure 5.4(a) can be enabled or disabled by the clock input. With the clock input held low, the [S] and [R] states are both held high. But these states are just inputs to the NAND *R–S* FF output stage. With [S] and [R] both high, the output FF remains unchanged, regardless of any changes in the *R* or *S* inputs. Thus no change in output can occur as long as the clock input is low. When the clock input goes high, then the output of the FF follows the *R* and *S* inputs according to the truth table. The output will be frozen or *latched* at the value it has when the clock input goes low.

■ EXAMPLE 5.2–6

Suppose *S* and *R* are both low and the clock input goes high and then low. While the clock is high, *S* and *R* determine [S] and [R]. With both *S* and *R* low, then [S] and [R] will be high. This keeps the output of the FF unchanged. When the clock goes low, there will have been no change in output. In the truth table, this is indicated by saying that the output following the clock pulse (Q_{n+1}) is the same as the output before the clock pulse (Q_n).   ■

■ EXAMPLE 5.2–7

Suppose that, while the clock pulse is high, *R* is low and *S* is high. This will force [R] to be high and [S] to be low, which forces *Q* to be high and $\overline{Q}$ to be low. If the clock input now goes low, this will force both [R] and [S] high, thus freezing (latching) the FF output in the set state. Thus the output following the clock pulse (Q_{n+1}) will be (1) regardless of what it was before the clock pulse.   ■

■ EXAMPLE 5.2–8

Consider what happens if both *S* and *R* are high when the clock input goes from high to low. With *S* and *R* both high, and the clock high also, then [S] and [R] will both be low. This will force both Q and $\overline{Q}$ high. However, when the clock goes low, one of the two NAND outputs [S] or [R] will go high a split second before the other. This will leave the FF output in a definite state, but there is no way to predict this state unless we know which NAND gate happens to be faster. Thus the truth table has a ? entered for the $S = 1$, $R = 1$ input state. ∎

Although the output FF in Figure 5.4(a) is a normal NAND-type FF with active-low inputs, the inverting action of the first set of NAND gates makes the S and R inputs to the main circuit active high. The FF can change state as soon as the clock input goes high. Thus changes occur on the *leading edge* of the clock input. Changes can continue to occur as long as the clock input stays high. The output is latched when the clock input goes low.

5.3 THE *D*-TYPE FLIP-FLOP

The *D*-type flip-flop remedies the indeterminate state problem that exists when both inputs to a clocked R–S FF are high. Figure 5.5(a) shows the circuit for a *D*-type FF. The D input is connected directly to what amounts to the S input of a clocked R–S FF. The D input inverted is connected to the R input of the clocked R–S FF. It is impossible for both the S and R inputs to simultaneously be high, so there is no indeterminate state.

The truth table in Figure 5.5(b) summarizes how the circuit works. When T is low, the output remains unchanged, regardless of the state of the D input (shown as an X, signifying the input state doesn't matter). When T is high, Q simply follows D. The output is latched when T goes low. Since this circuit is essentially a clocked R–S FF, it works the same way an R–S FF does, except for the absence of the

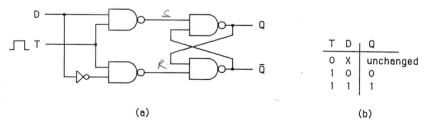

(a)

T	D	Q
0	X	unchanged
1	0	0
1	1	1

(b)

FIGURE 5.5 (a) *D*-type flip-flop. (b) Truth table for the circuit in (a).

indeterminate state. The simple *D*-type FF of Figure 5.5(a) is often referred to as a data latch since it latches the data on input *D* and holds it until the next clock pulse.

The 74100 dual 4-bit latch that we encountered in Chapter 4 is essentially two groups of four *D*-type FFs on one chip. Refer back to Figure 4.18 to recall how such latches are used. Latches are also available as tristate outputs. The 74363 and 74373 are both 8-bit tristate latches. This means that data can be latched into internal *D*-type FFs on a clock pulse, but the data will not be connected to the eight outputs until the tristate output is enabled. Such an arrangement is useful for connecting outputs of several latches to a common group of output lines (called a *bus*).

Edge-triggered *D*-type Flip-flop

It is possible to enhance the *D*-type FF in several ways. One potential problem with the *D*-type FF is that it will change state as long as the clock input is high. By adding some circuitry to the *D*-type FF, it is possible to make the FF latch the data that are present at input *D* *at the time the clock goes from low to high*. That is, data are latched at the precise instant of the leading edge of the clock. The FF is then referred to as an edge-triggered FF. Changes in *D* that occur following the leading edge of the clock have no effect.

Other enhancements include direct-set and direct-reset inputs. These inputs allow the FF to be set or reset independently of the clock input or data input. The circuit in Figure 5.6(a) incorporates both the edge-triggered and direct set–reset features. Since the circuit in Figure 5.6(a) has grown a little complicated, a simpler functional diagram is used to represent this circuit. This diagram is shown in Figure 5.6(b). Notice that the direct set ($\overline{S}$) and reset ($\overline{R}$) inputs are active low. The small arrow on the *T* input pulse indicates that the circuit is triggered on the leading edge of the pulse.

As a practical matter, for reliable operation the data input (*D*) must be stable for a short time before *T* goes from low to high (the *setup* time), and *D* must remain stable for a short time after *T* goes high (the *hold* time). When edge-triggered *D*-type FFs are fabricated on MSI ICs, the setup and hold times are specified by the manufacturer of the IC and depend on the TTL family involved. For example, for LS TTL circuits, the setup time is typically given as 25 nanoseconds and the hold time as 5 nanoseconds.

The detailed operation of the circuit in Figure 5.6(a) is a little complicated. The easiest part to understand is the operation of the $\overline{R}$ and $\overline{S}$ inputs. Normally, these inputs are held inactive (high). If $\overline{S}$ goes low, this low is fed directly to NAND gate 5 and the gate is forced high, setting the FF. The $\overline{R}$ input is likewise connected directly to NAND gate 6, and a low on $\overline{R}$ will force $\overline{Q}$ high, resetting the FF.

To understand the edge-triggered function of the circuit, we begin by assuming that the FF is set, *T* is low, $\overline{S}$ and $\overline{R}$ are high, and *D* is low. This sets up the set

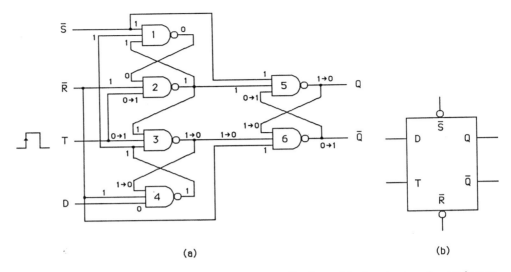

FIGURE 5.6 (a) An edge-triggered *D*-type flip-flop with direct-set and reset inputs. (b) Standard symbol for circuit in (a).

of states shown as 1's and 0's in Figure 5.6(a). You should study the figure closely to be sure that you understand why each input and output has the value shown. Now, what happens at the instant that *T* goes high? When *T* goes high, this will make all three inputs to NAND 3 high. Thus its output is forced low. This immediately forces the output of NAND 6 high, which in turn forces the output of NAND 5 low. All the changes forced by *T* going high are shown on Figure 5.6(a) by a small arrow leading from the state before the clock pulse to the state after the clock pulse. If no arrow is shown, the state does not change. The important point is that the circuit output changes as soon as *T* goes high. Notice also that, once the output of NAND 3 has gone low, this seals NAND 4 in a high-output state *regardless of the state of D following the clock transition*. This is what prevents any further changes in the output until the next leading edge of the clock.

Uses of Edge-triggered D-type Flip-flops

Edge-triggered *D*-type FFs are quite useful for capturing data at a predetermined instant of time. Since data are captured at the leading edge of the clock pulse, it is necessary only to control the time at which this leading edge occurs in order to capture the data. Edge-triggered *D*-type FFs are available as MSI ICs in groups of four (74175), six (74174), or eight (74273) and in tristate outputs in groups of four (73173) and eight (74374). Near the end of the book we will encounter tristate latches and edge-triggered *D*-type flip-flops when we discuss interfacing digital circuits to microprocessors.

5.4 THE *J–K* FLIP-FLOP

One of the most versatile types of flip-flops is the *J–K* flip-flop. This type of FF is usually constructed as a master–slave type FF, which internally consists of two clocked *R–S* FFs. However, not all *J–K* FFs are of the master–slave type. In either case, the operation is logically almost identical.

Figure 5.7(a) shows one way to construct a *J–K* master–slave FF from SSI gates and one diode. First, let us consider the basic function of the circuit. Then we will look at the circuit operation in detail. The clock is normally held low. In this condition, the states of the *J* and *K* inputs do not matter. The *Q* and $\overline{Q}$ outputs will remain unchanged as long as the clock remains low. To guarantee reliable performance, the *J* and *K* inputs should be stable prior to the time that the clock goes high. When a clock pulse occurs, the circuit output will not change state until the *falling* edge of the clock pulse. At this time, the outputs will change according to the truth table. In the table, Q_{n+1} refers to the state of the *Q* output following the clock pulse, while Q_n refers to the state of *Q* prior to the clock pulse. Let us review each entry in the truth table.

J = 0, *K* = 0: With this set of inputs, the output will remain unchanged by a clock pulse.

J = 0, *K* = 1: With this set of inputs, the FF will be reset following the clock pulse, regardless of the state of the FF prior to the clock pulse.

J = 1, *K* = 0: With this set of inputs, the FF will be set following the clock pulse, regardless of the state of the FF prior to the clock pulse.

J = 1, *K* = 1: With this set of inputs, the FF will reverse its state following the clock pulse. That is, if the FF was set prior to the clock pulse, it will be reset following the clock pulse, and vice versa.

The versatility of the *J–K* FF stems from these four different modes of operation. We will see in Chapter 6 how several *J–K* FFs can be interconnected to make many functionally different types of circuits.

Now let us consider the detailed operation of the circuit in Figure 5.7(a). Essentially, the circuit is comprised of two clocked *R–S* FFs, the first called the master FF and the second called the slave FF. Recall that for NAND-type clocked *R–S* FFs, the *R–S* inputs, and the clock input are active high. With this in mind, we will first consider the state where the clock is low. In this state, the outputs of both NAND 1, [*S*], and NAND 2, [*R*], will be high. These are inputs to NANDs 3 and 4, which comprise a simple *R–S* FF. Since the [*S*] and [*R*] inputs to this FF

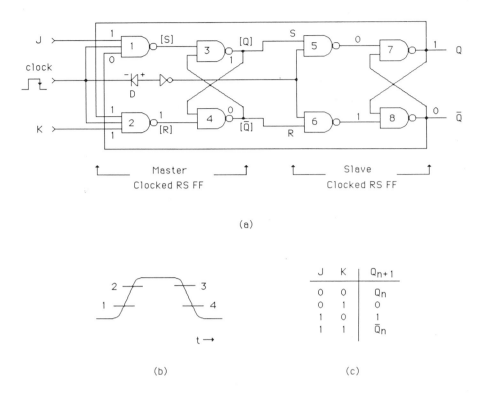

FIGURE 5.7   (a) *J–K* master–slave flip-flop. (b) Detailed profile of clock pulse. (c) Circuit truth table.

are both high, the FF is in its stable (remains unchanged) state. Thus the outputs of this FF, $[Q]$ and $[\overline{Q}]$, will be in some stable state, say 1 and 0, respectively. These outputs are the S and R inputs to the slave R–S clocked FF. Now recall that our assumption was that the clock input to the circuit was low. But the inverter following the clock input means that the clock input to the slave-clocked R–S FF is high. The slave clock is thus in its active state. This means that the state of S will be passed directly out as the Q output (and the state of R will be output as the $\overline{Q}$ output). Summarizing, then, when the clock input is low, the output of the master FF is passed directly through the slave FF and is output as Q and $\overline{Q}$.

Now let us consider what happens as the clock starts to go high. The diode comes into play at this point since, when the cathode ($-$) of the diode is held low, the anode ($+$) will be about 0.6 volt higher in voltage than the cathode. Thus, as the clock pulse begins to rise, the input to the inverter will be about 0.6 volt above the voltage of the clock pulse. This means that the inverter input will reach a logical high value before the clock pulse itself (which is an input to NANDs 1 and 2) does. Once the inverter input has reached logical high, the output will switch to logical low, *disabling the slave-clocked R–S FF.* Thus the slave outputs become latched before the clock pulse rises high enough to affect NANDs 1 or 2. This happens at point 1 in Figure 5.7(b).

To progress further, let us assume that the FF is initially set and that the *J* and *K* inputs are both 1. With the clock low, these circumstances lead to the states shown as 1's and 0's in Figure 5.7(a). As noted previously, as the clock starts to go high, the clock input to NANDs 5 and 6 will go low first, and the output of both of these gates will go high, latching the *Q* and $\overline{Q}$ outputs in their current state. This happens at point 1 in Figure 5.7(b). At point 2 in Figure 5.7(b), the clock gets high enough to be recognized as a logical high by NANDs 1 and 2. When this happens, NAND 1 will have two highs and a low as inputs. Thus its output will remain high. NAND 2 will have three highs as inputs, and thus its output will go low. This will force the output of NAND 4, $[\overline{Q}]$, to go high and the output of NAND 3, $[Q]$, to go low. The master FF has been reset. However, since the slave FF is disabled by the low on its clock input, the state of the slave FF remains unchanged. Summarizing, at point 2 in Figure 5.7(b), the master FF becomes enabled to the *J–K* inputs, and in our example this causes the master FF to be reset.

The clock pulse now begins to drop back toward zero. At point 3 in Figure 5.7(b), the pulse drops enough to be recognized as a logical low by NANDs 1 and 2. This will force the output of both these gates high and latch the state of the simple *R–S* FF comprised of NANDs 3 and 4. That is, the master FF will be in its latched state. Because of the 0.6-volt drop across the diode, however, the inverter is still seeing a logical high at its input, so the inverter output is still low. Thus the slave FF is also still latched. Finally, at point 4 in Figure 5.7(b), the clock input gets low enough so that the inverter sees a logical low at its input. The inverter output then goes high, enabling the slave-clocked *R–S* FF. The output of the master FF is now passed directly through the slave as the *Q* and $\overline{Q}$ outputs. But recall that the master FF had been reset. Thus the slave output will also now be reset. The FF has changed its state from set to reset at point 4 on the falling edge of the clock pulse. This is just what the truth table indicates should happen.

It is important to note that the preceding analysis assumes that *J* and *K* are stable during the high clock pulse. If *J* or *K* is allowed to change during the clock pulse, some strange things can happen. Therefore, sound practice demands that the designer guarantee that the *J* and *K* inputs be held stable during the clock pulse.

TTL *J–K* Flip-flops

J–K flip-flops are available as several TTL ICs. One common example is the 7476 shown in Figure 5.8. This IC has two *J–K* FFs with direct set and reset inputs. The connection diagram is shown in Figure 5.8(a). This circuit is available as the 7476, 74H76, and 74LS76. The 74 and 74H versions are positive-pulse-triggered

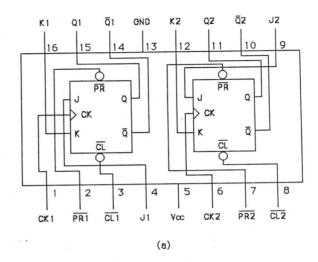

(a)

Mode	PR	Cl	CK	J	K	Q	Q̄
Asynchronous set	L	H	X	X	X	H	L
Asynchronous reset	H	L	X	X	X	L	H
Undetermined	L	L	X	X	X	H	H
Toggle	H	H	⎍	h	h	q̄	q
Reset	H	H	⎍	l	h	L	H
Set	H	H	⎍	h	l	H	L
Hold	H	H	⎍	l	l	q	q̄

q, q̄ = state of Q, Q̄ prior to clock pulse

⎍ = clock pulse

l, h = low, high voltage one set-up time prior to
        clock transistion

(b)

FIGURE 5.8 (a) Connection diagram for 7476 dual $J-K$ flip-flop. (b) Truth table for each $J-K$ flip-flop.

FFs. That is, JK information is loaded into the master FF while the clock is high and transferred to the slave at the falling edge of the clock. It is therefore necessary to keep the JK inputs stable during the entire high phase of the clock. The 74LS version is an edge-triggered JK FF. Thus it is necessary that the JK inputs be stable only one setup time prior to the falling edge of the clock pulse. The setup time for the 74 and 74H versions is listed as 0 nanoseconds. The setup time for the 74LS version is listed as 20 nanoseconds. The hold time for all three versions is listed as 0 nanoseconds.

Other 74XX *JK* FFs include the 7473, 74101-103, 74106-109, and 74112-114. For example, the 74103 is a dual edge-triggered *JK* FF with a direct-reset (but not direct-set) input. In this case, the *JK* inputs may change while the clock is high (like the 74LS76). The output depends only on the value of the *JK* inputs one setup time prior to the falling edge of the clock. As another example, the 74102 is a single edge-triggered *JK* FF with direct set and reset inputs. The *J* and *K* inputs are actually built-in 3-input AND gates. Thus there are three "*J*" inputs, all of which must be high in order for *J* to be high, and likewise for *K*.

PROBLEMS

1. Refer to Figure 5.2(a). Assume that R and S are both low. Show that the two sets of output states $Q = 1, \overline{Q} = 0$ and $Q = 0, \overline{Q} = 1$ are both stable output states.

2. In Figure 5.2(a), assume that R and S are both low and the FF is set. Show that when R is taken high this will reset the FF.

3. Design a switch debounce circuit similar to that of Figure 5.3(b) using NOR gates.

4. Design a clocked $R-S$ FF that uses NOR gates rather than NAND gates. Make sure you identify the RS inputs and the Q and $\overline{Q}$ outputs. Are the RS inputs active high or low? Is the clock active high or low?

5. Redesign the circuit of Figure 5.4(a) to include direct set and reset inputs. You may alter the number of inputs on some of the NAND gates.

6. The circuit in Figure 5.4(a) is subjected to the inputs shown in Figure 5.9. Make a plot of Q and $\overline{Q}$ as a function of time. Assume that the FF is initially set.

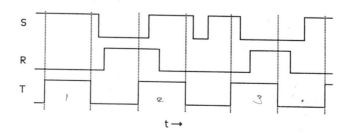

FIGURE 5.9 Use with Problem 6.

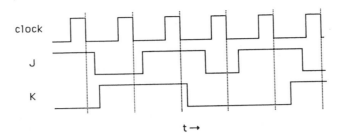

FIGURE 5.10   Use with Problem 10.

7. Redesign the circuit of Figure 5.5(a) to include direct set and reset inputs.

8. Make a copy of Figure 5.6(a) without the input and output states labeled with 1's and 0's. Now assume that the FF is reset, D is 1, the clock is low, and the RS inputs are high. Determine and label the state of every input and output for all the gates in the circuit. The clock now goes high. Redetermine the state of every input and output. Is the circuit now set or reset? The clock now goes low again. Redetermine the state of every input and output. Has the Q output changed as a result of the clock going low?

9. In the circuit in Figure 5.7, the J input is low and K input is high. The circuit is set ($Q = 1$). The clock is low. Determine the state of all gate inputs and outputs at this time. A clock pulse is now input. Determine the state of all gate inputs and outputs at each of the four times labeled on Figure 5.7(b). At the end of the pulse, is the circuit set or reset?

10. A $J-K$ FF is initially reset. It is subjected to the inputs shown in Figure 5.10. Make a plot of Q versus time for the FF.

11. Redesign the circuit of Figure 5.7(a) to include direct set and reset inputs.

12. Assuming standard TTL behavior, how would a $J-K$ FF behave if its J and K inputs were left unconnected?

13. The circuits shown in Figure 5.11 are both initially set. A series of five clock pulses is input to each. Make a table showing the state of the J and K inputs and the Q and $\overline{Q}$ outputs before the first clock pulse occurs. Using the values of J and K in your table, determine the state of the Q and $\overline{Q}$ outputs following the first clock pulse, and use this information to determine the new values of the J and K inputs following the first clock pulse. Continue this procedure to determine the state of Q and $\overline{Q}$ following each clock pulse.

14. Show that an edge-triggered D-type FF can be made from a $J-K$ FF by connecting a D line to the $J-K$ inputs in the manner of Figure 5.5.

15. One common type of burglar alarm involves placing metal foil on windows. As long as the foil is not broken, a circuit input remains grounded. If the foil

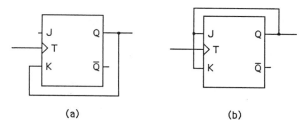

(a)   (b)

FIGURE 5.11 Use with Problem 13.

is broken, the circuit input goes high. Show how a FF could be used to indicate the breakage of a window, even if the break is quickly repaired.

16. The states of a series of devices are to be monitored with LEDs. The LEDs will go on when the devices are in a high state and off otherwise. Since the devices might change rapidly, it is desired to have a control switch that can be thrown to temporarily latch the state of the devices at the time the switch is thrown to the hold position. When the switch is returned to the monitor position, the lights should resume, indicating the instantaneous state of the devices. What type of FF would be best for this job? Design a circuit that monitors three lights in the manner described.

Chapter 6

FLIP-FLOP APPLICATIONS

Learning Objectives

After completing this chapter you should know:

- How to construct a simple asynchronous up or down counter with $J-K$ flip-flops.

- The difference between an asynchronous and a synchronous counter.

- How to add reset and preset functions to a counter.

- How the versatile 74193 binary up/down counter functions.

- What a BCD counter is and how it functions.

- Examples of using $J-K$ flip-flops to fabricate modulus-N counters.

- How to use MSI counters to fabricate modulus-N counters.

- What a shift register is and how it can be constructed from $J-K$ flip-flops.

- Examples of MSI shift registers.

- What a ring counter is and how it can be fabricated from $J-K$ flip-flops.

6.1 INTRODUCTION

In Chapter 5, we became acquainted with a number of kinds of flip-flop circuits. In this chapter we will explore some of the most important applications of such circuits. Among these are a wide variety of counters, shift registers, and ring counters. Applications of this type can be implemented by using discrete flip-flops and SSI gates. As before, however, we will also find that MSI circuits are available that implement these functions.

6.2 THE BASIC BINARY COUNTER

A *digital counter* is a circuit that accepts a sequence of clock pulses (a complete sequence of logical 0, 1, 0) on an input, and produces a multibit output that represents in some binary code the number of pulses that has been received. A *binary counter* is a counter that uses a straight binary code for its output. It can be made quite easily by simply connecting together a number of $J-K$ flip-flops. Such a circuit is shown in Figure 6.1(a). This circuit is a 4-bit binary counter. It is designed to count the number of pulses that arrive at the clock input to the first $J-K$ FF.

In Figure 6.1(a), the outputs of each FF have not been labeled Q and $\overline{Q}$, but rather A, $\overline{A}$, and so on. This is simply to make it easier to identify the different FFs. Notice that the J and K inputs to all the FFs in the circuit are left open. Assuming TTL circuitry, this means that all the J and K inputs will be high. Recall from Chapter 5 that when J and K are left high the FF simply toggles back and forth between its set and reset states at the *falling edge* of each clock pulse. With this in mind, let us analyze the operation of the circuit.

We assume a series of clock pulses, such as those shown in Figure 6.1(c), are input to FF A. At the falling edge of each clock pulse, FF A will reverse its state. Thus FF A will toggle indefinitely through the series of states 01010101. . . .Notice that on *every other* clock pulse FF A will change state from 1 to 0. In Figure 6.1(a), we see that the set output of FF A is connected directly to the clock input of FF B. Thus, each time the output of FF A goes from 1 to 0, FF B will toggle its state. FF B is likewise connected to the clock input of FF C. Thus, everytime FF B changes from 1 to 0, FF C will toggle its state, and similarly for FF D. If, for convenience, we assume that all four FFs are initially reset (so the beginning state is 0000), then the count will proceed as indicated in Figure 6.1(b). This information is repeated in a more graphic fashion in Figure 6.1(c). Notice that each FF only changes state on the falling edge of its clock input.

The binary counter in Figure 6.1(a) obviously counts upward in standard binary code. It is therefore called an up counter. It is also called an *asynchronous counter* because the FFs do not change their state simultaneously. FF B cannot

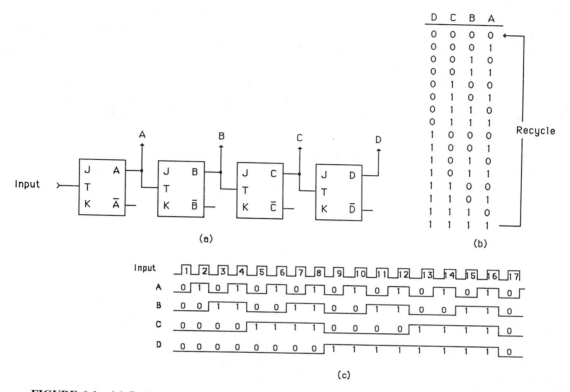

(a)

(b)

(c)

FIGURE 6.1 (a) Basic 4-bit asynchronous binary counter. (b) Sequence of states for the 4-bit counter. (c) Waveforms associated with the 4-bit counter.

change until FF *A* has already changed state, and FF *C* cannot change until FF *B* has changed, and so on. Thus the count "ripples" through the counter over a short, but finite, period of time. For this reason, the type of counter shown in Figure 6.1(a) is also called a *ripple counter*. Because of the ripple effect, there will be short periods of time when the count output will be incorrect because one of the earlier FFs has changed state, but the later FFs have not. We commented on this before in Chapter 4 when we discussed binary codes and intermediate states. Here we see how these intermediate states arise. It is obvious that the more bits the counter has, the worse the problem of intermediate states becomes. There is a definite delay time from one FF to the next. The last FF will thus change state only after all other previous FFs have changed. If the input clock rate is sufficiently high, it is quite possible to have the first FF changing to a new state before the last FF has even adjusted to the previous clock pulse. In this situation, the count on the counter is essentially meaningless. Asynchronous binary counters can be useful if the clock rate is not too high and if intermediate states cause no problem.

The Down Counter

It is quite easy to design an asynchronous binary down counter. We simply make use of the reset outputs of each FF, rather than the set outputs. Such a circuit is shown in Figure 6.2(a). Notice that the J–K inputs are still left unconnected, but now it is the reset outputs that are connected to the clock input of the next FF. Each FF will still toggle on the falling edge of its clock, but the falling edge now occurs when the reset output of the previous FF goes from 1 to 0, that is, when the *set* state goes from 0 to 1. To repeat, *each FF of the down counter will toggle when the set input of the previous FF goes from 0 to 1.* To satisfy ourselves that the counter indeed counts down, let us examine a few specific situations.

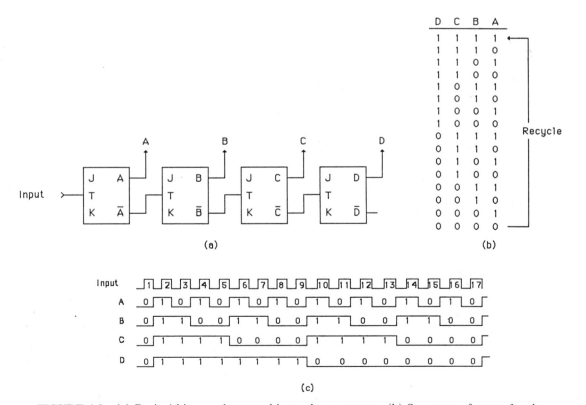

FIGURE 6.2 (a) Basic 4-bit asynchronous binary down counter. (b) Sequence of states for the down counter. (c) Waveforms associated with the 4-bit down counter.

■ EXAMPLE 6.2—1

First, suppose that the state of the counter is 1111. A new pulse arrives that toggles FF A. The counter thus changes state to 1110. Since FF A went from 1 to 0, this does not change FF B. Now another pulse arrives. FF A goes from 0 to 1, but this change also changes FF B (from 1 to 0). Thus the new count is 1101. The next pulse will toggle FF A from 1 to 0 (which will not affect FF B). Thus the new count is 1100. You can see that the count is proceeding downward.   ■

■ EXAMPLE 6.2—2

Suppose that the count output of the down counter has finally reached 0000. The next clock pulse will toggle FF A to 1. This 0 to 1 transition will toggle FF B from 0 to 1. But this 0 to 1 transition will toggle FF C from 0 to 1, which will toggle FF D also from 0 to 1. Thus the count state will have gone from 0000 to 1111, which will complete the cycle. Figures 6.2(b) and (c) show the count sequence in table and graphic form. Notice that FFs B, C, and D always change state on the leading edge of the *set* output of the previous FF.   ■

The Up—Down Counter

It is entirely possible to have a single counter that will count either up or down, depending on the state of a control input. It is only necessary to arrange circuitry that will *either* connect the set output of one FF to the clock input of the next, *or* connect the reset output of one FF to the clock input of the next. The circuit in Figure 6.3 accomplishes this.

In the circuit in Figure 6.3, there is a single count-control input. This count control is connected directly to each AND gate numbered 1, and the inverted count control is connected to each AND gate numbered 2. Let us assume that the count

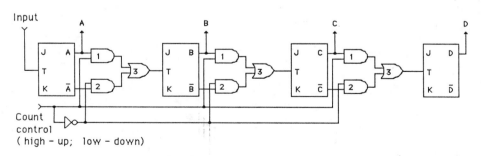

FIGURE 6.3   A 4-bit asynchronous binary up/down counter.

control is high. This means that all the number 1 AND gates will be enabled. By enabled, we mean that the output of each of these AND gates is now under control of the set output from FF A, B, or C. For example, if FF A is set, then the output of the number 1 AND gate to which it is connected will be high. This high will pass through the OR gate directly to the clock input of FF B. If FF A is reset, then the output of the AND gate will be zero, and this will keep the output of the OR gate zero. Thus the state of the set output of FF A is passed directly to the clock output of FF B. A similar situation exists for the other FFs. But this is just the connection necessary to produce an up counter. Notice that if the count control is high then the inverted count control must be low, disabling all the number 2 AND gates (forcing all their outputs low).

If the count control is made low, then all the number 1 AND gates will be disabled, but the number 2 AND gates will be enabled. This will now connect the reset output of each FF to the clock input of the next FF. A down counter is thus produced.

Synchronous Binary Counters

The counters that we have seen so far have been asynchronous. That is, each FF must wait until the previous FF has changed before it too can change. Because the resulting intermediate states can be a problem, it would be useful if we could design a counter where all the FFs changed at the same time. A counter designed so that all the count outputs change at the same time is called a *synchronous counter*. Such a counter is shown in Figure 6.4.

The synchronous counter has the same clock connected to the clock input of all the J–K FFs that make up the counter. The trick is to make sure that each FF changes state only under the correct circumstances. If the counter is an up counter, then what are these circumstances? First, of course, FF A should toggle with each clock pulse since FF A is the least significant bit of the count. FF B should only toggle when FF A is already high. Thus we connect the set output of FF A directly

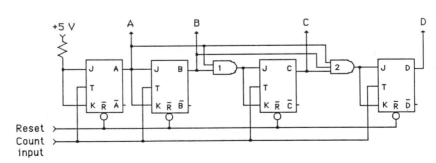

FIGURE 6.4 A synchronous binary counter with reset input.

to both the J and K inputs of FF B. If FF A is high, FF B will toggle on the next pulse (along with FF A). If FF A is low, then FF B will not change (when J and K are both 0, no change occurs). What about FF C? It should change only if *both* FF A and FF B are high. That is, the count is something like 1011 and must now change to 1100. Thus we AND the set outputs of FF A and FF B and input the result to the J and K inputs of FF C. If FFs A and B are both set, the output of AND 1 will be high, and FF C will toggle at the next clock pulse (as will FFs A and B). Otherwise, FF C will remain unchanged. Finally, FF D should change only if FFs A, B, and C are all set. Thus we AND the set outputs of FFs A, B, and C and input the result to the J and K inputs of FF D. This is done with AND gate 2 in Figure 6.4.

Figure 6.4 also introduces one additional control feature. There is a direct reset line connected to each FF. It is an active-low line that is normally held high. If this line is made low at any time, all four FFs will be reset and the count will become 0000. This is a very useful feature for conveniently keeping track of the number of counts that have arrived following a given start pulse.

Other Counter Controls

In addition to being able to reset a counter, it is sometimes convenient to be able to preset the counter to a given value. The circuit in Figure 6.5 shows one way to do this.

The scheme used in Figure 6.5 is not a true preset approach. That is, to produce correct results, the circuit must first be reset and then preset. Let us see how it works. The reset line is clearly active-low and is usually held high. The

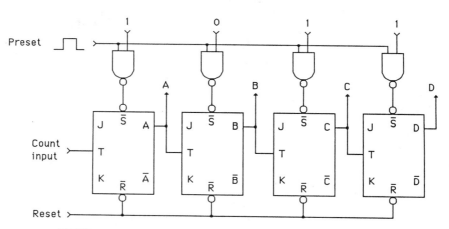

FIGURE 6.5 An asynchronous counter with reset and preset.

preset line is active-high. As long as the preset line is held low, the four NAND gates will have high outputs. Since the direct-set inputs to each FF are active-low, as long as the output of the NAND gates remains high, the FFs will not be affected. Now suppose that we wish to introduce a preset count to the counter. We place that count on the preset input to the four NAND gates and then enable the NAND gates with a high pulse on the preset input. For any NAND gate with a preset input of 1, the output of the NAND gate will go low, *setting the corresponding FF*. For any NAND gate with a preset input of 0, the output of the NAND gate will stay high, and the corresponding FF will be unaffected. The only problem with this scheme is that the preset control has no power to *reset* a FF. That is, suppose in Figure 6.5 that FF *B* happened to be set. The preset inputs shown in the figure are such that we desire FF *B* to become reset following the preset pulse. But the 0 on the preset input to NAND gate 2 will leave the output of this gate high and thus will make no change in the state of FF *B*. This is why it is first necessary to reset all the FFs using the reset input and then use the preset line to set only those FFs that are desired to be set.

It is possible to design circuitry that will preset each FF to whatever state is desired without first having to reset each FF. A scheme to do this is shown in Figure 6.6. In this circuit, the input labeled Preset is active-high. Thus it is normally kept low. When the Preset line is low, the outputs of NAND gates 1 and 2 are both forced high, disabling both the direct-set and reset inputs, which are active-low. When the Preset line is made high, there are two possibilities. The Value In line will be either high or low. Suppose it is high. Then the output of NAND 1 will be forced low (activating the set input), and the output of NAND 2 will be forced high (by the low from NAND 1), thus keeping the reset input inactive. The FF will thus be set by the 1 on the Value In line when the Preset line goes active. The other possibility is that the Value In line contains a 0 when the Preset line goes high. In this case, NAND 1 will be forced high by the 0 on the Value In line (thus keeping the set input inactive), and NAND 2 will be forced low (by the high from NAND 1 and the high on the Preset line). This low will reset the FF. Thus a 0 on the Value In line will reset the FF when the Preset line goes active.

We can add a direct reset capability to the circuit in Figure 6.6. This is shown in Figure 6.7. The Reset line is connected to one input of NAND gate 3 (shown as a low-level input OR). The other input to NAND gate 3 comes from the reset

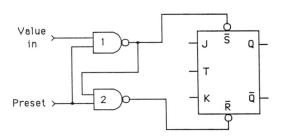

FIGURE 6.6 Preset without the necessity of reset first.

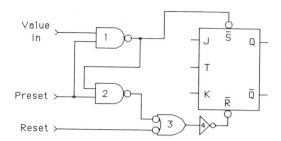

FIGURE 6.7 Preset with reset added.

portion of the preset circuitry. If either input to NAND gate 3 goes low, its output will go high. Inverter 4 will invert this high to a low, which will reset the FF. The Reset input is thus active-low and is normally held high. Likewise, the output of NAND 2 is normally high since the Preset input is normally held low (recall it is active-high). The FF is thus reset either by a low on the Reset line or by a 0 on the Value In line when the Preset line goes high.

6.3 SOME MSI COUNTERS

In Section 6.2 we developed all the tools to assemble a very versatile counter. The various options, such as up/down control, reset, and preset, could all be added to a counter. This would require a fair amount of SSI circuitry. Fortunately, some MSI ICs contain all the options we have discussed.

The 74193 Counter

One very versatile counter is the 74193 synchronous binary counter. The logic diagram for this circuit is shown in Figure 6.8. A careful examination of the logic in Figure 6.8 shows that the 74193 incorporates all the features we have discussed in the previous sections. It is a *very* useful personal exercise to try to follow the logic of Figure 6.8 (you might want to try it as a test of your understanding). Also shown in this figure are the pin-connection numbers.

The 74193 contains a number of features. There are two count inputs labeled UP COUNT and DOWN COUNT. Both are normally held high. Counts are applied to one or the other of the two count inputs. The counts are actually registered on the rising edge of the external count input.

There is a direct-reset input labeled CLEAR, which is active-high. This line is normally held low. There is also a preset input labeled LOAD, which is active-low. Thus it is normally held high. When the LOAD input is made low, the data

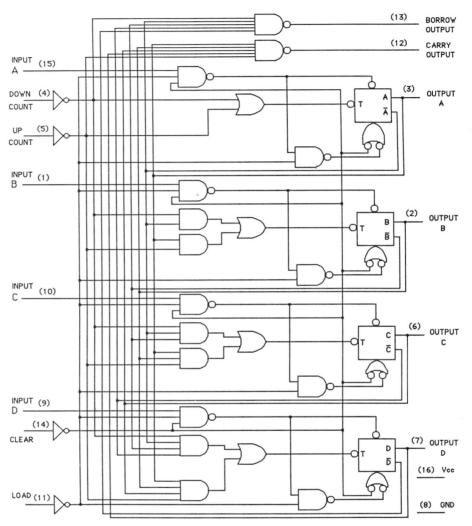

FIGURE 6.8 Logic diagram and pin-outs for the 74193 binary counter.

on the data inputs A to D will be loaded into the counter. The LOAD input would then normally be made high again. The count shows up on outputs A to D.

Finally, there are two more outputs on the 74193 to be discussed. These are the CARRY OUTPUT and the BORROW OUTPUT. Normally, CARRY OUTPUT will be high. However, when the counter reaches 1111 AND the UP COUNT input is low, then the CARRY OUTPUT will go low for the low period of the UP COUNT input. This brief negative pulse from the CARRY OUTPUT can be used as an UP COUNT input to another 74193 counter. That is, counters can be ganged together to form 8, 12, 16 or higher bit counters. The BORROW OUTPUT works in the same way as the CARRY OUTPUT. Thus, when the count reaches 0000 AND the DOWN COUNT input is low, then the BORROW OUTPUT will go low for the low period of the DOWN COUNT input. Again, this output can be connected to the DOWN COUNT input of another 74193 to form a counter with more bits.

Timing Diagram for the 74193 Counter

Figure 6.9 is a timing diagram showing different hypothetical inputs to a 74193 counter and the resulting outputs. Let us follow this diagram through from beginning to end. It begins with the CLEAR input low (inactive), the LOAD input high (inactive), and both count inputs high (inactive). The DATA inputs are also low at this time, although this is really irrelevant. The CARRY and BORROW outputs are high, as they are most of the time. The count output happens to be 0101. First, we see that the DATA inputs are changed to 1101 (13). This has no effect at this time. Then a CLEAR pulse arrives (the CLEAR line goes high temporarily). This resets the counter to 0000. Next the LOAD line goes temporarily low. This loads the data on the DATA inputs into the counter. The count is now set at 1101 (13). Next a series of five pulses arrives on the COUNT UP input. On the rising edge

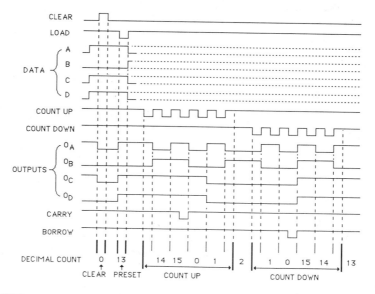

FIGURE 6.9 A hypothetical input sequence for the 74193 binary counter.

of each pulse, the count increases by 1. Notice what happens when the count reaches 15 (1111). At this point there is a brief negative pulse on the CARRY output during the time the counter is at 15 AND the COUNT UP input is low. The count then changes to 0000 and the CARRY output returns high again. After the count reaches 2 (0010), the COUNT UP input stays high, and the COUNT DOWN input begins to toggle. It receives five pulses in succession. On the rising edge of each pulse, the count decreases by 1. Again note what happens when the count reaches 0000. At this point there is a brief negative pulse on the BORROW output during the time the counter is at 0000 AND the COUNT DOWN input is low. The count then changes to 1111 and the BORROW output returns high again.

The 74193 as a Modulus-N Counter

The 74193 can be made into a modulus-N counter with little trouble. When we say modulus-N, we mean that the counter will count only through N distinct states before recycling to the beginning of its count. The 74193 normally counts through the cycle 0 to 15 and thus has 16 states. This makes it a modulus-16 counter under normal circumstances. Suppose we connect the 74193 as shown in Figure 6.10. Notice that we have permanently wired the data inputs with the binary number 0110 (6). Once the counter reaches 1111 (15), a short negative pulse will be produced on the CARRY output. This pulse will be sent, because of the direct-wire connection, to the LOAD input. The negative pulse will reload the counter with the number 0110, and the count will start again from this point. What will actually happen is that the count 1111 (15) will be reached and held for half a clock cycle of the UP COUNT input while the input is high. As soon as the UP COUNT input goes low, the CARRY output will also go low and this will reload the counter with 0110. This load is completed slightly into the low clock cycle. As soon as the load is complete, the CARRY output will go high again, since the count is no longer 1111. When the count input goes high, the count of 0110 will increment to the next count. Thus the count of 1111 will be held for about one-half of the clock cycle, and the count of 0110 will also be held for about one-half of the clock cycle. The

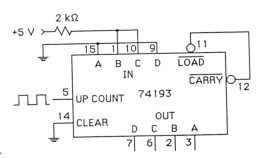

FIGURE 6.10   A modulus-10 counter.

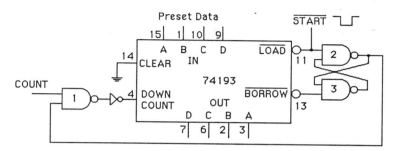

FIGURE 6.11   A down counter that stops at 0000.

net effect, however, is that the counter cycles through the states 6 to 15, a total of ten states. The counter is thus a modulus-10 counter.

By wiring a different binary number into the data-in inputs, we could construct a counter of any modulus from 2 to 15. It is also possible to achieve the same kind of result by using the DOWN COUNT input for the clock pulses and connecting the BORROW output to the LOAD input. Then each time the counter reaches 0000 it will be reloaded with the same number and begin to count down again.

The 74193 as One-cycle Counter

As a final example of using the 74193, consider the circuit shown in Figure 6.11. We assume that some preset data are permanently connected to the data-in inputs. When a negative START pulse is received on the LOAD input, this will load the counter with the preset number. It will also set the FF made up of NAND gates 2 and 3. This high will enable NAND gate 1 so that counts can pass to the DOWN COUNT input. The circuit will start to count down until it reaches 0000. At this point, a short negative pulse on the BORROW output will reset the FF made of NAND gates 2 and 3. This will disable NAND gate 1 and stop any further counting. The set output from NAND gate 2 could also be connected to a light or other signal device to signal when the counter has reached zero. It would also be no problem to have several counters ganged together to form a multibit counter. The FF comprised of NANDs 2 and 3 would only be at the end of the chain following the last counter.

BCD Counters

A BCD counter is a sequential counter that counts through a series of 10 distinct states before recycling; in other words, a modulus-10 counter. Usually, when we speak of a BCD counter, we also mean that it counts through the standard BCD

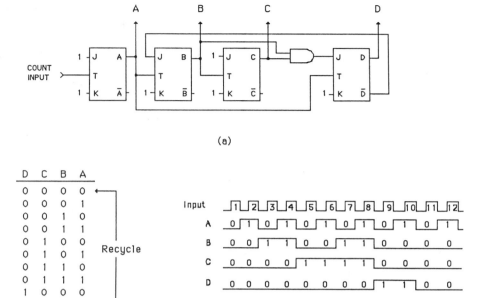

(a)

(b)   Recycle

(c)

FIGURE 6.12 (a) An asynchronous BCD counter. (b) The sequence of states for the counter. (c) Waveforms associated with the counter.

sequence of states. It is possible to design a modulus-10 counter that counts through any series of 10 binary states. Thus we could design a counter that cycles through the 10 XS3 Gray states or the 10 2421 states. In Chapter 8, we will see how to complete such designs. For the moment, however, we will concern ourselves with standard BCD code.

Figure 6.12(a) shows a design for a BCD counter. In this case it is an asynchronous counter since the count input is not connected to the clock input of all the J–K flip-flops in the circuit. What makes this counter different from the standard binary counter is the use of feedback from the outputs of later FFs to the inputs of earlier FFs. Let us examine how the circuit works. The first FF has both J and K inputs tied high, so this FF toggles with every input pulse to its clock input. The second FF is more complicated. Its K input is tied high, but its J input comes from the reset output of FF D. This means that, as long as the count is 0111 or less, the J input of FF B will be high. Since the clock input of FF B comes from the set output of FF A, FF B will toggle everytime FF A resets until the count gets to 1000. FF C has both its J and K inputs tied high, and its clock input comes from the set output of FF B. Thus FF C will toggle everytime FF B resets. Finally, we come to FF D. Its clock input comes from the set output of FF A.   The K input to FF D is tied high, but the J input comes from the AND of the set outputs of FFs B and C. Thus FF D can only be set when both FF B and C are also set.

If we assume that the counter in Figure 6.12(a) begins with all the FFs reset, then we can follow the count as pulses are received. As long as the count is below 1000, the first three FFs form a simple binary counter. That is, all three will have both their J and K inputs high, and the clock of each is fed by the set output of the previous FF. Thus the count will proceed normally until it reaches 0111. At this point, the first three FFs will toggle normally on the next clock pulse, but we must examine FF D. Since FFs B and C are both set when the count is 0111, the J input to FF D will be high. Thus, the next time FF A resets, FF D will toggle to 1. With the count at 0111, the next clock pulse will reset FF A. So once the count reaches 0111, the next pulse will cause the count to go to 1000. At this point, the J input to FF B will be low since FF D is set. This means that FF B must remain reset until FF D toggles its output. Since FF C gets its clock input from the set output of FF B, FF C is also stuck in the reset state until FF D toggles. With FFs B and C both reset, the J input to FF D is low. Thus the next clock pulse to FF D will cause it to reset. This clock pulse will come from the set output of FF A the next time FF A toggles from 1 to 0. But with the count at 1000, FF A is currently reset. Thus the next pulse to FF A will first set FF A. The count will become 1001. Following this, the next pulse to FF A will reset FF A, thus sending a falling edge to the clock input of FF D. This resets FF D. With FFs A and D being reset and FFs B and C already reset, the count returns to 0000 and the cycle begins anew. Figure 6.12(b) shows the count sequence, and Figure 6.12(c) shows in graphic form the state of each FF as the count input toggles. Notice that each FF toggles on the falling edge of the count input. This ignores the gate delays that are present from one FF to the next. A more precise graphing of the waveforms would show that each FF changes state after the falling edge of the clock input, with FF A changing first, followed by FFs B and C. FF D will toggle at the same time as FF B since both these FFs toggle on the falling edge of the set output of FF A.

It is common to gang several BCD counters together to form a many-digit decimal counter. The output from FF D of one counter is tied to the count input of the next higher counter. This works fine unless the count rate gets so high that the propagation delays become important.

Frequency Dividers

Consider the clock input to a BCD counter. Assume that the clock input repeats itself in a regular fashion [such a signal is shown as the input in Figure 6.12(c)]. This repeating clock signal is said to have a *period* and a *frequency*.

The period is the length of time necessary for the signal to repeat itself. The unit of measure for the period is time, such as seconds, milliseconds, or microseconds.

The frequency is the number of periods that occur in 1 second. The unit of measure for the frequency is the number of periods per second, which is called the hertz *(Hz).*

■ EXAMPLE 6.3–1

A clock signal repeats itself every 2 milliseconds. What is its period and its frequency?

The period is just the repeat time, which has been given as 2 milliseconds. The frequency is obtained by dividing 1 second by 2 milliseconds or 0.002 seconds. The result is

$$\frac{1}{0.002} = 500 \text{ Hz} \qquad ■$$

It is often useful to reduce the frequency of a clock signal. Usually, this is accomplished using a circuit that divides the frequency of the initial signal by some integer, and this integer is often a multiple of 10. A circuit that accomplishes frequency division is called a *frequency divider*. BCD counters, used individually or in groups, are often used as frequency dividers. For example, when one counter is used, FF *D* toggles once every 10 input pulses, in effect dividing the input frequency by 10. Two counters ganged together produce a divide-by-100 circuit. That is, the *D* FF of the second counter toggles once every 100 clock cycles. It is thus easy to produce a circuit that divides by any power of 10. The only limiting factor is that the frequency must not get so high that the first counter cannot keep up because of invalid intermediate states.

■ EXAMPLE 6.3–2

A clock signal exists that has a frequency of 1 megahertz (1 million Hz or 1 MHz). It is desired that this frequency be reduced to 1 kilohertz (1 thousand hertz or 1 kHz). How many BCD counters are needed to accomplish this?

Each BCD counter will reduce the frequency of the signal by a factor of 10. Since the desired reduction is from 1 megahertz to 1 kilohertz, a factor of 1000, three BCD counters are required. ■

It is not always desired to divide a clock frequency by factors of 10. It may be of interest to divide the frequency by 2 or 3 or some other number. The frequency of a clock signal can be divided by *N* by using a modulus-*N* counter.

■ EXAMPLE 6.3—3

A clock signal of 6 megahertz is input into the clock input of a $J-K$ FF for which both the J and K inputs are held high. The Q output of the $J-K$ FF is input to the clock input of a modulus-3 counter. What is the frequency of the signal observed at the Q output of the $J-K$ FF and at the msb output of the modulus-3 counter?

The Q output of the $J-K$ FF will change state each time the input clock completes a cycle. Thus the input clock must complete two cycles to cause the Q output to complete one cycle. The Q output therefore has a frequency of half the input clock signal. The single $J-K$ FF is a divide-by-2 frequency divider. Thus the frequency of the Q output is 3 megahertz.

The 3-megahertz signal from the Q output of the $J-K$ FF is the clock input to the modulus-3 counter. The msb output of a modulus-3 counter will complete one cycle for every three clock cycles. That is, the frequency of the msb output of a modulus-3 counter will be one-third that of the clock input. Therefore, the MSB output of the modulus-3 counter will be 1 megahertz. ■

MSI BCD Counters

A number of MSI BCD counters are available. The most commonly used is perhaps the 7490A, which is asynchronous, and is actually a modulus-2 counter and a modulus-5 counter on the same IC. Other counters are the 74160 and 74162, which are high-speed decade counters that feature look-ahead capability. We will discuss these last two ICs more fully shortly.

The 7490A BCD Counter

First, let us examine the 7490A BCD counter. Figure 6.13 shows a logic diagram of this IC, as well as pin number identifications. FF A has no internal connection to the other FFs on the chip. Thus FF A can be used separately as a single $J-K$ FF if necessary. The other three FFs together form a modulus-5 counter. The verification of this is left as a problem. If the IC is to be used as a decade counter, then the output of FF A must be tied to the input of FF B. That is, pin 12 must be tied directly to pin 1. The circuit then becomes a BCD counter. Notice that there is a reset-to-zero pair of inputs through a NAND gate. When both of these inputs are made high, the circuit is reset to 0000. Often the two NAND inputs, pins 2 and 3, are tied together and made into a single input. There is also a reset-to-9 pair of inputs as well. When these inputs are both made high, the circuit is reset to 1001.

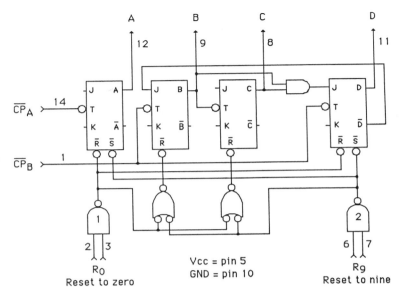

FIGURE 6.13 The 7490A asynchronous BCD counter.

The 74160 Counter

For high-speed counting, the 7490A is unsuitable. It is then that a circuit such as the 74160 becomes useful. The connection diagram for this circuit is shown in Figure 6.14(a). We will not go into all the operational aspects of this circuit. However, the basic operation is straightforward. The circuit counts input pulses on the CP input in BCD code on the Q outputs. Counts are registered on the rising edge of the CP input. There are also parallel load inputs labeled $D0$ to $D3$. These inputs are enabled by a low on the $\overline{PE}$ input. This line is normally kept high. There is an active-low, direct-reset input labeled $\overline{MR}$. There are two count enables labeled CEP and CET. Both of these must be high in order for counting to be enabled. There is also a look-ahead output labeled TC. This output goes high whenever the count has reached 1001 AND CET is high. The TC output is used when several counters are ganged together. Let us see how this works.

Figure 6.14(b) illustrates how several counters are usually connected. Note that the count signal goes to all counters simultaneously. However, only the first counter is permanently enabled since its CET and CEP inputs are left unconnected or tied high. The TC output of the first counter is sent to the CEP input of all the other counters. Thus the count on the first counter must reach 1001 in order for any other counter to be enabled. The second counter has its CET input left open or tied high. Thus the second counter will become enabled as soon as the first

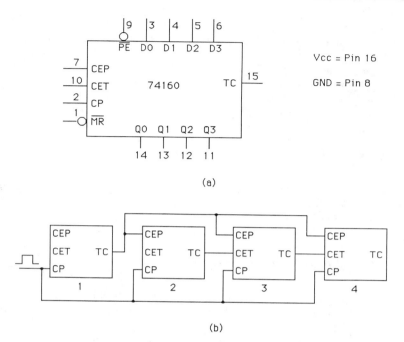

FIGURE 6.14   (a) Connection diagram for a 74160 decade counter.
(b) Typical method of ganging counters together.

counter reaches 1001. However, the third counter gets its *CET* input from the *TC* output of the second counter. Thus the third counter will not be enabled unless both the first and second counters are at 1001. The fourth counter gets its *CET* input from the *TC* output of the third counter. This output will not go high unless the *CET* input of the third counter is high, which, in turn, requires that the *TC* output of the second counter also be high. Thus the fourth counter will not be enabled unless the first three counters have a count of 1001. When the first three counters reach 1001, then all four counters will be enabled and will increment at the next clock pulse. You might ask, why not simply leave the *CEP* inputs of all counters high, and just connect the *TC* output of one counter to the *CET* input of the next counter instead of the connection method used in Figure 6.14(b)? If this were done, then we would have essentially a ripple counter. That is, suppose that the counts on counters 2 to 4 had reached 9, but the count on counter 1 was at 8. Then none of counters 2 to 4 would be enabled. When counter 1 reached 9, it would enable counter 2, which, after a short gate delay, would enable counter 3, and so on. If the counting rate was high, counter 4 might not get enabled in time to register the next count. In the method used in Figure 6.14(b), when the count reaches 9990, counters 2 to 4 will then have their *CET* inputs enabled. As soon as counter 1 reaches 9, counters 2 to 4 will be simultaneously enabled by having their CEP inputs taken high as the TC output of counter 1 goes high. Thus there will be no ripple effect, since the CET inputs of counters 2 to 4 have long since been enabled.

6.4 MODULUS-N COUNTERS

It is possible to connect several $J–K$ FFs in such a way as to create virtually any count sequence that is desired. The modulus of a counter is simply the number of distinct states in the count cycle. The actual code that is counted through can be straight binary or any other sequence of binary states. We will now illustrate these concepts with a few examples.

Modulus-3 Counter

Figure 6.15(a) shows a modulus-3 counter. There are only three states in the counting sequence. These are shown in Figure 6.15(b). To understand the operation of the counter, let us begin by assuming that both FFs are reset (the count is 00). Because the reset output of FF B is connected to the J input of FF A, this means that both the J and K inputs of FF A are now high. Thus the first clock pulse will

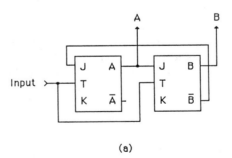

(a)

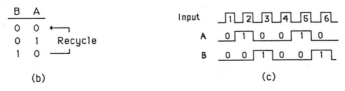

(b)   (c)

FIGURE 6.15 (a) A modulus-3 counter. (b) Sequence of states for the counter. (c) Waveforms associated with count sequence.

toggle FF *A*. The *J* input of FF *B* is connected to the set output of FF *A*. Thus the *J*–*K* inputs of FF *B* are now 0–1 respectively. Since FF *B* is already reset, the first pulse will leave it unchanged. We therefore see that following the first clock pulse the count will be 01 (where *A* is the lsb). Once FF *A* has become set, the inputs to FF *B* will be 1–1. Thus the second clock pulse will toggle FF *B*. Since FF *B* is still reset, however, the inputs to FF *A* are still 1–1. So the second pulse will also toggle FF *A*. Thus, following the second clock pulse, the count will be 10. The inputs to FF *A* will now be 0–1, and the inputs to FF *B* will be 0–1 also. Therefore, the third clock pulse will reset FF *B* and leave FF *A* reset. The count will now be 00, and the cycle is complete.

State Tables

One way to follow the changing states of a counter such as shown in Figure 6.15(a) is to make a table that includes the present output state of each FF and the present *J*–*K* input states. Such a table is shown in Figure 6.16. To make such a table, we begin with the first counter state, 00. We then determine the resulting state of each input and output and list these in the table. The states of *J* and *K* thus obtained tell us what the next state of each FF will be. This determines the next count listed under the Counter State heading. The process is repeated until the count cycle is determined.

Modulus 3/6/12 Counter

Once we have fashioned a modulus-3 counter, we can combine it with other FFs to produce other counter types. Figure 6.17(a) shows how this could be done.

We simply add two *J*–*K* FFs at the end of the modulus-3 counter. Thus, each time the output of the modulus-3 counter goes from 1 to 0, FF *C* will register a count. The counting sequence up through FF *C* is shown in Figure 6.17(b) as the modulus-6 count sequence. Notice that although the sequence contains six distinct states, they do not follow the standard binary sequence. A modulus-6 counter can be designed to follow the standard binary sequence, but the circuit in Figure 6.17(a) will not accomplish this. The second FF produces a modulus-12 counter. The count

Counter State	FF B J	K	B	FF A J	K	A
00	0	1	0	1	1	0
01	1	1	0	1	1	1
10	0	1	1	0	1	0
00	0	1	0	1	1	0

FIGURE 6.16 Table of inputs and outputs for each state of the counter in Figure 6.15(a).

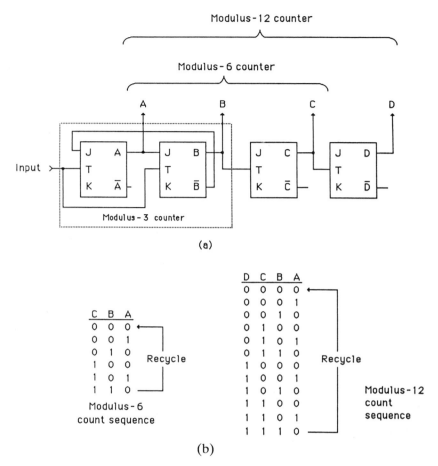

FIGURE 6.17   (a) A modulus-3/6/12 counter. (b) Count sequence for the modulus-6 and modulus-12 counter outputs.

sequence for the complete circuit is shown in Figure 6.17(b). Again, the sequence does not follow the standard binary sequence.

Modulus-5 Counter

Figure 6.18 shows a modulus-5 counter constructed from three FFs. Two FFs will not suffice since they could produce at most four distinct states. Thus we must use at least three FFs, which could produce up to eight distinct states. It would be possible to use even more FFs and still end up with some type of modulus-5 counter, although it would be inefficient to do so.

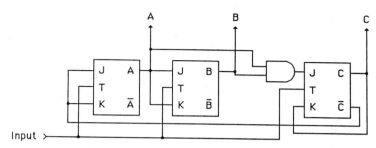

FIGURE 6.18 A synchronous modulus-5 counter.

Rather than analyze the operation of the circuit in Figure 6.18 in step by step fashion, let us simply refer to the table in Figure 6.19. This table is constructed in the same fashion as the table in Figure 6.16. We begin with the count state 000. From this state, we can determine the $J-K$ inputs of each FF. These are listed in the table. Once these $J-K$ inputs are known, we can determine how each FF will react to the next clock pulse. This determines the next count state, which is then entered in the Counter State column and in the output column of each FF. We continue this procedure until the count cycle is determined.

This procedure allows us to determine the count cycle of any sequence of $J-K$ FFs, however they are interconnected. Thus we can analyze the behavior of any existing counter design that uses $J-K$ FFs. We have not yet developed the ability to reverse this procedure. That is, given a desired sequence of count states, how do we design the circuit that will produce the sequence? We will see how to do this in Chapter 8.

6.5 SHIFT REGISTERS

When data are transmitted over a telephone line, they are sent 1 bit at a time in serial fashion. The data, however, usually start out as a group of bits of some definite length, say eight. These data must be loaded into a register and then sent over the phone line 1 bit at a time. The type of register used is called a *shift register*.

Counter State	FF C J	K	C	FF B J	K	B	FF A J	K	A
0 0 0	0	0	0	0	0	0	1	1	0
0 0 1	0	0	0	1	1	0	1	1	0
0 1 0	0	0	0	0	0	1	1	1	0
0 1 1	1	0	0	1	1	1	1	1	1
1 0 0	0	1	1	0	0	0	0	0	0
0 0 0	0	0	0	0	0	0	1	1	0

FIGURE 6.19 Table of inputs and outputs for each state of the modulus-5 counter in Figure 6.18.

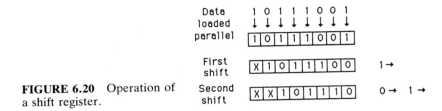

FIGURE 6.20 Operation of a shift register.

To visualize the operation of a shift register, refer to Figure 6.20. In this example, an 8-bit group of information is loaded into the register in parallel fashion. Then, at each ensuing cycle of the clock, 1 bit is shifted out to the right and sent to some other circuit. Inside the register, each bit shifts one position to the right. The leftmost position in the register is marked with an X following the first clock pulse to show that it is irrelevant what it now contains (usually a 0 will be shifted into this position). The figure shows the results of the first two clock pulses following the loading of the register.

A simple 4-bit shift register that is loaded in a serial, rather than parallel, fashion is shown in Figure 6.21. The operation of the circuit is very simple. The first FF is essentially a D-type FF because of the way the incoming data are connected to the $J-K$ inputs. Thus, at each pulse of the clock, the data on the serial input will be clocked into FF A. Because of the way the outputs of FF A are connected to the $J-K$ inputs of FF B, the current state of FF A will be passed to FF B at the next clock pulse, and likewise for FFs C and D. Thus bits come in from the left and pass out at the right, simply shifting through the register. The circuit also includes a direct-reset input so that the register can be cleared if desired.

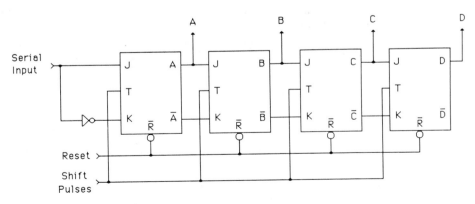

FIGURE 6.21 A simple shift-right register.

MSI Shift Registers

A number of shift registers are available in MSI form, some quite sophisticated. We will look at the operation of a few of these in some detail.

The 74165 Shift Register

Figure 6.22(a) is the connection diagram for the 74165 8-bit serial/parallel-in, serial-out shift register. Figure 6.22(b) shows the function table for this circuit. This is

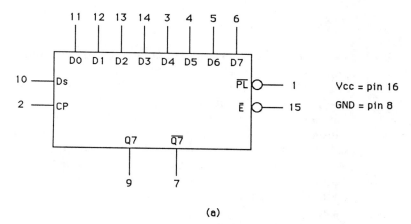

(a)

OPERATING MODES	INPUTS					Qn REGISTER			OUTPUTS	
	$\overline{PL}$	$\overline{E}$	CP	Ds	D0 -> D7	Q0	Q1 -> Q6		Q7	$\overline{Q7}$
Parallel Load	L	X	X	X	L	L	L -> L		L	H
	L	X	X	X	H	H	H -> H		H	L
Serial Shift	L	L	↑	l	X	L	q0 -> q5		q6	$\overline{q6}$
	H	L	↑	h	X	H	q0 -> q5		q6	$\overline{q6}$
Hold	H	H	X	X	X	q0	q1 -> q6		q7	$\overline{q7}$

H = HIGH voltage.
h = HIGH voltage level one setup time prior to LOW-to-HIGH clock transition.
L = LOW voltage level.
l = LOW voltage level one setup time prior to LOW-to-HIGH clock transition.
qn = Lowercase letters indicate the state of the referenced output one
      setup time prior to the LOW-to-HIGH clock transition.
X = Don't-care.
↑ = LOW-to-HIGH clock transition.

(b)

FIGURE 6.22 (a) Connection diagram for the 74165 serial/parallel-in, serial-out register. (b) Function table for the 74165.

the kind of circuit that would be used on the transmit end of a data-transmission line. Data can be loaded in parallel into the $D0$ to $D7$ inputs when the $\overline{PL}$ line is made low, or in a serial fashion on the Ds line when the $\overline{E}$ input is made low. Serial data are shifted in and out on the rising edge of the CP clock input when the $\overline{E}$ line is held low. The $Q7$ output is the normal output for shifted data. There is also a $\overline{Q7}$ line available for convenience. A typical use sequence of this circuit might be to present parallel data on the $D0$ to $D7$ inputs and then load these data by taking the $\overline{PL}$ input low. Once the data were loaded, the $\overline{PL}$ line would be taken high. Shifting out would be initiated by taking the $\overline{E}$ line low. At each rising edge of the CP input, a new bit would be shifted into the $Q7$ output for transmission to some other circuit.

The 74164 Shift Register

The 74165 IC is well suited for sending serial data. When such data are received, however, it is convenient to be able to ultimately read the data in a parallel fashion after all the data have been received. To do this a serial-in, parallel-out circuit is useful. The 74164 8-bit, serial-in parallel-out circuit shown in Figure 6.23(a) is just such a circuit. The truth table for the circuit is shown in Figure 6.23(b). There is a direct-reset input, $\overline{MR}$, which is active low. Data are clocked in on either of two data inputs, D_{sa} or D_{sb}. The usual procedure is to simply tie these two inputs together and use them as a single input line. Alternatively, one input can be tied high and the other used for data input. Parallel outputs are available at all times on the $Q0$ to $Q7$ outputs. A typical use sequence might include a reset pulse followed by eight clock pulses that clock in 8 bits of data. These data can then be read at a single time on the eight outputs lines.

The 7495 Shift Register

Other, somewhat more versatile, shift registers are available. One such circuit is the 7495 shift register. A logic diagram of this circuit, including pin assignments, is shown in Figure 6.24. Two main operating modes are available. If the MODE CONTROL is made high, this disables AND gates 1, 4, 7, and 10, effectively cutting off the output of each FF from the $J-K$ inputs of the next FF. AND gates 2, 5, 8, and 11 are enabled, allowing data from the parallel data inputs to pass directly into the FFs on the falling edge of clock $C2$. When the MODE CONTROL is made low, AND gates 2, 5, 8, and 11 are disabled, and AND gates 1, 4, 7, and 10 are enabled. Now the SERIAL INPUT is connected to the $J-K$ inputs of FF A, and the output of each FF is connected to the $J-K$ inputs of the next FF. Thus, on the falling edge of clock $C1$, new data will be clocked into FF A, and existing data will be shifted 1 bit to the right.

The 7495 can be connected in a variety of different ways to achieve different functions. For example, if output D is connected back to the serial input, the circuit

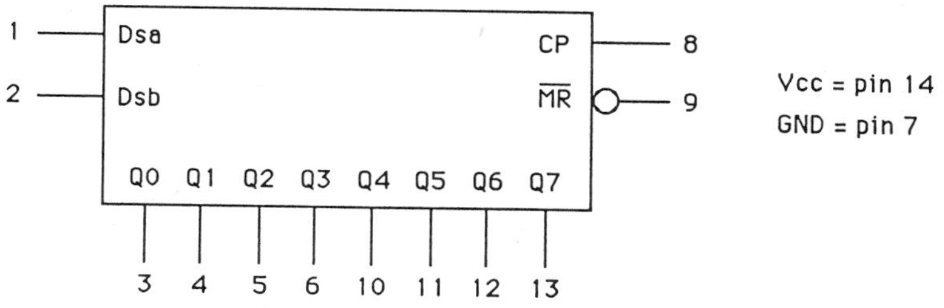

(a)

OPERATING	INPUTS				OUTPUTS	
MODE	$\overline{MR}$	CP	Dsa	Dsb	Q0	Q1 -> Q7
Reset	L	X	X	X	L	L -> L
Shift	H	↑	l	l	L	q0 -> q6
	H	↑	l	h	L	q0 -> q6
	H	↑	h	l	L	q0 -> q6
	H	↑	h	h	H	q0 -> q6

H = HIGH voltage.
h = HIGH voltage level one setup time prior to LOW-to-HIGH clock transition.
L = LOW voltage level.
l = LOW voltage level one setup time prior to LOW-to-HIGH clock transition.
qn = Lowercase letters indicate the state of the referenced output one
    setup time prior to the LOW-to-HIGH clock transition.
X = Don't-care.
↑ = LOW-to-HIGH clock transition.

(b)

FIGURE 6.23 (a) Connection diagram for the 74164 serial-in, parallel-out shift register. (b) Truth table for the 74164.

becomes a *ring counter*. That is, data simply circulate around the counter at each clock pulse on $C1$, assuming the mode control is set low (for serial operation).

Another variation has the D output connected to the C input, the C output to the B input, the B output to the A input, and new data coming in on the D input. If the mode control is set high (for parallel operation), then at each clock pulse on $C2$ new data will be clocked into FF D, what was in D will shift to C, and so on. Output A is where the data will be shifted out. We thus have a shift-left register. With these connections, the circuit can be changed between shift left and shift right simply by toggling the mode control from high to low.

Ring Counters

A ring counter is simply a shift register where the same bit pattern circulates endlessly through the register. This type of circuit has a number of uses. One type of use occurs when it is necessary to have the same pattern of bits arrive at a given

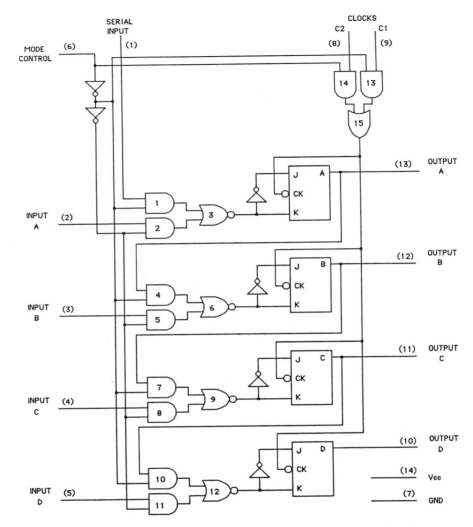

FIGURE 6.24   Pin-outs and logic diagram for the 7495 4-bit shift register.

output over and over again, perhaps to control the sequence of some machine cycle. A 4-bit ring counter would have a pattern of 4 bits that would cycle repeatedly at each output. A second use of ring counters is for frequency division.

Figure 6.25 shows a 4-bit ring counter. Ignoring the preset input for a moment, the remainder of the circuit is quite simple. It is just a normal shift register with the output of the last FF fed back to the input of the first FF. Shift pulses simply circulate whatever data are in the counter. The preset input sets the first FF and resets all the others. Thus this particular ring counter will normally have only 1 bit

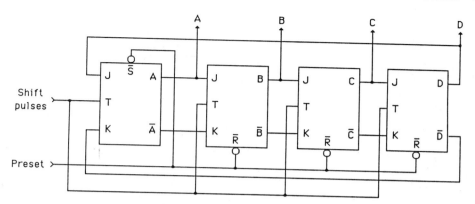

FIGURE 6.25 A simple 4-bit ring counter.

set and the other 3 reset. The single bit circulates around the counter as clock pulses arrive. This circuit could be used as a divide-by-4 circuit since only one pulse will arrive at FF D for each four pulses that arrive at the clock input. By simply extending the number of FFs, a divide-by-N circuit could be constructed.

The circuit of Figure 6.25 could be enhanced by adding the kind of preset inputs shown in Figure 6.7. If this were done, then it would be possible to preset any bit pattern whatever into the ring counter. Of course, a 7495 MSI circuit could be used for this task just as easily, since it can be connected as a ring counter and already has parallel load inputs.

PROBLEMS

1. What is the difference between an asynchronous and synchronous counter?
2. How many flip-flops are necessary to achieve 16 distinct states? How about 10 states? Twenty-four states?
3. A binary counter consisting of six FFs currently holds the count 010110. How many counts are necessary to bring the counter to a count of 110011?
4. What is the largest count that can be achieved with 10 FFs?
5. How many FFs would be necessary to hold a count of 600?
6. What is meant by the modulus of a counter?
7. Refer to Figure 6.4. The count ($DCBA$) currently stands at 0110. What are the states of the J–K inputs of each FF?

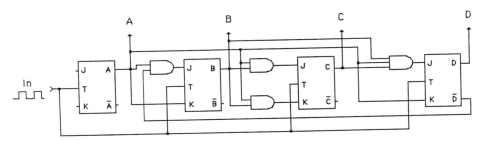

FIGURE 6.26 To be used with Problem 11.

8. Refer to Figure 6.5. The count (*DCBA*) currently stands at 1101. The preset inputs are at 1011. The PRESET line is taken high. What does the count stand at?

9. The output of a modulus-3 counter is connected to the input of a BCD counter. The output of the BCD counter is connected to the clock input of a single *J–K* FF whose *J–K* inputs are open. If a clock signal of frequency 1 kHz is input to the modulus-3 counter, what will the frequency of the output from the *J–K* be?

10. Refer to Figures 6.8 and 6.9. In Figure 6.9, there is a LOAD pulse that loads the counter with the decimal number 13. During this load pulse, what is the state of each gate in Figure 6.8?

11. Refer to the circuit in Figure 6.26. Assume this counter starts in the 0000 state. Make up a state table similar to that of Figure 6.19 and determine the count sequence for this counter. Is the counter synchronous or asynchronous?

12. In Figure 6.13, FFs *B*, *C*, and *D* were said to form a modulus-5 counter. Develop an argument that verifies this, and determine what the counting sequence is.

13. Suppose in Figure 6.14(b) that the *CEP* outputs of all the ICs are left open, and that the *TC* output of each IC is connected directly to the *CET* input of the next circuit. Suppose further that the count has reached 9999 on the rising edge of the clock input. The *TC* output of counter 1 now goes high. However, suppose that there exists a 10-ns delay before the TC output of counter 2 can follow suit, and likewise for counters 3 and 4. What limit would this place on the count frequency if counter 4 must be enabled before the rising edge of the next clock pulse?

14. It is desired that 4 bits be loaded in parallel into a 74160 IC. Describe the state that all the control inputs should assume to accomplish this.

15. Using the circuit diagram in Figure 6.24, explain why clock input *C*1 is used for clocking in serial data and *C*2 is used for clocking in parallel data.

16. Imagine the 7495 IC in Figure 6.24 connected in the following way. The serial mode is enabled. The output of FF *D* is connected to an inverter whose output

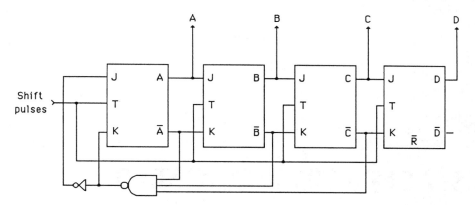

FIGURE 6.27   To be used with Problem 17.

is connected to the serial input. The circuit is assumed to be cleared initially (count at 0000). Determine the sequence of count states for the next 10 clock pulses. This type of arrangement is called a Johnson counter. How many unique count states does it have?

17. The circuit in Figure 6.27 is called a self-correcting ring counter. No matter what the initial state of the counter, after two clock pulses it will contain only one set bit, which thereafter will be circulated around the counter. Using initial counts of 1111 and then 0101, verify this statement by determining the count for the first five clock pulses.

18. What is the binary output of a 4-bit binary up counter just after the thirteenth clock pulse is received? Assume that the initial count is 0000. How would your answer be different if the initial count were 1011?

19. A 1-MHz square wave is input to the clock input of a 4-bit, binary up counter. Assuming that the count outputs are $DCBA$ from msb to lsb, what are the frequencies of the waveforms observed at the A, B, C, and D outputs?

20. A 74164 shift register has just been cleared (all outputs reset to zero). The 8-bit code 10011110 is to be clocked into the register. What is the output of the register after four clock pulses?

Chapter 7

ASTABLE AND MONOSTABLE MULTIVIBRATORS

Learning Objectives

After completing this chapter you should:

■ Know what an astable multivibrator is and how one can be fabricated from discrete electronic components.

■ Know how an astable multivibrator can be fabricated from logic gates and other circuit components, and how to calculate the frequency of the multivibrator.

■ Know what a monostable multivibrator is and how one can be constructed from discrete electronic components.

■ Know how a monostable multivibrator can be fabricated from logic gates and other circuit components, and how to calculate the on period of the multivibrator.

■ Be familiar with the use of TTL monostable ICs.

■ Know what a Schmitt-trigger input is.

- Know how to use a simple logic gate with a Schmitt-trigger input to produce an astable multivibrator.

- Know how to use an op-amp (or comparator) to produce an astable multivibrator.

- Be familiar with the versatile 555 timer IC and with many of its uses as an astable or monostable multivibrator.

7.1   INTRODUCTION

Multivibrator or clock circuits form an integral part of many digital circuits. A computer, for example, requires a system clock to synchronize its operations. Multivibrators are needed for time measurements. Often a single pulse of a specified length is required, or a certain delay is needed between an incoming pulse and an outgoing pulse. Sometimes an incoming pulse is supposed to trigger a specified number of outgoing pulses. All these tasks make an understanding of multivibrator circuits necessary.

We will examine two basic types of multivibrator circuits known as the *astable multivibrator* and the *monostable multivibrator*. The astable multivibrator is a continuously oscillating circuit that produces a signal that varies between two fixed voltages in a periodic fashion. The monostable multivibrator produces a single pulse when it is triggered.

7.2   ASTABLE MULTIVIBRATORS

Astable multivibrators are those that oscillate between two fixed voltages on a periodic basis. Usually, such a multivibrator spends equal amounts of time in each voltage state. However, this is not always the case. Figure 7.1 shows a typical output from an astable multivibrator. Note that the output spends an amount of time T_H in the high-voltage state and an amount of time T_L in the low-voltage state during each oscillation cycle. The *duty cycle* of the multivibrator is the fraction of time the output of the multivibrator spends in the high state. This time is given by

$$D = \frac{T_H}{T_H + T_L} \tag{7.1}$$

Thus if the high and low times are equal, the duty cycle is 0.50 or 50%. The duty cycle can be anywhere from virtually 0% to nearly 100%.

The *period* of the astable multivibrator is the time required for the multivibrator to complete one oscillation. Thus the period is given by

$$T = T_H + T_L \tag{7.2}$$

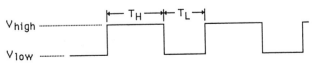

FIGURE 7.1   The output of a typical astable multivibrator.

The *frequency* of the astable multivibrator is the number of oscillations completed per second and is thus obtained by dividing 1 second by the period of the multivibrator. Thus

$$f = \frac{1}{T}$$

(7.3)

■ EXAMPLE 7.2–1

For a given astable multivibrator, $T_H = 0.4$ millisecond (ms) and $T_L = 0.6$ millisecond. Find the duty cycle, the period, and the frequency of the multivibrator.

$$D = \frac{0.4 \text{ ms}}{0.4 \text{ ms} + 0.6 \text{ ms}}$$

$$= \frac{0.4 \text{ ms}}{1.0 \text{ ms}} = 0.4 \quad \text{or} \quad 40\%$$

$$T = 0.4 \text{ ms} + 0.6 \text{ ms} = 1.0 \text{ ms}$$

$$f = \frac{1}{1 \text{ ms}} = 1000 \text{ Hz} = 1 \text{ kHz}$$

■

Astable multivibrators can be constructed in many ways. Figure 7.2(a) shows a discrete-component astable multivibrator. The frequency of oscillation depends on the RC combinations of R_2, C_2 and R_3, C_1. One RC pair controls the time the

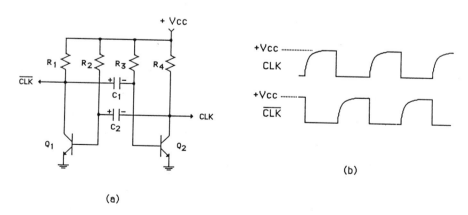

(a)

(b)

FIGURE 7.2 (a) A discrete-component astable multivibrator. (b) Waveforms associated with the multivibrator.

circuit is in the high state, and the other RC pair controls the time the circuit output is in the low state. These two times do not need to be equal; but if they are equal, then a 50% duty cycle results. If the RC pairs are identical, then the frequency of oscillation is given by

$$f = \frac{1}{1.4\ RC} \tag{7.4}$$

The values of R and C in formula (7.4) are in ohms and farads, respectively, and the resulting value of f is in hertz. One thing to notice about the output of the circuit as shown in Figure 7.2(b) is that the transition from low to high state is not sharp. This is because this transition involves charging a capacitor through a resistor, a process that requires time. The transition from high to low state is much sharper since it involves discharging a capacitor through a saturated transistor, a much faster process.

■ EXAMPLE 7.2–2

Suppose that in Figure 7.2(a) the RC pairs are equal and that R is 10 kilohms (kΩ) and C is 2 microfarads (μF). Find the RC product and the frequency of the oscillator.
The RC product will be

$$RC = (10\ \text{k}\Omega)(2\ \mu\text{F})$$
$$= (10 \times 10^3\ \Omega)(2 \times 10^{-6}\ \text{F})$$
$$= 2 \times 10^{-2}\ \text{s}$$

1.4 RC will then be 2.8×10^{-2} second

Then $f = 1/(1.4\ RC) = 1/(2.8 \times 10^{-2}\ \text{second}) = 36$ hertz.   ■

There are other discrete-component designs that produce results similar to the circuit in Figure 7.2(a). We will not pursue these, however, since it is much easier to construct multivibrators using IC logic gates.

IC Astable Multivibrators

An astable multivibrator can be constructed in a number of ways using IC logic gates to replace the discrete transistors that were used in the circuit in Figure 7.2(a). One way to do this is shown in Figure 7.3(a).

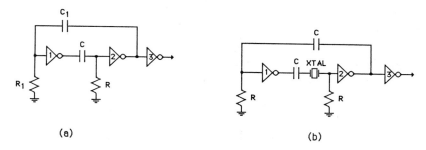

FIGURE 7.3 Two simple IC-based multivibrators.

The basic multivibrator in Figure 7.3(a) does not actually include inverter 3, which is present only as a buffer between the multivibrator and the circuit receiving the multivibrator output.

With the circuit in Figure 7.3(a), there is no exponential rounding of the clock signal edges. The state transitions are sharp. Assuming that $R_1 = R$ and $C_1 = C$, the frequency of oscillation of the circuit is given by

$$f = \frac{1}{2RC} \tag{7.5}$$

As in the earlier similar formula, R is in ohms, C in farads, and f in hertz.

One problem with any circuit that depends on RC values for frequency determination is the stability of component values. The values of R and C can change because of changes in temperature or humidity or simply change with time. More frequency stability can be achieved by introducing a crystal into the circuit. This is shown in Figure 7.3(b). The basic RC components are chosen to produce an oscillation frequency as close as possible to the natural frequency of the crystal. The introduction of the crystal keeps the output frequency very stable. Crystals are available over a wide range of frequencies.

7.3 MONOSTABLE MULTIVIBRATORS

On many occasions a single pulse must be produced rather than a continuous oscillation. A circuit that produces such a pulse is called a monostable multivibrator. As with astable multivibrators, monostable multivibrators can be designed from discrete components or IC gates. In addition, monostable multivibrators are available as single IC components. Let us first consider a discrete-component monostable.

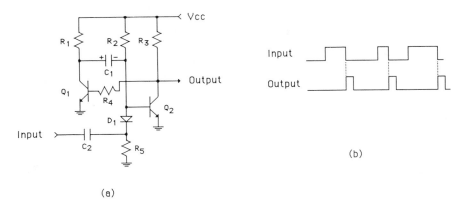

(a)

(b)

FIGURE 7.4 (a) A discrete-component monostable multivibrator. (b) Input and output waveforms for the multivibrator.

Figure 7.4(a) shows a discrete-component monostable. Both the inputs and outputs to this circuit normally stand low. This is a stable situation, which should persist indefinitely unless an external signal upsets the balance.

The circuit is triggered when a positive pulse arrives at the input. In particular, the circuit is triggered by the falling edge of the positive pulse, that is, by the high to low transition at the end of the positive pulse. At the falling edge of the positive pulse, the output of the circuit goes high and remains in the high state for a fixed period of time. The circuit output then returns low and remains this way until the next positive pulse. The time that the circuit spends in its unstable high state depends on how long it takes to charge C_1. This is determined by the R_2C_1 time constant. The actual time of the pulse produced by the one shot is approximately

$$T = 0.69R_2C_1 \tag{7.6}$$

Some care must be exercised in choosing R_2 since it must be small enough in value to ensure that Q_2 becomes completely saturated in the stable state. The value of C_1 is then chosen to produce the desired time constant.

■ EXAMPLE 7.3–1

Using the circuit in Figure 7.4(a), it is desired to produce an output pulse of duration 0.01 millisecond. The value of R_2 is set at 2 kilohms. What value should be chosen for C_1?

Equation (7.6) can be rewritten to give

$$C_1 = \frac{T}{0.69R_2}$$

$$= \frac{0.01 \times 10^{-3}S}{0.69 \times 2 \times 10^{3}\ \Omega}$$

$$= 0.72 \times 10^{-8}\ F = 0.0072\ \mu F \qquad\blacksquare$$

Figure 7.4(b) shows the input and output waveforms for the monostable multivibrator of Figure 7.4(a). Notice that it is the trailing edge of the input pulse that triggers the circuit, and the duration of the output pulse has nothing to do with the duration of the input pulse.

Monostable Multivibrators Produced from IC Logic Gates

It is relatively easy to produce a monostable multivibrator using logic gates. A typical circuit is shown in Figure 7.5(a). Both the inputs and outputs of this circuit normally stand high. With this in mind, let us analyze the circuit.

Let us first investigate the situation where the signal input to NAND 1 and the output from NAND 2 are both high and inquire as to whether this situation is

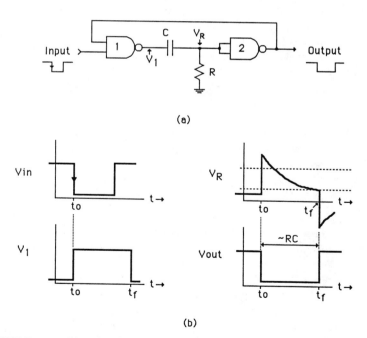

(a)

(b)

FIGURE 7.5   (a) A monostable multivibrator constructed from IC gates. (b) Waveforms at various points in the circuit of (a).

stable. The assumed state of NAND 2 implies that both inputs to NAND 1 must be high. It also implies that the output of NAND 1 and the input to NAND 2 are low. This further implies that both sides of capacitor C are at a low voltage, and thus the capacitor is uncharged. The voltage across R would also be zero. This set of conditions is entirely static and therefore stable.

Now imagine that a negative input pulse is applied to the signal input of NAND 1. This causes the output of NAND 1 to immediately go high. This high is transmitted through the uncharged capacitor to the input of NAND 2, so the output of NAND 2 goes low. The high at the input of NAND 2 is actually maintained by current flowing through C and particularly R. As C charges, this current diminishes, reducing the voltage drop across R. When this voltage drops low enough to be recognized as a logical low at the input of NAND 2, the output of NAND 2 will go high and the output pulse will terminate. The voltages at various points in the circuit during a trigger pulse are shown in Figure 7.5(b).

IC Monostable Multivibrators

Several monostable multivibrators are available in the 74XX series. Among these are the 74121, 74122, and 74123. The major differences between IC monostables involves whether or not they are *retriggerable*. This concept can be explained as follows. Normally, we think of an input pulse triggering the monostable. The result is a single output pulse of fixed duration. But what if a new triggering pulse arrives *before the end of the output pulse*? There are two possibilities. If the new input pulse causes the output pulse to start over (in effect, extending the output pulse), the monostable is said to be retriggerable. If the new input pulse has no effect, the monostable is not retriggerable. Clearly, if a monostable is retriggerable, then the output pulse can be extended by additional input pulses that occur while the output pulse is still in progress. Let us examine one particular IC monostable.

The 74121 Monostable Multivibrator

Figure 7.6 shows the logic diagram and pin connections for the 74121 nonretriggerable monostable multivibrator. The external components R and C determine the output pulse duration. The circuit contains, in addition to the actual monostable, two gates that can be used for triggering the circuit. One of these is a normal NAND gate. The other is an AND gate with a Schmitt-trigger input (we will talk about this type of input shortly). The monostable itself triggers on a low to high transition. However, from the user's point of view, the external trigger pulse can be either a rising edge or a falling edge. For example, suppose A_1, A_2, and B are all held high. Then the output of the NAND gate is low, and so is the output of the AND gate. If either A_1 or A_2 goes low (falling edge), the NAND output will

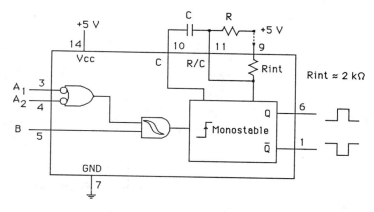

FIGURE 7.6 A 74121 monostable multivibrator. External components R and C determine the output pulse duration.

go high, and so will the AND output. This will trigger the monostable. On the other hand, suppose that either A_1 or A_2 (or both) is held low, and also B is held low. Then the output of the NAND gate is high, but the output of the AND gate is low. Now if B goes high (leading edge), then the output of the AND gate will go high, triggering the monostable. Thus the external inputs can be set up so that the triggering signal can be either a leading edge or a falling edge.

The external components R and C determine the duration of the output pulse of the monostable according to the equation

$$t = (\log_e 2)RC \tag{7.7}$$

In this equation, R is in ohms, C in farads, and t in seconds.

Output pulse durations from 30 nanoseconds up to 40 seconds can be achieved by properly selecting R and C. R is permitted to range from 2 to 40 kilohms, while C may range from 10 picofarads to 10 microfarads. It is also possible to make use of the internal 2-kilohm resistor, in which case no external resistor is required.

More than one monostable can be used at a time to achieve various design aims. For example, sometimes it is necessary to produce a delay between an incoming trigger pulse and the resulting output pulse. The circuit in Figure 7.7 shows one way to do this. The first monostable is triggered by the rising edge of the input pulse. The resulting pulse is connected to the second monostable in such a way that the falling edge of the pulse triggers the second monostable. Thus the output of the second monostable is delayed by the length of the output pulse of the first monostable. The relevant waveforms are also shown in Figure 7.7.

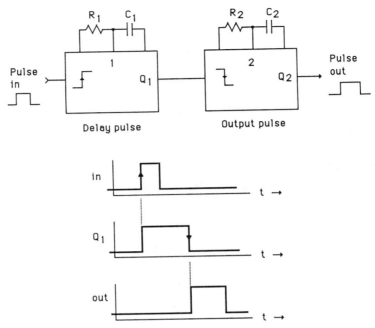

FIGURE 7.7 Two monostable multivibrators connected to produce an output pulse of length determined by R_2 and C_2 following a delay determined by R_1 and C_1. Note that the first monostable is triggered on the leading edge of the input pulse, while the second monostable is triggered on the falling edge of Q_1.

7.4 SCHMITT-TRIGGER INPUTS

One very common problem in digital electronics is to detect and count pulses on a given input. Sometimes, however, the input signal is not a particularly clean electrical signal. For example, suppose a given input is designed to detect and count low to high transitions on a particular line. Standard TTL inputs have a critical voltage below which the input is recognized as a logical low and above which the input is recognized as a logical high. As the input crosses this threshold, the circuit will respond to the changing input.

Consider the input signal shown in Figure 7.8. This is a greatly magnified and expanded view of a noisy digital input signal. Notice that, as the signal approaches and then crosses the critical threshold voltage for the input circuit, the voltage actually crosses this threshold three times in quick succession before completing the crossing. The input circuit will actually count three input pulses instead of just one. This type of problem is overcome by the *Schmitt-trigger* input.

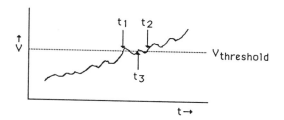

FIGURE 7.8 A magnified view of a noisy digital input.

The Schmitt-trigger input has two threshold voltages instead of just one. When a digital circuit has a Schmitt-trigger input, the logic symbol is modified by the addition of a small hysteresis curve placed inside the logic symbol. Figure 7.9(a) shows two examples of this convention. To understand the function of the Schmitt-trigger input, let us refer to Figure 7.9(b), which is an input/output voltage curve for an inverter with a Schmitt-trigger input. Assume that the input voltage is zero, and thus the inverter output is high (3.4 volts in this case). As the input voltage is raised, the output remains unchanged until the critical *low to high* voltage is reached. This occurs at just below 1.7 volts. As the voltage crosses this threshold, the output of the inverter switches from high to low (to about 0.2 volt). As the input voltage is raised further, the circuit output remains low. Now suppose we begin to lower the input voltage. Notice that this voltage can drop below the original low to high threshold and *not produce any change of output*. It is not until the voltage drops to nearly 0.9 volt, the *high to low* threshold, that the inverter output switches back to high. There is about a 0.8-volt margin between the two thresholds. We can see how this helps with a noisy input signal. Referring back to Figure 7.8, the noise on the signal would need to be over 0.8 volt before any problem would arise.

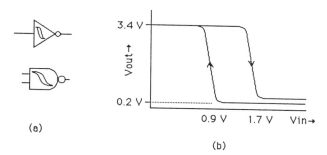

FIGURE 7.9 (a) Logic symbols for an inverter and a NAND gate with Schmitt-trigger inputs. (b) Voltage response for an inverter with a Schmitt-trigger input.

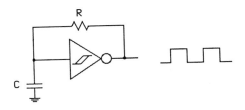

FIGURE 7.10   A simple astable multivibrator constructed from an inverter with a Schmitt-trigger input.

The Schmitt trigger input can be used to produce a very simple multivibrator. Consider the circuit in Figure 7.10. Suppose that the output of the circuit has just switched from low to high. If we use the voltage curves of Figure 7.9(b), this means that the input voltage must have just reached 0.9 volt (on the way down) and the output has just become 3.4 volts. There will thus be 2.5 volts across the resistor, and current will flow through this resistor into the capacitor. The capacitor will thus begin to increase its charge and its voltage. When this voltage increases to 1.7 volts, the output of the inverter will switch low (to 0.2 volt), and the current will reverse its direction, discharging the capacitor. This charging and discharging of the capacitor cause the inverter output to switch back and forth endlessly.

The R and C values determine the frequency of oscillations in the circuit in Figure 7.10. Unfortunately, the value of R must be kept quite small (less than about 400 ohms) to provide sufficient input current for the inverter. If the resistor is any larger than this, the inverter input will get all the current, and the capacitor will not charge. The small value required for R is no problem as long as low-frequency oscillations are not required. For low-frequency oscillations (which would require an inordinately large capacitor for a 400-ohm resistor), an emitter follower can be used. This is shown in Figure 7.11. In this case, the voltage between the base and emitter of the transistor will remain essentially fixed at about 0.6 volt. The current from the inverter output will mostly be used to charge the capacitor. Only a small amount of current will flow across the base–emitter junction. The current to drive the inverter input will come from the 5-volt supply through the collector–emitter junction. To sustain the few milliamperes of input current required by the inverter input, very little base current is required.

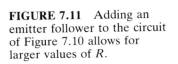
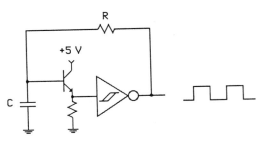

FIGURE 7.11   Adding an emitter follower to the circuit of Figure 7.10 allows for larger values of R.

Op-Amp/Comparator Schmitt-trigger Circuits

What amounts to a Schmitt-trigger multivibrator can be easily constructed using an operational amplifier (op-amp) or a comparator (which is essentially a fast op-amp). We will not go into the details of op-amps here, but only note the essential characteristics.

Basically, an op-amp is just an amplifier with a very large gain (usually several thousand at least). The circuit symbol for the op-amp is shown in Figure 7.12. The op-amp has two inputs called the noninverting input ($+$) and the inverting input ($-$). There are also two voltage supply inputs, which usually are connected to two equal voltages of opposite polarity (for example, $+$ and -12 volts). In operation the op-amp tries to amplify the voltage difference between its two inputs by its very large gain. As a practical matter, this means that, if the voltage at the ($+$) input is even a few millivolts higher than at the ($-$) input, the output of the op-amp will go positive as far as it can (to its positive saturation voltage). This positive saturation voltage is usually just slightly less than $+V_s$. Likewise, if the voltage at the ($-$) input is slightly greater than the voltage at the ($+$) input, the op-amp's output will go as far negative as it can (to its negative saturation voltage). This negative saturation voltage is usually just slightly above $-V_s$. In effect, the basic op-amp or comparator has two stable output states unless negative feedback is introduced.

To see how an op-amp or comparator can be used to produce a Schmitt-trigger multivibrator, consider the circuit shown in Figure 7.13(a). Let us consider the effect of the two possible output voltages (positive and negative saturation voltages). When the output is high, voltage-divider action will make the ($+$) input have a positive voltage as shown in Figure 7.13(b). When the output is low, the ($+$) input will have a negative voltage of approximately the same magnitude. Now suppose that the output has just switched from low to high. Just before the switch, the output was negative and the voltage at ($+$) was negative. Since a switch has occurred, the voltage at ($-$) must have just dropped below that of ($+$), which was *negative at the time of the switch*. Thus, at the switch from negative to positive output, the capacitor is negatively charged. Once the switch has occurred, the voltage at ($+$) will be positive, several volts above the negative voltage at ($-$).

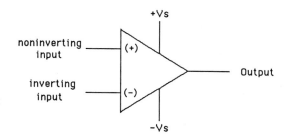

FIGURE 7.12 Basic circuit symbol for an operational amplifier or comparator.

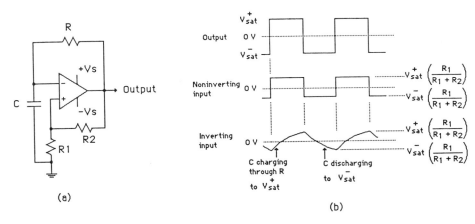

FIGURE 7.13 (a) An op-amp multivibrator. (b) Relevant waveforms.

Thus the positive output is quite stable for the time being. However, the positive output will cause current to flow through resistor R and into the capacitor. This will first discharge the capacitor and then begin to charge it in the opposite direction. When the voltage on the charging capacitor reaches the magnitude of the voltage at ($+$), the output will switch from positive to negative and the process will reverse itself. This can be seen by examining the waveforms in Figure 7.13(b).

The Schmitt-trigger part of the circuit consists of just the two resistors $R1$ and $R2$ and the op-amp. The ($-$) input is the input to the Schmitt trigger. The R and C components can then be seen to be connected in a manner identical to that of Figure 7.10.

7.5 THE 555 TIMER

One of the most versatile IC clock circuits is the 555 timer. Among the variety of functions that this IC can serve, the two that will concern us are the monostable and astable multivibrator functions. The logic diagram and pin connections for the 555 timer are shown in Figure 7.14.

Let us examine the 555 timer more closely. The circuit contains two comparators, the outputs of which are connected to the active-high inputs of an RS flip-flop. The *reset* output, $\overline{Q}$, of the FF is connected to two places. One is the input of an inverter whose output is the "output" of the 555. The other is the base of an internal *NPN* transistor. When $\overline{Q}$ is low, this transistor will be cut off, and when $\overline{Q}$ is high, the transistor will be saturated.

Now let us look at the three resistors labeled $R1$, $R2$, and $R3$. These internal resistors are all 5 kilohms and form a voltage-divider chain. That is, assuming the

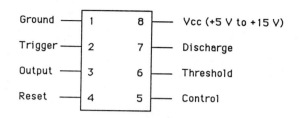

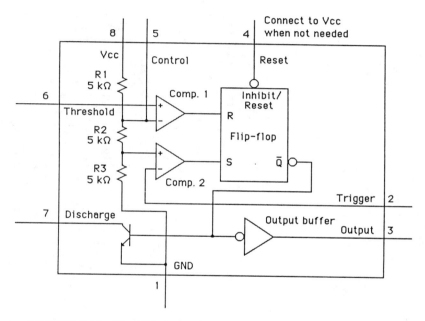

FIGURE 7.14 The 555 timer pin-outs and internal equivalent circuit.

control input is left open, the three resistors divide *Vcc* into three equal parts. The
(−) input to comparator 1 will be held at $\frac{2}{3}Vcc$, and the (+) input to comparator
2 will be held at $\frac{1}{3}Vcc$. There is a second input to each comparator. The (+) input
to comparator 1 is available as an external input at pin 6 called the threshold input.
The (−) input to comparator 2 is available at pin 2 as the trigger input. These
external inputs can be used to change the outputs of comparators 1 and 2 and thus
set or reset the internal FF. It is obviously necessary to avoid a situation where
the outputs of both comparators go high simultaneously. This would make both
the *R* and *S* inputs to the FF active at the same time, an ambiguous situation.

There is an active-low, direct-reset input available at pin 4 of the 555. This
is often tied to *Vcc* unless it is specifically needed. The discharge input normally
has a capacitor tied to it. This capacitor will be discharged whenever the FF $\bar{Q}$

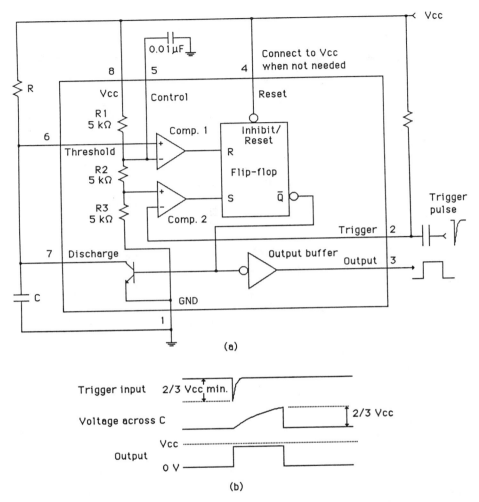

FIGURE 7.15 (a) 555 timer connected as a monostable multivibrator.
(b) Relevant waveforms initiated by triggering.

output goes high, causing the transistor to saturate. Finally, the control input at pin 5 can be used to place any desired voltage directly on the ($-$) input of comparator 1 and one-half this voltage on the ($+$) input of comparator 2.

Let us see how the 555 timer could be used as a monostable multivibrator. Figure 7.15(a) shows the required circuit. Normally, the output stands low, which means that the input to the inverter must be high. This high will keep the transistor in saturation, so the capacitor is completely discharged. This also makes the voltage at pin 6 (threshold input) equal to zero. The trigger input is being held high, so the ($-$) input to comparator 2 is at *Vcc*. If we examine the inputs to comparators

1 and 2, we see that both comparators should be producing a low output since in each case the $(-)$ input is at a higher voltage than the $(+)$ input. Thus the RS inputs to the FF are both inactive, and the FF is in the reset state (since $\overline{Q}$ is high).

Now suppose a negative pulse appears at the trigger input (of magnitude at least $-\frac{2}{3}Vcc$). This pulse will temporarily pull the $(-)$ input of comparator 2 below the $(+)$ input (which is at $\frac{1}{3}Vcc$). This will cause the output of comparator 2 to go briefly positive, setting the FF. With the FF set $(\overline{Q} = 0)$, two things happen. First, the output at pin 3 will go high. In addition, the transistor will be cut off, and the capacitor C will begin to charge through R. As the capacitor charges, the voltage at pin 6 will slowly rise. This is shown in Figure 7.15(b). When this voltage reaches $\frac{2}{3}Vcc$, the voltage at the $(+)$ input of comparator 1 will exceed the voltage at its $(-)$ input. This will cause the comparator output to go positive, resetting the FF. This reset action will cause the circuit output to go low again, and also cause the transistor to saturate, discharging the capacitor. The cycle is now complete.

The time required for the 555 internal FF to switch states is about 100 nanoseconds, placing a lower limit on the useful pulse duration. Above this limit, the pulse duration is given by

$$t = (1.1)RC \tag{7.8}$$

R can range in value from 1 kilohm to about $(1.3)Vcc$ megohms. That is, if Vcc is 5 volts, R can range from 1 kilohm to 6.5 megohms.

■ **EXAMPLE 7.5–1**

Using the circuit of Figure 7.15(a), choose values of R and C to achieve an output pulse of 1-millisecond duration. We are free to choose R and then use Eq. (7.8) to determine C. If the resulting value of C is an inconvenient value, then the value of R could be adjusted to yield a more suitable value for C.

Let us choose R equal to 10 kilohms. Equation (7.8) can then be rewritten to yield

$$C = \frac{t}{1.1\,R} = \frac{10^{-3}\,\text{s}}{1.1 \times 10^4\,\Omega}$$
$$= 9.1 \times 10^{-8}\,\text{F} = 0.091\,\mu\text{F}$$

This is a physically convenient size of capacitor, although not one of the standard values available. We would probably choose a 0.1-microfarad capacitor, which is a commonly available size. This would require a slight change in the value of R. Thus

$$R = \frac{t}{1.1C} = \frac{10^{-3} \text{ s}}{1.1 \times 10^{-7} \text{ F}}$$

$$= 9.1 \times 10^3 \ \Omega = 9.1 \text{ k}\Omega$$

This value of R is a standard value, and thus would be readily available. The convenient solution to our problem is thus to choose a value for R of 9.1 kilohms and a value for C of 0.1 microfarad. ∎

■ EXAMPLE 7.5–2

Using the circuit of Figure 7.15(a), choose values for R and C to achieve a pulse duration of 10 seconds. Again, let us begin with a value for R of 10 kilohms. This yields a value for C of

$$C = \frac{t}{1.1R} = \frac{1 \text{ s}}{1.1 \times 10^4 \ \Omega}$$

$$= 9.1 \times 10^{-5} \text{ F} = 91 \ \mu\text{F}$$

This value is not a commonly available size. The closest convenient value is 100 microfarads, which is available as a fairly small electrolytic capacitor. Choosing this value, R must be adjusted slightly. Thus

$$R = \frac{1 \text{ s}}{1.1 \times 10^{-4} \text{ F}} = 9.1 \text{ kilohms}$$

∎

The 555 monstable configuration can be used in any situation where a monostable is required. One interesting use is in the orderly start-up of a digital circuit when power is first applied or interrupted. It is often advantageous, in this situation, to have the circuit disabled for a period of time until the supply voltage reaches its stable value. Two circuits that achieve this are shown in Figures 7.16(a) and (b). In Figure 7.16(a), the capacitor C holds pins 2 and 6 low for a period of time after power is established, keeping the FF set and the output high. This is useful if a high is needed to keep the digital circuit disabled. In Figure 7.16(b), the capacitor keeps the voltage at pins 2 and 6 high for a period of time after power is established. This keeps the FF reset and the output low. The configuration in Figure 7.16(b) would be used if a low were needed to keep a digital circuit disabled for a short time following power-up.

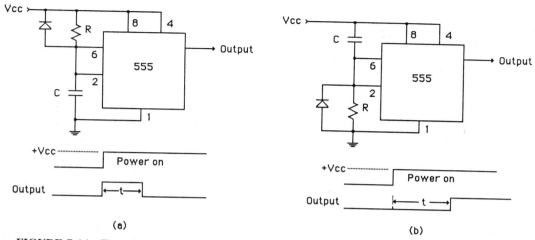

FIGURE 7.16 Two circuits that produce (a) a temporary high output at power on, (b) a temporary low output at power on.

The 555 as an Astable Multivibrator

The 555 can also be used as an astable multivibrator. The typical configuration is shown in Figure 7.17. The operation of the circuit is not too different from that of the monostable configuration.

In this case, however, pin 6 and pin 2 (threshold and trigger) are tied together. In addition, when the transistor goes into saturation, the capacitor cannot discharge quickly, but instead must discharge through R_B. Let us pick up the circuit operation just as the output has gone low. At this time, the voltage at pin 2, 6 must have just reached $\frac{2}{3}Vcc$ in order to cause the output of comparator 1 to go high and reset the FF. When the FF resets, the transistor saturates and the capacitor begins to discharge through R_B. The voltage at pin 2, 6 will now go down exponentially as the capacitor discharges. When this voltage drops below $\frac{1}{3}Vcc$, the output of comparator 2 will go positive, setting the FF. This will turn off the transistor, and the capacitor will now begin to charge through both R_A and R_B. As the capacitor charges, the voltage at pin 2, 6 will rise again and will eventually reach $\frac{2}{3}Vcc$. This will again cause the output of comparator 1 to go positive, resetting the FF. The cycle is complete.

The circuit output is an oscillation with unequal high and low times. This is because the capacitor discharges through R_B, but charges through both R_A and R_B. The high and low time intervals are shown in Figure 7.16. If R_A is made much smaller than R_B, then the two times will be nearly equal, and the duty cycle will be close to 50%.

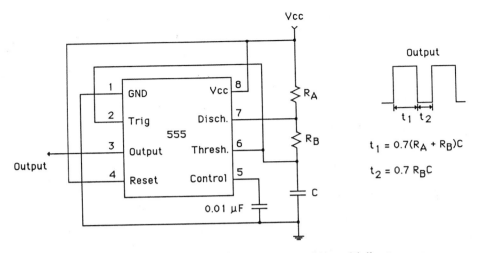

FIGURE 7.17 A 555 timer used as an astable multivibrator.

The output of the 555 astable multivibrator shown in Figure 7.17 is high for a period of time given by

$$t_1 = 0.7(R_A + R_B) \tag{7.9}$$

and low for a period of time given by

$$t_2 = 0.7R_B \tag{7.10}$$

Some care must be taken in choosing the values of R_A and R_B. When the capacitor discharges, it discharges through R_B. If the value of R_B is chosen too small, the discharge current could damage the transistor inside the 555 timer. Also, when the capacitor is discharging, there is a current through R_A that is also flowing through the transistor in the 555 timer. This current is in addition to the capacitor discharge current. Thus care must be taken that both R_A and R_B are chosen large enough to limit the current through the 555 transistor during the discharge phase of the astable cycle. These resistors should be a few hundred ohms at least.

■ **EXAMPLE 7.5–3**

Using the circuit of Figure 7.17, choose values of R_A, R_B, and C so that the frequency of the output is 10 kHz and the duty cycle is 55%.

The frequency requirement can be translated into the statement that the *period* of the output, which is $1/f$, is 10^{-4} second. This can be expressed as

$$t_1 + t_2 = 0.7(R_A + 2R_B)C = 10^{-4} \text{ second}$$

The duty-cycle requirement can be expressed as

$$\frac{t_1}{t_1 + t_2} = \frac{R_A + R_B}{R_A + 2R_B} = 0.55$$

A little algebraic rearrangement of this last equation yields

$$\frac{R_B}{R_A} = 0.45$$

Let us take $R_A = 10$ kilohms. Thus $R_B = 4.5$ kilohms. There is no standard resistor of this value, but let us set this issue aside for the moment. With the values of the resistors determined, the equation involving the period can be used to determine the value of C.

$$0.7(R_A + 2R_B)C = 10^{-4} \text{ s}$$

$$C = \frac{10^{-4} \text{ s}}{0.7(10^4 + 9 \times 10^3 \ \Omega)}$$

$$= 7.5 \times 10^{-9} \text{ F} = 0.0075 \ \mu\text{F}$$

This is a nonstandard value for capacitors. We would probably choose a value or 0.01 microfarad. This would require a recalculation of the values of R. In particular, although the *ratio* of the two values of R would remain the same, the individual values would need to be decreased to compensate for the increase, for convenience, in the calculated value of C. Thus, since C has been increased by multiplying by a factor of $\frac{4}{3}$, the values of the resistors must be decreased by multiplying each by a factor of $\frac{3}{4}$ (note $\frac{4}{3} \times \frac{3}{4} = 1$). The resulting values for the resistors then become

$$R_A = 7.5 \text{ k}\Omega, \qquad R_B = 3.37 \text{ k}\Omega$$

The 7.5-kilohm value is standard, and the 3.37 value is quite close to a standard value of 3.3 kilohms, which should be sufficiently accurate for our purposes. If real precision is required in both duty cycle and frequency, variable resistors could be used so that adjustments could be made to achieve the exact results desired.■

PROBLEMS

1. An oscillator has a normally low output. When triggered, the oscillator produces a single high pulse of 4-ms duration. Is this a monostable or astable oscillator?

2. An oscillator produces a series of periodic pulses that consist of 4 ms in the high state followed by 6 ms in the low state, and so on. What is the duty cycle of the signal?

3. In Figure 7.2(a), suppose that resistors R_1 and R_4 are reduced in value. Will this make the rising edge of the clock signals in Figure 7.2(b) rise more quickly or be more rounded?

4. In Figure 7.2(a), the values of R_2 and R_3 are 2 kΩ, while the values of C_1 and C_2 are 0.2 μF. What is the *period* of the oscillation of the circuit?

5. In Figure 7.3(a), suppose inverter 3 were not present and that the output of inverter 2, in addition to being fed back through the capacitor, was connected to the input of the next circuit. Explain how this might change the oscillation frequency of the circuit.

6. In the circuit of Figure 7.3(a), R and R_1 are chosen as 5 kΩ. Assuming C and C_1 are equal in value, what should this value be if the period of oscillation is supposed to be 15 ms?

7. In Figure 7.5(a), does the duration of the input pulse have any effect on the operation of the circuit? Specifically, is there any requirement that the input pulse be longer or shorter than the output pulse?

8. In Figure 7.6, A_2 and B are high. What must occur on input A_1 for an output pulse to be produced?

9. Use the 74121 to design a circuit that will produce a 40-μs pulse each time input A_2 goes high. Show how inputs A_1 and B must be connected, and use the internal resistor rather than an external resistor.

10. In Figure 7.6, notice that the internal AND gate has a Schmitt-trigger input. This being the case, which circuit input would you use if you wished to trigger the monostable with a signal that might be somewhat noisy?

11. Using the circuit of Figure 7.7, produce a design that will accept an input pulse, produce a 10-ms delay, and then produce an output pulse of 25-ms duration.

12. The circuit shown in Figure 7.10 could just as easily use a Schmitt-triggered NAND gate or NOR gate instead of an inverter. Show how this might be done in two different ways for each gate.

13. The output of the circuit in Figure 7.11 has just gone high. Assuming that the threshold voltages of Figure 7.9(b) are correct, what does this imply about the voltage at the input to the inverter? What about the voltage at the base of the transistor? If the high output voltage of the inverter is 3.4 V, what voltage

drop exists across the resistor? If the resistor has a value of 1 kΩ, how much current flows through the resistor? If the capacitor has a value of 0.033 μF, how long will it take for this capacitor to increase its voltage by 0.8 V, assuming the current through the resistor stays essentially constant? This time is one-half of the period of oscillation. Therefore, what is the frequency of oscillation?

14. In Figure 7.13(a), assume that $+Vs$ and $-Vs$ are $+12$ V and -12 V, respectively. Assume also that the saturation voltages of the op-amp are $+10$ V and -10 V. $R1$ and $R2$ are each 5 kΩ. R is 10 kΩ and C is 0.01 μF. Just before the output switches positive, what will the voltage be at the $(+)$ input of the op-amp? Just as the output switches positive, what will the voltage be at the $(-)$ input of the op-amp? Just after the output has switched positive, what will the voltage drop across R be? When the capacitor finally charges to the point where the circuit is about to switch negative, what will the voltage at the $(-)$ input of the op-amp be? What will the voltage across R be at this time?

15. In the 555 logic diagram of Figure 7.14, Vcc is 12 V. The input at pin 6 is 10 V, and the input at pin 2 is 5 V. Is the FF set or reset? What is the circuit output? Is the transistor saturated or cut off?

16. In Figure 7.15(a), why does the negative trigger pulse need to be at least $\frac{2}{3}Vcc$ in magnitude?

17. Choose appropriate values of R and C so that the circuit in Figure 7.15(a) will produce an output pulse of 10-s duration.

18. Using the circuit in Figure 7.17, choose values of R_A, R_B, and C such that the duty cycle is 60% and the period of oscillation is 0.01 s.

19. One method of controlling a dc motor is to vary the dc voltage applied to the motor windings. An efficient way of doing this is to apply a signal such as that shown in Figure 7.17. The *effective* dc voltage that the motor "sees" is Vcc multiplied by the duty cycle. Thus, to control the *effective* voltage, we control the duty cycle. Suppose that *one* of the resistors, R_A or R_B, could be replaced by a variable resistor. And suppose that you wished to use this variable resistor to vary the duty cycle between 60% and 85%. Which resistor would you choose to make the variable resistor? Based on your choice, design a circuit that produces a 60% duty cycle at 1 kHz (the frequency at the 85% duty cycle will, of course, be somewhat different than 1 kHz). Notice that your variable resistor must be chosen large enough so that it can be adjusted to cover the range of resistances needed to produce any duty cycle between 60% and 85%.

Chapter 8

KARNAUGH MAPPING AND DIGITAL DESIGN

Learning Objectives

After completing this chapter you should:

■ Know what Karnaugh mapping is as applied to combinational logic circuits.

■ Know how to plot a two-, three-, or four-variable truth table on a Karnaugh map.

■ Know how to use a Karnaugh map to minimize a sum-of-products Boolean function derived from a truth table.

■ Know how to plot any Boolean function in a sum-of-products form on a Karnaugh map.

■ Know how to use a Karnaugh map to reduce a Boolean function.

■ Understand the concept of don't-care states and know how to plot these states on a Karnaugh map.

- Know how to use don't-care states on a Karnaugh map to aid in reduction.

- Know how to construct Karnaugh maps for sequential logic derived from JK flip-flops.

- Know how to use sequential Karnaugh maps to design counters that count in any desired sequence.

8.1 INTRODUCTION

When an engineer or technician sets out to design a digital circuit, one of the first steps is to develop a truth table (for combinational logic designs) or a next-state table (for sequential logic designs). These tables define what the circuit is supposed to do. The problem is, how does one get from the table to the actual circuit design?

In the case of combinational logic designs, we have already seen that the truth table can be used to form a sum-of-products Boolean expression for the logic function or functions to be implemented. We can then use various Boolean identities to try to reduce the Boolean expression to what we hope is its minimum form. It is difficult in many instances, however, to be sure that the expression is actually in a minimum form. What is needed is a technique that allows us to reduce sum-of-products Boolean expressions to a minimum form in a systematic way. Several such techniques are available. Among these, one of the simplest to use, provided the number of input variables is not too large, is the technique of Karnaugh mapping. In the next section we will develop this technique and apply it to a number of design problems.

In addition to developing the basic Karnaugh mapping technique for combinational circuits, we will consider a variety of ways to implement the reduced Boolean function once it is obtained. In some cases involving multiple outputs, we will see that greater efficiency can be achieved if we do not insist on using the minimum Boolean expression.

Different techniques are required when dealing with sequential logic circuits. Here we generally know what series of states we want the circuit to cycle through. The problem is how to interconnect flip-flops to achieve the desired sequence. Many techniques are available, depending on what type of FF we want to use in the basic design. We have many types of FFs to choose from, including $R-S$, D-type, $J-K$, and T flip-flops (a T FF is one that reverses state with each clock pulse, for example, a $J-K$ with both J and K tied high). Once we have decided which type of FF to use, a design technique can be chosen. In this chapter, we will develop a Karnaugh mapping technique for circuit design using $J-K$ FFs. We will use this technique to complete a number of design examples.

8.2 COMBINATIONAL LOGIC KARNAUGH MAPPING

We begin by considering a 2-input digital circuit. The two inputs will be labeled A and B. With two inputs, there are four possible input combinations, or product terms, that can be formed. These are shown in Figure 8.1. Each product term in Figure 8.1 is a product of two *literals*, the name given to the individual terms such as A, B, $\overline{A}$, and $\overline{B}$. Furthermore, each product term is referred to as a *minterm*. Since there are four product terms, there are therefore four minterms, which we

Decimal	Binary A B	Product Term	Minterm Designation
0	0 0	$\overline{A}\ \overline{B}$	m_0
1	0 1	$\overline{A}\ B$	m_1
2	1 0	$A\ \overline{B}$	m_2
3	1 1	$A\ B$	m_3

FIGURE 8.1 Minterm designations for each input combination of a 2-input circuit.

label m_0 to m_3. The subscripts correspond to the decimal equivalent of each product term.

When we use a truth table to construct a sum-of-products Boolean expression, we are simply writing the function as the sum of all minterms for which the function is 1. If there are two inputs, each minterm automatically contains two literals, one for each input. Consider the Boolean function

$$F = \overline{A}B + A\overline{B}$$

From Figure 8.1 we see that this could also be written as

$$F = m_1 + m_2$$

These two expressions are equivalent. A Karnaugh map is a graphical way of presenting the same information that is found in a table such as that of Figure 8.1. Figure 8.2 shows a Karnaugh map for a 2-input logic circuit. The two variables A and B and their inverses form labels along each side of the map. The labels are arranged in such a way that each square of the map represents one of the minterms for a 2-input circuit. There are four squares on the map, one for each minterm.

The map in Figure 8.2 is essentially a blank form. It shows only the minterms that represent a 2-input circuit. To use this map for a specific Boolean expression, we must have the expression in a sum-of-products form, where each product term contains all the input literals. For example, in the expression for F given earlier, both terms in the expression contain a literal for A and B. An expression in this form can be immediately plotted on a Karnaugh map. We simply identify each

	$\overline{A}$	A
$\overline{B}$	m_0 $\overline{A}\ \overline{B}$	m_2 $A\ \overline{B}$
B	m_1 $\overline{A}\ B$	m_3 $A\ B$

FIGURE 8.2 A 2-input Karnaugh map.

$$\begin{array}{cc} \overline{A} & A \end{array}$$

FIGURE 8.3 Karnaugh map for the expression $F = \overline{A}B + A\overline{B}$.

minterm in the Boolean expression and then put a 1 in the corresponding square on the Karnaugh map. Figure 8.3 shows the Karnaugh map for the expression F that was introduced earlier.

An expression such as $F = \overline{A}\overline{B} + A$ cannot be directly plotted on a Karnaugh map since each product term in this expression does not contain all the input literals. The A term has no literal for the B input (either B or $\overline{B}$). Before the preceding expression could be plotted on a Karnaugh map, it would need to be expanded. This could be done by multiplying A by $(B + \overline{B})$. Then we would have $F = \overline{A}\overline{B} + AB + A\overline{B}$. This new version of the expression could be plotted since each term does contain all input literals.

Three-variable Karnaugh Maps

Karnaugh maps for two input variables are not particularly useful since the techniques of Boolean algebra make reduction a very simple matter. However, Karnaugh maps can be constructed for any number of inputs. Figure 8.4 shows the input combinations that are possible for a 3-input circuit with inputs A, B, and C. Again, each product term is given a minterm designation corresponding to the decimal number of the product term. A Karnaugh map representing the information in Figure 8.4 is shown in Figure 8.5(a). Again, there is a square for each minterm. Notice how the labels have been arranged at the sides of the map. Each square

Decimal	Binary A B C	Product Term	Minterm Designation
0	0 0 0	$\overline{A}\ \overline{B}\ \overline{C}$	m_0
1	0 0 1	$\overline{A}\ \overline{B}\ C$	m_1
2	0 1 0	$\overline{A}\ B\ \overline{C}$	m_2
3	0 1 1	$\overline{A}\ B\ C$	m_3
4	1 0 0	$A\ \overline{B}\ \overline{C}$	m_4
5	1 0 1	$A\ \overline{B}\ C$	m_5
6	1 1 0	$A\ B\ \overline{C}$	m_6
7	1 1 1	$A\ B\ C$	m_7

FIGURE 8.4 Product terms and minterm designations for a 3-input circuit.

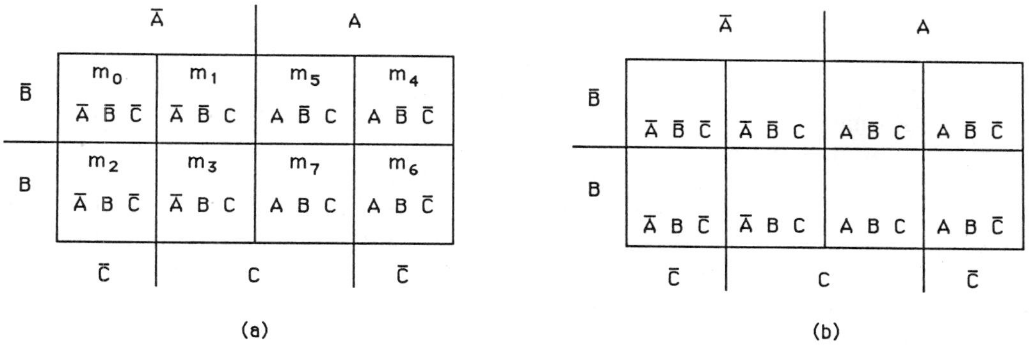

	$\overline{A}$		A	
$\overline{B}$	m_0 $\overline{A}\ \overline{B}\ \overline{C}$	m_1 $\overline{A}\ \overline{B}\ C$	m_5 $A\ \overline{B}\ C$	m_4 $A\ \overline{B}\ \overline{C}$
B	m_2 $\overline{A}\ B\ \overline{C}$	m_3 $\overline{A}\ B\ C$	m_7 $A\ B\ C$	m_6 $A\ B\ \overline{C}$
	$\overline{C}$	C		$\overline{C}$

(a)

	$\overline{A}$		A	
$\overline{B}$	$\overline{A}\ \overline{B}\ \overline{C}$	$\overline{A}\ \overline{B}\ C$	$A\ \overline{B}\ C$	$A\ \overline{B}\ \overline{C}$
B	$\overline{A}\ B\ \overline{C}$	$\overline{A}\ B\ C$	$A\ B\ C$	$A\ B\ \overline{C}$
	$\overline{C}$	C		$\overline{C}$

(b)

FIGURE 8.5 Karnaugh maps for a 3-input circuit: (a) with minterm designations; (b) without minterm designations.

has a unique combination of labels, which means that each square represents a unique product and therefore a unique minterm.

As a matter of convenience, the minterm designations are not usually written on the map. Thus we will use maps of the type shown in Figure 8.5(b), which shows only the product terms. You might ask if the method of labeling shown in Figure 8.5 must be followed exactly. The answer is no. We could interchange letters at will (for example, A and $\overline{A}$ with B and $\overline{B}$, and so on). This would make the map look different, but would not affect its basic function (which we will get to shortly).

The three-variable Karnaugh map is used in the same manner as the two-variable map. Thus consider the Boolean function

$$F = AB\overline{C} + \overline{A}BC + \overline{A}\,\overline{B}C$$

We simply place a 1 in the square on the Karnaugh map that represents each product term in this expression. This is shown in Figure 8.6. It is very easy to reverse this process. Suppose we had been given the map in Figure 8.6. The Boolean expression it represents (with no reduction) is simply the OR of each product term that has a 1 in its square.

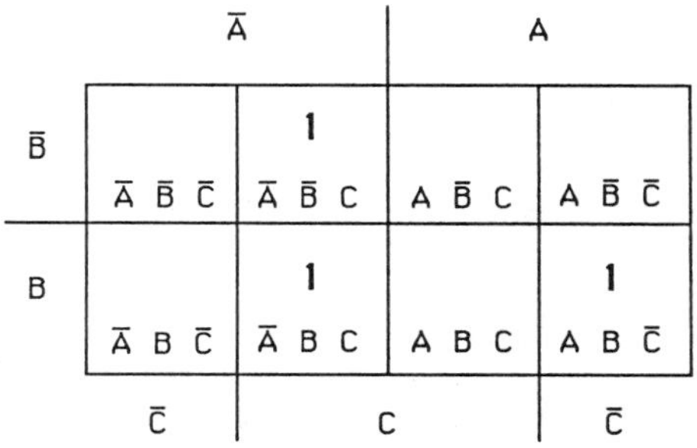

FIGURE 8.6 Karnaugh map of the function $F = AB\overline{C} + \overline{A}BC + \overline{A}\,\overline{B}C$.

	$\overline{A}$		A	
$\overline{B}$		1		
	$\overline{A}\ \overline{B}\ \overline{C}$	$\overline{A}\ \overline{B}\ C$	$A\ \overline{B}\ C$	$A\ \overline{B}\ \overline{C}$
B		1		1
	$\overline{A}\ B\ \overline{C}$	$\overline{A}\ B\ C$	$A\ B\ C$	$A\ B\ \overline{C}$
	$\overline{C}$	C		$\overline{C}$

Four-variable Karnaugh Maps

By extending the principles developed so far, it is easy to see how a four-variable Karnaugh map can be constructed. Figure 8.7 lists all the product terms and min-term designations associated with a 4-input circuit. The corresponding Karnaugh map is shown in Figure 8.8(a). Again, we will dispense with the minterm designations within each square and therefore use the map shown in Figure 8.8(b).

A four-variable Karnaugh map is used in the same way as a three- or two-variable map. A Boolean expression to be mapped must be in a sum-of-products form, with each product containing four literals. Each term therefore corresponds to a square on a four-variable Karnaugh map. On the map a 1 is placed in each square that is represented by a product in the Boolean expression. As an example, consider the expression

$$F = \overline{A}\,\overline{B}\,\overline{C}\,\overline{D} + \overline{A}\,\overline{B}\,C\,D + \overline{A}\,B\,\overline{C}\,\overline{D} + A\,\overline{B}\,\overline{C}\,\overline{D} + A\,\overline{B}\,\overline{C}$$

The resulting Karnaugh map is shown in Figure 8.9. Notice that there is a 1 in each square that corresponds to one of the products in the Boolean expression.

Reduction Using Karnaugh Maps

Boolean algebra reduction, in the final analysis, is largely a matter of applying the law of complements $(A + \overline{A} = 1)$. Even in cases where this does not seem to be

Decimal	Binary A B C D	Product Term	Minterm Designation
0	0 0 0 0	$\overline{A}\ \overline{B}\ \overline{C}\ \overline{D}$	m_0
1	0 0 0 1	$\overline{A}\ \overline{B}\ \overline{C}\ D$	m_1
2	0 0 1 0	$\overline{A}\ \overline{B}\ C\ \overline{D}$	m_2
3	0 0 1 1	$\overline{A}\ \overline{B}\ C\ D$	m_3
4	0 1 0 0	$\overline{A}\ B\ \overline{C}\ \overline{D}$	m_4
5	0 1 0 1	$\overline{A}\ B\ \overline{C}\ D$	m_5
6	0 1 1 0	$\overline{A}\ B\ C\ \overline{D}$	m_6
7	0 1 1 1	$\overline{A}\ B\ C\ D$	m_7
8	1 0 0 0	$A\ \overline{B}\ \overline{C}\ \overline{D}$	m_8
9	1 0 0 1	$A\ \overline{B}\ \overline{C}\ D$	m_9
10	1 0 1 0	$A\ \overline{B}\ C\ \overline{D}$	m_{10}
11	1 0 1 1	$A\ \overline{B}\ C\ D$	m_{11}
12	1 1 0 0	$A\ B\ \overline{C}\ \overline{D}$	m_{12}
13	1 1 0 1	$A\ B\ \overline{C}\ D$	m_{13}
14	1 1 1 0	$A\ B\ C\ \overline{D}$	m_{14}
15	1 1 1 1	$A\ B\ C\ D$	m_{15}

FIGURE 8.7 Product terms and minterm designations for a 4-input circuit.

(a)

	Ā	Ā	A	A	
B̄	m_0 $\overline{A}\,\overline{B}\,\overline{C}\,\overline{D}$	m_2 $\overline{A}\,\overline{B}\,C\,\overline{D}$	m_{10} $A\,\overline{B}\,C\,\overline{D}$	m_8 $A\,\overline{B}\,\overline{C}\,\overline{D}$	D̄
	m_1 $\overline{A}\,\overline{B}\,\overline{C}\,D$	m_3 $\overline{A}\,\overline{B}\,C\,D$	m_{11} $A\,\overline{B}\,C\,D$	m_9 $A\,\overline{B}\,\overline{C}\,D$	D
B	m_5 $\overline{A}\,B\,\overline{C}\,D$	m_7 $\overline{A}\,B\,C\,D$	m_{15} $A\,B\,C\,D$	m_{13} $A\,B\,\overline{C}\,D$	
	m_4 $\overline{A}\,B\,\overline{C}\,\overline{D}$	m_6 $\overline{A}\,B\,C\,\overline{D}$	m_{14} $A\,B\,C\,\overline{D}$	m_{12} $A\,B\,\overline{C}\,\overline{D}$	D̄
	C̄	C	C̄		

(b)

	Ā	Ā	A	A	
B̄	$\overline{A}\,\overline{B}\,\overline{C}\,\overline{D}$	$\overline{A}\,\overline{B}\,C\,\overline{D}$	$A\,\overline{B}\,C\,\overline{D}$	$A\,\overline{B}\,\overline{C}\,\overline{D}$	D̄
	$\overline{A}\,\overline{B}\,\overline{C}\,D$	$\overline{A}\,\overline{B}\,C\,D$	$A\,\overline{B}\,C\,D$	$A\,\overline{B}\,\overline{C}\,D$	D
B	$\overline{A}\,B\,\overline{C}\,D$	$\overline{A}\,B\,C\,D$	$A\,B\,C\,D$	$A\,B\,\overline{C}\,D$	
	$\overline{A}\,B\,\overline{C}\,\overline{D}$	$\overline{A}\,B\,C\,\overline{D}$	$A\,B\,C\,\overline{D}$	$A\,B\,\overline{C}\,\overline{D}$	D̄
	C̄	C	C̄		

FIGURE 8.8 Karnaugh maps for a 4-input circuit: (a) with minterm designations, (b) without minterm designations.

true, it actually is. For example, consider the expression

$$F = \overline{A}B + A$$

Theorem (T–8) from Chapter 3 says that this expression reduces directly to $B + A$. We don't think of Theorem (T–8) as having anything to do with the law of complements. However, consider the following development:

$$F = \overline{A}B + A = \overline{A}B + A(B + \overline{B})$$
$$= \overline{A}B + AB + A\overline{B} = \overline{A}B + AB + A\overline{B} + AB$$

$$\text{(since } AB = AB + AB\text{)}$$

	Ā	Ā	A	A	
B̄	1 $\overline{A}\,\overline{B}\,\overline{C}\,\overline{D}$	$\overline{A}\,\overline{B}\,C\,\overline{D}$	$A\,\overline{B}\,C\,\overline{D}$	1 $A\,\overline{B}\,\overline{C}\,\overline{D}$	D̄
	$\overline{A}\,\overline{B}\,\overline{C}\,D$	1 $\overline{A}\,\overline{B}\,C\,D$	1 $A\,\overline{B}\,C\,D$	$A\,\overline{B}\,\overline{C}\,D$	D
B	$\overline{A}\,B\,\overline{C}\,D$	$\overline{A}\,B\,C\,D$	$A\,B\,C\,D$	$A\,B\,\overline{C}\,D$	
	1 $\overline{A}\,B\,\overline{C}\,\overline{D}$	$\overline{A}\,B\,C\,\overline{D}$	$A\,B\,C\,\overline{D}$	$A\,B\,\overline{C}\,\overline{D}$	D̄
	C̄	C	C̄		

FIGURE 8.9 Karnaugh map of the function $F = \overline{A}\,\overline{B}\,C\,D + A\,\overline{B}\,C\,D + A\,\overline{B}\,\overline{C}\,\overline{D} + \overline{A}\,B\,\overline{C}\,\overline{D} + \overline{A}\,\overline{B}\,\overline{C}\,\overline{D}$.

$$= (\overline{A} + A)B + A(\overline{B} + B)$$

$$= B + A$$

Thus once the Boolean expression is written in a sum-of-products form with each product containing all the literals, the reduction is seen as several applications of the law of complements. This fact is what makes reduction using Karnaugh maps work.

If you look at any of the Karnaugh maps presented so far, you will see that the product terms in adjacent squares (either horizontally or vertically) differ by only one literal. But two terms in a Boolean expression that differ by only one literal can be reduced by using the law of complements. That is, the two terms $\overline{A}\,\overline{B}CD + A\overline{B}CD$ can be reduced in the following way:

$$\overline{A}\,\overline{B}CD + A\overline{B}CD = (\overline{A} + A)\overline{B}CD$$

$$= \overline{B}CD$$

You will notice in Figure 8.9 that the two squares representing the preceding two terms are adjacent, and both have 1's in them. These two squares can be grouped together, and only the variables $\overline{B}CD$, common to both, are retained. That is, the variables A and $\overline{A}$ are dropped because they are different in the two squares. This is essentially how Karnaugh map reduction proceeds.

The basic Karnaugh map reduction procedure is then as follows. First, we plot the Boolean expression on a Karnaugh map. Then we look for *rectangular* groups of 1's of size 2, 4, 8, or 16. Within each group, we keep only those variables that are common to the entire group. The rest are thrown out. Every 1 on the Karnaugh map must be accounted for by being included in at least one group (even if it is a group of one). A 1 may be included in as many groups as desired in order to facilitate grouping. However, the aim is to minimize the number of groups. Thus a new group should not be created if it does not use any *unused* 1's. Although it may not be immediately obvious, for purposes of grouping, squares along the top row of a Karnaugh map are adjacent to squares along the bottom row of the map. Likewise, squares along the left column are adjacent to squares along the right column. You can verify this by noting that these squares differ by only one variable. The procedure is best illustrated by example.

■ EXAMPLE 8.2–1

Figure 8.10 shows a Karnaugh map that has been constructed for the Boolean function listed below the map. This function starts with 10 products of four literals each. Each product term is represented by a 1 on the map. We now try to include every 1 in as large a group as possible. The map shows four groups can be found that include all the 1's. In two instances, 1's have been used twice. To determine

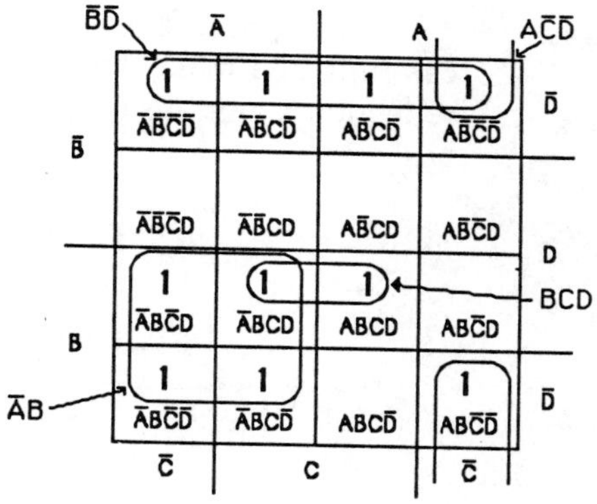

FIGURE 8.10 $F = \overline{A}\,\overline{B}\,\overline{C}\,\overline{D}$
$+\overline{A}\,\overline{B}\,C\overline{D} + A\overline{B}\,C\overline{D} +$
$A\overline{B}\,\overline{C}\,\overline{D} + \overline{A}\,B\underline{C}\,\overline{D} + \overline{A}BCD$
$+ \underline{A}BC\underline{D} + \overline{A}\,B\underline{C}\underline{D}$
$+\underline{A}B\overline{C}\overline{D} + ABC\overline{D} =$
$\overline{A}\,CD + BCD + \overline{B}\,\overline{D} + \overline{A}\,D$

the Boolean expression for each group, we keep only the variables that are constant within the group. For example, in the group of four along the top, only the $\overline{B}\,\overline{D}$ terms remain unchanged within the group. Thus this group reduces to the Boolean term $\overline{B}\,\overline{D}$. The original 10-term expression is seen to reduce to only four terms, and these terms each have less than four literals. You might want to try this reduction by using standard Boolean techniques. ∎

■ EXAMPLE 8.2–2

As a second example, consider Figure 8.11. Here the original expression maps into the four corners of the map. But these four corners are adjacent to each other both horizontally and vertically. Thus they can be combined into one group of four. The only common terms are $\overline{C}\,\overline{D}$. Thus the expression reduces to $\overline{C}\,\overline{D}$. ∎

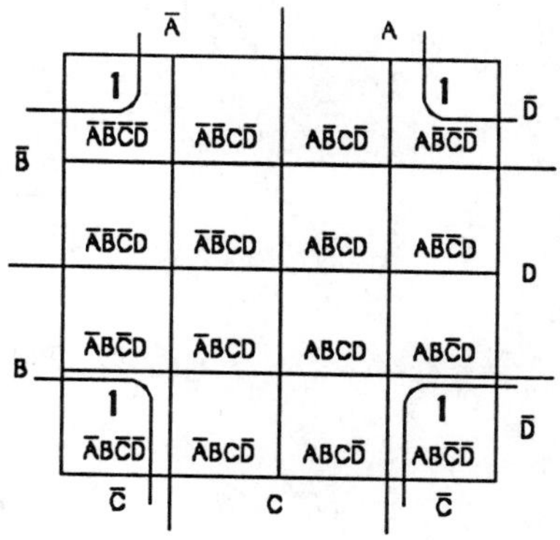

FIGURE 8.11
$F = \overline{A}\,\overline{B}\,\overline{C}\,\overline{D} + A\overline{B}\,\overline{C}\,\overline{D} + \overline{A}\,B\overline{C}\,\overline{D} + AB\overline{C}\,\overline{D}$
$= \overline{C}\,\overline{D}$

Figure 8.12 shows several examples of map reduction. Study these examples carefully and make sure you understand them. Notice, as in Figure 8.12(d), that reduction is not always possible.

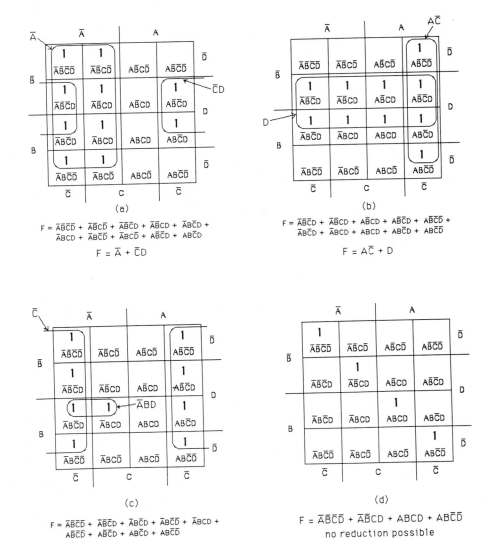

$F = \overline{A}\overline{B}\overline{C}\overline{D} + \overline{A}\overline{B}\overline{C}D + \overline{A}\overline{B}C\overline{D} + \overline{A}\overline{B}CD + \overline{A}B\overline{C}D + \overline{A}B\overline{C}\overline{D} + \overline{A}B\overline{C}D + AB\overline{C}\overline{D} + AB\overline{C}D$

$$F = \overline{A} + \overline{C}D$$

(a)

$F = \overline{A}\overline{B}\overline{C}D + \overline{A}\overline{B}CD + \overline{A}B\overline{C}D + \overline{A}BCD + A\overline{B}\overline{C}\overline{D} + A\overline{B}\overline{C}D + A\overline{B}CD + AB\overline{C}D + AB\overline{C}\overline{D}$

$$F = A\overline{C} + D$$

(b)

$F = \overline{A}\overline{B}\overline{C}\overline{D} + \overline{A}\overline{B}\overline{C}D + \overline{A}B\overline{C}D + \overline{A}B\overline{C}\overline{D} + \overline{A}BCD + AB\overline{C}\overline{D} + A\overline{B}\overline{C}D + AB\overline{C}D + AB\overline{C}\overline{D}$

$$F = \overline{A}BD + \overline{C}$$

(c)

$$F = \overline{A}\overline{B}C\overline{D} + \overline{A}\overline{B}CD + ABCD + AB\overline{C}\overline{D}$$

no reduction possible

(d)

FIGURE 8.12   Examples of Karnaugh map reductions.

Don't-care States

In the examples so far we have assumed that every input combination results in a predefined output that is either 0 or 1. That is, in the truth table for the Boolean function, we assume that in the function column there are only 0's and 1's. There are occasions, however, where this is not the case. Suppose we are to design a circuit that has four inputs. This circuit could in principle have 16 different input combinations. Suppose, however, that we know that certain input combinations will never occur. How do we list these in a truth table? These states are called *don't-care* states since we don't care how the circuit would react to them (they will never occur). We list these states as X's in our truth table. Let's consider a specific example.

■ **EXAMPLE 8.2—3**

We are told to design a 4-input circuit (*ABCD*) where the binary inputs 1010 to 1111 will never occur. That is, only the first 10 binary combinations will occur. For those 10 combinations, we are given the desired circuit outputs. These outputs are shown in the truth table in Figure 8.13. We see in the truth table that the six don't-care states have been marked by X's. We now must transfer the information in the truth table to a four-variable Karnaugh map. The 1's and 0's are easy enough. What about the X's? These are transferred directly to the map as shown in Figure 8.14. The reduction rule is this. X's may be used in grouping with 1's *if this is convenient*. That is, use the X's if they help to include any 1's in a larger group. You don't need to use the X's. In Figure 8.14, the X's have proved very useful in

Inputs	Output
A B C D	F
0 0 0 0	0
0 0 0 1	0
0 0 1 0	0
0 0 1 1	0
0 1 0 0	1
0 1 0 1	1
0 1 1 0	1
0 1 1 1	1
1 0 0 0	1
1 0 0 1	1
1 0 1 0	X
1 0 1 1	X
1 1 0 0	X
1 1 0 1	X
1 1 1 0	X
1 1 1 1	X

Don't–care states

FIGURE 8.13 Truth table to be implemented.

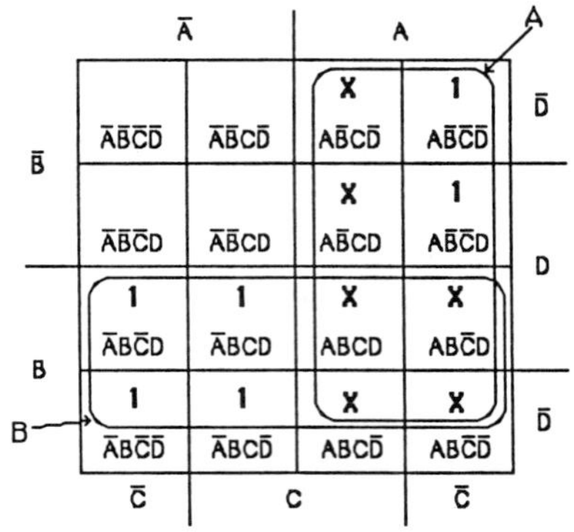

FIGURE 8.14 Karnaugh map for the truth table of Figure 8.13. $F = A + B$.

enlarging the groups of 1's. The final reduction turns out to be very simple when the X's are included. ∎

Even though the original truth table does not specify the state of the circuit for any of the don't-care inputs, once we complete a reduction, the resulting Boolean expression will have these states specified. Thus in Figure 8.14, the result $F = A + B$ is unambiguous about what happens if one of the don't-care inputs *should* occur.

The Karnaugh mapping technique helps with obtaining a minimized Boolean function from the original truth table. It does not tell you how to implement this function. Usually, there are many things to consider when implementing a design. If there will be thousands of copies of the circuit ultimately produced, then it pays to take extra time to produce a design that will minimize things like parts count, wiring complexity, and parts cost. On the other hand, if very few copies will be made, then it pays to reduce the design time. This may involve using more expensive MSI ICs, which can provide a quick design solution.

■ **EXAMPLE 8.2–4**

To illustrate the range of design choices, we will proceed with the design of a code converter. The circuit will accept BCD code as inputs and produce XS3 Gray code as outputs. We will assume that only valid BCD inputs will ever occur. There will thus be six don't-care inputs. Figure 8.15 shows the truth table for the circuit. Since there are four outputs to this circuit, there will be four Karnaugh maps, one for each output. The resulting maps are shown in Figure 8.16. The maps are labeled W, X, Y, and Z, respectively, to identify them with the appropriate output.

Figure 8.16 shows the groupings of 1's and X's that are possible and the resulting Boolean expressions. Once we have these reduced expressions, we must

Inputs BCD				Outputs XS3 Gray			
A	B	C	D	W	X	Y	Z
0	0	0	0	0	0	1	0
0	0	0	1	0	1	1	0
0	0	1	0	0	1	1	1
0	0	1	1	0	1	0	1
0	1	0	0	0	1	0	0
0	1	0	1	1	1	0	0
0	1	1	0	1	1	0	1
0	1	1	1	1	1	1	1
1	0	0	0	1	1	1	0
1	0	0	1	1	0	1	0
1	0	1	0				
1	0	1	1				
1	1	0	0		Don't-care		
1	1	0	1				
1	1	1	0				
1	1	1	1				

FIGURE 8.15 Truth table for BCD to XS3 Gray code converter.

decide how to implement them. The direct approach is to use SSI gates. The resulting circuit is shown in Figure 8.17. If we implement this circuit exactly as shown, it will be somewhat awkward, since AND gates and OR gates are not the preferred type of SSI gates. We can easily convert this circuit to a NAND gate circuit. First we replace AND gates 1 to 7 with NAND gates. We then replace OR gates 8 to 10 with NAND gates drawn as low-level-input OR gates. This will require attention to the direct inputs to the three ORs (inputs not from other NANDs). For example, if OR gate 8 is replaced by a NAND gate, it will essentially become an OR gate with inverters at the input. The A input to OR gate 8 would then need to be inverted, since the inverter at the input to the NAND (low-level-input OR) would produce the wrong logic otherwise. But this can be taken care of by simply using the output from the inverter at the A input. That is, instead of inputting A into gate 8, we will simply input $\overline{A}$. If these changes are made, the circuit in Figure 8.17 will require five 2-input NAND gates, three 3-input NAND gates, two 4-input NAND gates, and four inverters. Since 2-input NAND gates come four on a chip (7400), we need two of these chips. Three-input NAND gates come three on a chip (7410), so we need one of these. Four-input NAND gates come two on a chip (7420), so we need one of these. Finally, inverters come six to a chip (7404), so we need one of these. The final parts count is five ICs.

  There are other ways we could proceed. For example, we saw in Chapter 4 that multiplexers could be used to implement Boolean expressions. Figure 8.18 shows how three 16-input 74150 multiplexers could be used to implement our code converter. ∎

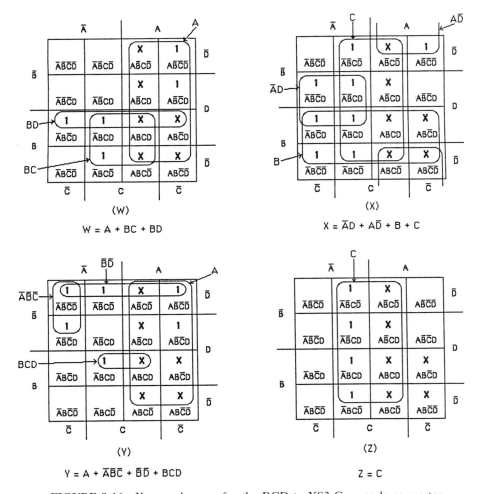

FIGURE 8.16   Karnaugh maps for the BCD to XS3 Gray code converter.

■ **EXAMPLE 8.2–5**

Consider the truth table shown in Figure 8.19. The Karnaugh maps for the three outputs X, Y, and Z are shown in Figure 8.20. If we minimize each map, the results are as shown in the figure and summarized as follows:

$$X = B\overline{C} + CD + A\overline{C}$$
$$Y = B + \overline{A}CD$$

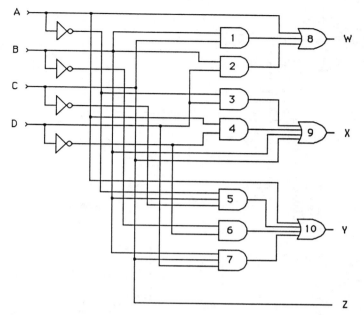

FIGURE 8.17 Circuit implementation of the BCD to XS3 Gray code converter.

$$Z = BC + A\overline{B}\,\overline{C} + ACD$$

Direct SSI implementation of these equations, not counting inverters, requires seven AND gates and three OR gates, a total of ten gates. However, notice that in the maps for Y and Z there are groups of 1's that exactly cover some of the minterms in the map for X. The dotted-line groupings in the map for X in Figure 8.19 show which 1's these are. The gates that produce the terms $\overline{A}CD$ (from Y) and $A\overline{B}\,\overline{C}$ and ACD (from Z) can be used directly to implement the terms covering the corresponding 1's on the X map. If we rewrite the X expression in a nonminimum form by using the dotted-line groupings of 1's on the X map, the result for X is

$$X = \overline{A}CD + ACD + A\overline{B}\,\overline{C} + B\overline{C}$$

The first three terms already exist in the Y and Z expressions and thus do not need to be implemented a second time. This reduces the number of AND gates required to five rather than seven. Thus two gates have been saved by using a nonminimum form for X. When working with multiple-output Karnaugh reductions, you need to be alert to the possibility of using terms from one map that can cover minterms in another map. ∎

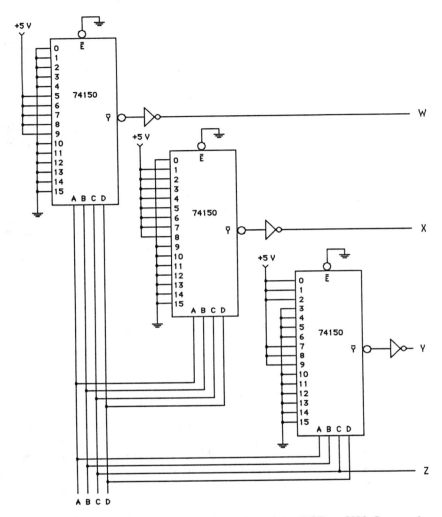

FIGURE 8.18 Multiplexer implementation of the BCD to XS3 Gray code converter.

Modifying the Rules

It was stated earlier that, for a Boolean expression to be in a form that could be plotted on a Karnaugh map, the expression had to be a sum of products, with each term containing a literal for all the input variables. That is, if the equation contained four input variables (A, B, C, and D), then each term in the Boolean expression

	Inputs			Outputs		
A	B	C	D	X	Y	Z
0	0	0	0	0	0	0
0	0	0	1	0	0	0
0	0	1	0	0	0	0
0	0	1	1	1	1	0
0	1	0	0	1	1	0
0	1	0	1	1	1	0
0	1	1	0	0	1	1
0	1	1	1	1	1	1
1	0	0	0	1	0	1
1	0	0	1	1	0	1
1	0	1	0	0	0	0
1	0	1	1	1	0	1
1	1	0	0	1	1	0
1	1	0	1	1	1	0
1	1	1	0	0	1	1
1	1	1	1	1	1	1

FIGURE 8.19 Truth table for Example 8.2-5.

had to contain a literal for each of these variables. Another way of saying this is that each term had to be a minterm.

If some thought is given, however, to the Karnaugh reduction procedure we have developed, it can be seen that the preceding restriction is not necessary. For example, consider the lengthy Boolean expression listed under Figure 8.10. This expression reduces to $A\overline{C}\overline{D} + BCD + \overline{B}\overline{D} + \overline{A}D$. Clearly, the Karnaugh map represents both the original expression, consisting only of minterms, and the reduced expression, which does not consist of true minterms (since each product lacks one or more literals). Each product in the reduced expression comes from a definite area of 1's on the Karnaugh map. It is therefore possible to reverse the reduction procedure. That is, given the reduced expression, it is a simple matter to fill in the corresponding squares on the Karnaugh map represented by this expression.

Consider the form that a reduced expression from a Karnaugh map will always take. Each term in the reduced expression is in product form. However, the number of literals in the term may be anywhere from the number of variables represented by the map down to zero (a term like 1). Even though each term in the reduced expression may not be a true minterm, each term will be mintermlike. That is, it will be a simple product of literals with no parentheses. We can now modify the earlier restriction requiring a Boolean expression to consist entirely of minterms and say instead that, for a Boolean expression to be in a form to be plotted on a Karnaugh map, the expression must consist entirely of mintermlike terms (some of which, of course, may be minterms). The advantage of this is that Karnaugh reduction can now be applied directly to Boolean expressions that are already somewhat reduced, but not yet in minimum form.

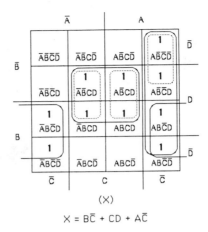

(X)

$X = B\bar{C} + CD + A\bar{C}$

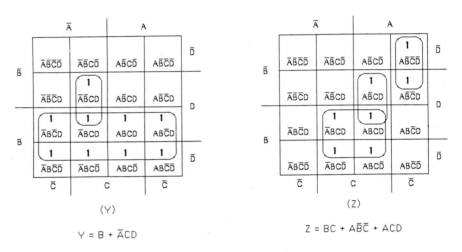

(Y)

$Y = B + \bar{A}CD$

(Z)

$Z = BC + A\bar{B}\bar{C} + ACD$

FIGURE 8.20 Karnaugh maps for Example 8.2-5.

■ **EXAMPLE 8.2−6**

Consider Figure 8.21(a). Below this figure is a Boolean expression that consists of mintermlike terms, but is not completely reduced. Each term can be plotted on the Karnaugh map since we know which area of the map each term represents. For example, the *ABC* term represents that area of the map where *A*, *B*, and *C* are common to each square. This is indicated in the figure. Each term is plotted on the Karnaugh map in turn (do not add a second 1 in a square if one is already present). We now examine the resulting map to see if it can be further reduced. Figure 8.21(b)

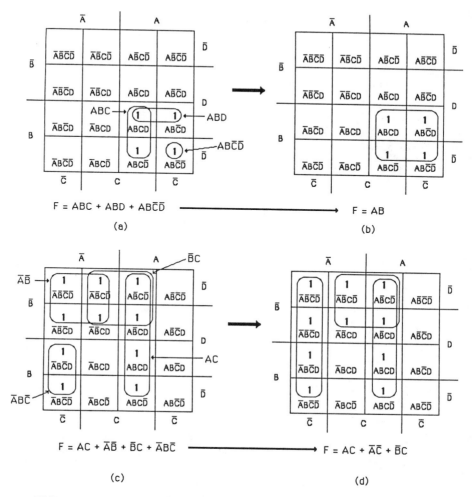

FIGURE 8.21 Karnaugh mapping applied to expressions containing min-termlike terms.

shows the result. Clearly, the single four-square area can be reduced to the term AB. Thus the original Boolean expression was not in a minimum form.

■ **EXAMPLE 8.2—7**

Consider the Boolean expression below Figure 8.21(c). It is not immediately obvious whether this expression is in minimum form or not. However, when it is plotted on a Karnaugh map [Figure 8.21(c)] and then reexamined [Figure 8.21(d)], it becomes obvious that further reduction is possible. ■

More Than Four Inputs

When a circuit has more than four inputs, the Karnaugh mapping technique becomes progressively more awkward. The fifth input variable cannot easily be added to a two-dimensional Karnaugh map. In effect, what must be done is to place two 4-input maps on top of each other in a three-dimensional structure. Other methods such as the Quine–McCluskey procedure now become more efficient. Nevertheless, since a great many circuits of practical interest involve four or fewer inputs, the Karnaugh mapping technique is a very useful method.

8.3   SEQUENTIAL KARNAUGH MAPPING

Sequential digital circuits can be used to carry out a number of tasks. For example, it is often necessary to examine a serial stream of data on an input line to determine when a particular sequence of bits occurs. Another very common task is the production of a particular sequence of counts on the multiple outputs of a digital circuit. We will examine the problem of designing a specialized counter in this section and take up the detection of code sequences in the next section.

Counting circuits use flip-flops of one kind or another. In implementing a specialized counter, the design procedure depends on the type of flip-flop to be used. We will develop methods that utilize the clocked $J-K$ type of flip-flop.

The design problem begins with a specification of the number of states in the sequence to be cycled through. The number of states in this sequence will determine how many FFs must be used in the design. Recall that N FFs can take on 2^N distinct states. Thus three FFs can take on eight unique states, four FFs can take on sixteen states, and so on. For example, suppose it is decided that an application requires a six-state counter. We know that two FFs is not enough since they will produce at most four distinct states. Three FFs will work since eight states are now possible.

Once we have decided on the number of states, say six, we must decide exactly what the count sequence will be. With three FFs, each state will be defined by three literals, say A, B, and C. Suppose we decide on the series of states in Figure 8.22(a). This figure shows how the counter is supposed to cycle through a specific series of six states. An alternative way of showing this information is the *next-state* table of Figure 8.22(b). In a next-state table, all the possible states are listed sequentially under the State column. To the right of each state in the State column is listed the next state the counter will cycle to. This second set of entries forms the Next State column. For states that never occur in the sequence, an X is placed to their right, in the Next State column.

Flip-flop outputs

State	A	B	C	
0	0	0	0	
1	1	0	0	
2	0	0	1	
3	1	1	1	
4	1	1	0	
5	0	1	1	
6	1	0	1	Not
7	0	1	0	used

(a)

State A B C	Next State A B C
0 0 0	1 0 0
0 0 1	1 1 1
0 1 0	X
0 1 1	0 0 0
1 0 0	0 0 1
1 0 1	X
1 1 0	0 1 1
1 1 1	1 1 0

(b)

FIGURE 8.22 (a) Desired count sequence for a counter. (b) Next-state table for the counter.

Since we will be using clocked *J–K* FFs, the counter will change states on the falling edge of each clock pulse. The information in either Figure 8.22(a) or (b) is sufficient to tell us how the counter will change at each clock pulse. We assume that the two states 101 and 010 can never occur and are not part of the count sequence. This may require that the counter be reset at start-up so that it begins in one of the known states of the sequence. Otherwise, depending on the final design, it is possible that, if the counter comes on in one of the not-used states, it will hang up in some kind of loop involving these states.

Let us now see how we can develop a Karnaugh map for the counter specified by the tables in Figure 8.22. Since there are three FFs, we require a standard Karnaugh map appropriate for three input variables. With this map we simply label each square that is part of the count sequence with the number that represents the state. That is, from Figure 8.22(a) we have states labeled 0 to 5, each corresponding to a definite product of the three literals *A*, *B*, and *C*. We simply place the state number in the square on the Karnaugh map representing the appropriate product. This is shown in Figure 8.23. Arrows are used to graphically illustrate the count

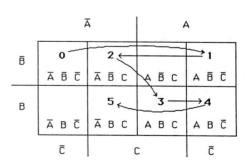

FIGURE 8.23. Karnaugh map for the six-state counter defined by Figure 8.22.

sequence, although they are not really necessary. Notice that two squares on the map have no number in them since they are not part of the count sequence.

The map in Figure 8.23 simply shows the count sequence in a graphic way. The next step is to use this map to produce three more maps, one for each flip-flop. These new maps will have special marks in each square depending on how each FF changes from one state to the next. There are five possibilities:

1. FF changes from reset to set.
2. FF changes from set to reset.
3. FF is set and remains set.
4. FF is reset and remains reset.
5. State does not occur, a don't-care state.

We now introduce five symbols to be used for each of these circumstances. These symbols are shown in Figure 8.24 and listed here. The symbols are to be placed in the square representing the *current* state and are determined by the change from the *current* state to the *next* state.

1   FF changes from reset to set $(0 \rightarrow 1)$.
+   FF is set and remains set $(1 \rightarrow 1)$.
−   FF changes from set to reset $(1 \rightarrow 0)$.
0   FF is reset and remains reset $(0 \rightarrow 0)$.
X   State never occurs; don't-care state.

We now create a Karnaugh map for each FF using these symbols. This is done by examining either the state table (such as Figure 8.22) or the corresponding Karnaugh map (such as Figure 8.23). Let us do this for each of the three FFs that make up our six-state counter. The results are shown on the three maps of Figure 8.25. Each map is labeled A, B, or C corresponding to the FF that it represents. We will now go through the development of the A FF map in detail.

Refer to Figure 8.23. In this figure, notice that there are two states that never occur. These are don't-care states and are marked with an X on all three maps in Figure 8.25. Now let us start with state 0, and look at how FF A changes as we go

Symbol to represent state change	Current state	Next state
1	0	1
+	1	1
−	1	0
0	0	0
X	Don't-care	

FIGURE 8.24 Symbols used to represent flip-flop state changes.

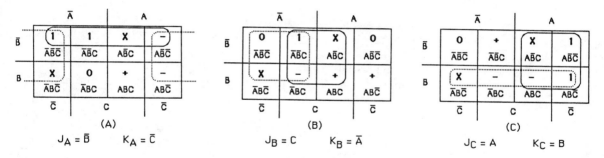

FIGURE 8.25 Karnaugh maps for the six-state counter.

from state 0 to state 1. We see that FF A changes from reset to set, so we put a 1 in the square representing state 0. Now we move to state 1 and see that, in going from state 1 to 2, FF A changes from set to reset. Thus we put a $-$ in the square representing state 1. Next we move to state 2 and note that, in going from state 2 to 3, FF A changes from reset to set. So we put a 1 in the square representing state 2. In going from state 3 to 4, the change is set to set. Thus the square for state 3 gets a $+$. From state 4 to 5, the change is set to reset, so the square for state 4 gets a $-$. From state 5 back to 0, the change is reset to reset, so the square for state 5 gets a 0. This completes the map for FF A. You should study the other two maps carefully to make sure how they were arrived at.

Now that we have completed a Karnaugh map for each FF, these maps must be used to determine the various interconnections between FF outputs and FF inputs. Not surprisingly, the map for FF A will be used to determine the logic that must be connected to the J and K inputs of FF A, and likewise for FFs B and C. The procedure is somewhat similar to that used for reducing combinational Karnaugh maps, except that the symbols that must be grouped are different. There are, in fact, two sets of grouping rules, one for the J input and one for the K input. These rules are as follows:

J Input Equation

    1. Each square with a 1 must be used in a group.
    2. Each square with a 0 may not be used in a group.
    3. All other symbols are optional.

K Input Equation

    1. Each square with a $-$ must be used in a group.
    2. Each square with a $+$ may not be used in a group.
    3. All other symbols are optional.

With these rules, we can complete two sets of groupings on the map for each FF. To help keep the two sets of groupings distinct from each other, we will use solid lines for the J groupings and dotted lines for the K groupings. The maps in Figure 8.25 have the groupings indicated on them.

The groupings are used just as they were in reducing the combinational Karnaugh maps. Thus the logical expression represented by a given group consists of the product of those literals that are identical in every square within the group. Below each map in Figure 8.25 are listed the logical expressions for the J and K inputs for the corresponding FF. For example, on the map for FF A we see that there is one solid-line group, and that in this group only the term $\overline{B}$ is common to every square. Thus the J input of FF A should be connected to the reset output of FF B (that is, to output $\overline{B}$). Likewise, on the map for FF A we see that there is one dotted-line group, and that in this group only the term $\overline{C}$ is common to every square. Thus the K input of FF A should be connected to the reset output of FF C. The other two maps in Figure 8.25 have been reduced in the same manner. Study these results carefully to be sure you understand how they were arrived at.

Once the J and K input equations are known, the circuit design can be completed. This is done in Figure 8.26. As a check, it is useful to imagine the counter in its initial state and then trace the logic to determine what the next state will be. Thus, if we assume that the counter is reset, the J, K inputs for the three FFs will be

$$A: J = 1, K = 1; \qquad B: J = 0, K = 1; \qquad C: J = 0, K = 0$$

With this set of inputs, at the next clock pulse FF A will toggle, FF B will remain reset, and FF C will remain reset. The count will thus change to 100, which is

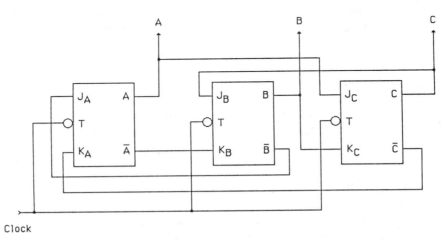

Clock

FIGURE 8.26 Circuit implementation of the six-state counter defined by the tables in Figure 8.22.

correct. You should continue this analysis through all the counting states to assure yourself that the counter works correctly.

We now return to the two don't-care states in the original count table. How would the counter behave if it somehow were placed into one of these states? First, consider the state 101. If the counter is in this state, the J, K inputs of the three FFs will be

$$A: J = 1, K = 0; \qquad B: J = 1, K = 0; \qquad C: J = 1, K = 0$$

On the next clock pulse, FF A will remain set, FF B will set, and FF C will remain set. The new count will be 111, which is part of the desired count sequence. Thus, after being placed in the 101 don't-care state, the counter will fall immediately into the correct count sequence. What about the 010 don't-care state? You should verify for yourself that if the counter is placed in the 010 state the next clock pulse will place it in the 000 state.

■ EXAMPLE 8.3–1

As a further illustration of the sequential Karnaugh mapping procedure, let us design a counter that counts directly in XS3 Gray code. Figure 8.27 illustrates the count sequence and identifies the don't-care states. (This counter could be implemented by using a BCD counter and then converting the BCD code into XS3 Gray code using a circuit like that of Figure 8.17.)

With 10 states, it is obvious that we must use four FFs. Thus we will require four 4-variable Karnaugh maps, one for each FF. These maps are shown in Figure

	Outputs				
State	A	B	C	D	
0	0	0	1	0	
1	0	1	1	0	
2	0	1	1	1	
3	0	1	0	1	
4	0	1	0	0	
5	1	1	0	0	Recycle
6	1	1	0	1	
7	1	1	1	1	
8	1	1	1	0	
9	1	0	1	0	
10	0	0	0	0	
11	0	0	0	1	
12	0	0	1	1	Don't-care
13	0	1	0	0	
14	1	0	0	1	
15	1	0	1	1	

FIGURE 8.27 The count sequence for an XS3 Gray code counter.

8.28. To facilitate reading the maps, the state number has been placed in the upper-left corner of each square on each map. The appropriate symbols have been added to each map according to how the FF associated with the map changes from one state to the next. Then the symbols have been grouped according to the J or K rules. Finally, the Boolean expression for each J and K input is determined from the groupings. All this information is shown in Figure 8.28.

Once we have the J and K input equations for each FF, all that remains is to introduce the appropriate logic between the outputs and inputs of the four FFs. Figure 8.29 shows the circuit implementation. Two points need to be made about this circuit diagram. First, the combinational logic has been implemented using all

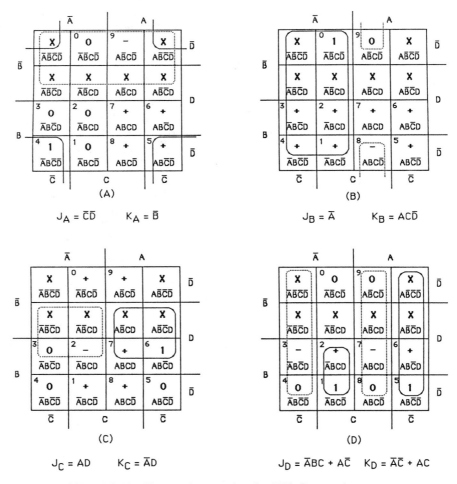

FIGURE 8.28   Karnaugh maps for the XS3 Gray code counter.

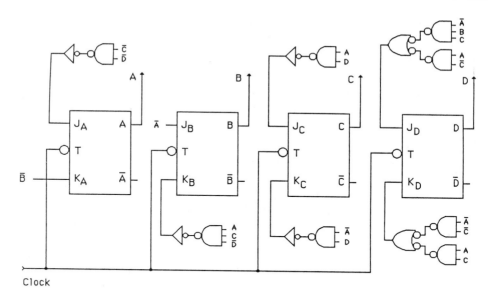

FIGURE 8.29 Circuit implementation of the XS3 Gray counter.

NAND gates and inverters. This is common practice. Second, on the diagram we have simply labeled the inputs to the gates rather than actually drawing lines from FF outputs to gate (or J, K) inputs. This greatly reduces the difficulty of reading the diagram. ∎

We will now further illustrate the combinational and sequential Karnaugh mapping techniques with an example that combines both techniques.

∎ EXAMPLE 8.3–2

Suppose we wanted to set up a circuit that would simulate the random throw of a single die. One way to do this would be to create a high-frequency clock to drive a six-state counter. The outputs from this counter would then cause a "digital die" to quickly cycle through the six standard die states 1 to 6. To "throw" the die, a person simply starts and stops the clock driving the counter. Since the clock frequency is high, the "thrower" will not know what state the die will be in when the clock is stopped.

First, let us identify the digital die. Figure 8.30(a) shows what it looks like. Each dot on the die is actually an LED that can be turned on by a logical high. Figure 8.30(b) shows one type of driver circuit for lighting each LED.

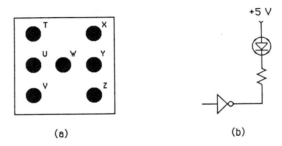

(a)   (b)

FIGURE 8.30 (a) Layout of a digital die. (b) Driver circuitry for each LED in the die.

We can break our design task into a number of parts. First, we need to design a clock that runs at some suitably high frequency. Anything over a few hundred counts per second would be adequate. Next we need to design a six-state counter. The actual count sequence can be any six-state sequence, since the code converter that comes next can be assigned to convert the code of any sequence chosen. Next we must design the code converter that takes the count produced as an input and produces the outputs needed to light up the die in the correct fashion. Finally, there is the circuitry to drive the die. Figure 8.31 schematically illustrates the different parts of the design. We will actually restrict ourselves to the design of the six-state counter and the code-converter circuit.

To design the six-state counter, we first realize that a minimum of three FFs is needed. Having decided on three FFs, we must decide on a count sequence. The obvious approach would be to select the first six binary states 000 to 101 and use them in numerical order. This choice turns out to require two AND gates for its implementation. Recall that earlier we designed a six-state counter that required no logic gates at all. The J–K inputs for each FF were taken directly from the set or reset outputs of other FFs. Thus we will simply use this earlier design.

The problem of determining the sequence of states that minimizes the logic required to implement an N-state counter is not easy to solve. For a six-state counter, 420 distinct combinations of 3-input states are possible. There are methods

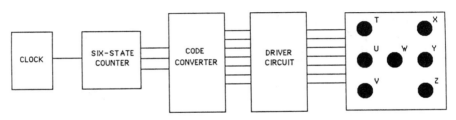

FIGURE 8.31 Schematic of the solution to Example 8.3-3.

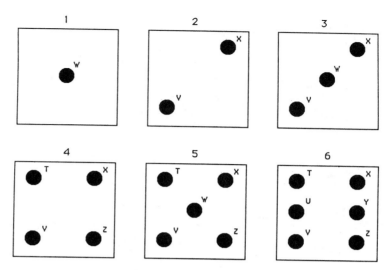

FIGURE 8.32 Standard die format.

that can be used to find the best sequence of states ("best" from the viewpoint of minimizing the gates required), but they are beyond the scope of this text.

Our next task is to design the code converter. To do this, we must ascertain what die segments must be illuminated to produce each number. Figure 8.32 illustrates the standard die format. With this figure we can determine which segments need to be illuminated at each count in the sequence. This information is summarized in Figure 8.33. The outputs T to Z will be used to drive the seven LEDs on the die. These seven outputs constitute seven Boolean functions, which must be determined and then implemented. However, a great deal of duplication is involved. In particular, the functions T, U, and V are identical to the functions Z, Y, and X, respectively. Thus actually only four independent Boolean functions must be determined.

Die State	Counter State A B C	Die Segment States T U V W X Y Z
1	0 0 0	0 0 0 1 0 0 0
2	1 0 0	0 0 1 0 0 1 0 0
3	0 0 1	0 0 1 1 1 0 0
4	1 1 1	1 0 1 0 1 0 1
5	1 1 0	1 0 1 1 1 0 1
6	0 1 1	1 1 1 0 1 1 1
	1 0 1	Don't-care
	0 1 0	

FIGURE 8.33 State table and output table for the die problem.

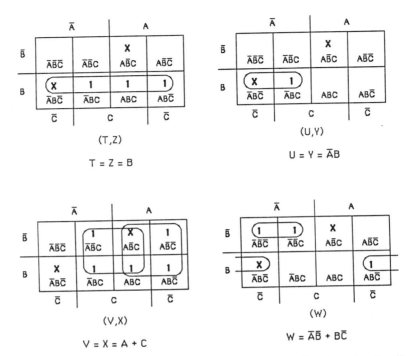

FIGURE 8.34 Karnaugh maps and reduced Boolean functions for the digital die problem.

The information from Figure 8.33 is used to complete four Karnaugh maps, one each for the four independent outputs required. These maps are shown in Figure 8.34. From the maps it is a simple matter to determine the reduced Boolean functions for each of the circuit outputs. These functions can now be implemented with SSI gates. The circuit implementation, including the counter, is shown in Figure 8.35. An implementation using NAND gates (and one inverter) has been chosen. There is one somewhat subtle point that should be noted. NAND gate 2 (shown as a low-level input OR gate) is being used to implement the function (A + C). This is done by inputting $\overline{A}$ and $\overline{C}$ to this gate. The implied inverters at the inputs to the gate result in the effective inversion of $\overline{A}$ and $\overline{C}$ and the ORing of the results. Since four 2-input NAND gates come on a single IC (7400) and since an inverter can be made from a NAND gate, the combinational part of the circuit requires only two SSI ICs. As a matter of comparison, had we chosen to use the most direct count sequence (000 to 101), the complete circuit (including counter) would have required three inverters, five 2-input NAND gates, and one 3-input NAND gate. This implementation would require an extra IC and several additional wiring connections. ∎

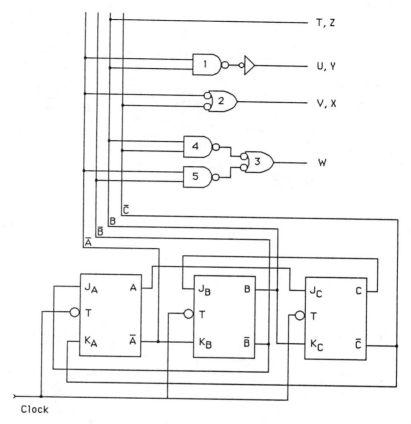

FIGURE 8.35   Circuit implementation of the digital die.

■ EXAMPLE 8.3–3

There are many ways we might solve the die problem. For example, even though a six-state counter can be implemented with three FFs, there is no reason that four FFs could not be used. One reason for doing this is that it then becomes possible to make the counter directly drive the die. That is, we simply design a six-state counter using four FFs, where the six states are just the values of T, U, V, and W for each die state. The Karnaugh maps in Figure 8.36 are drawn for the six count states that are needed. The many don't-care states help simplify the Boolean reduction. Thus the resulting Boolean equations for the J and K inputs for the four FFs require only two AND gates (or two NAND gates and two inverters, which can be obtained from a single 7400). Thus, by adding one FF to the circuit, we eliminate the need for code conversion entirely. This also reduces the parts count,

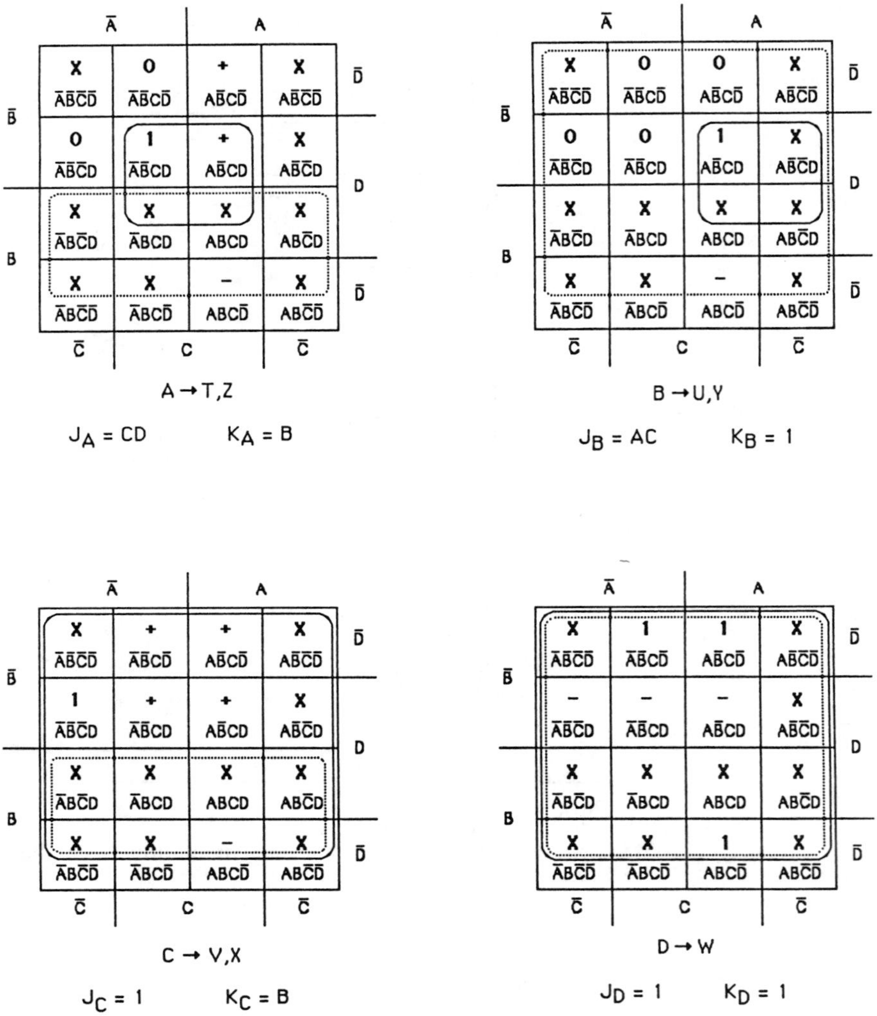

$A \rightarrow T, Z$

$J_A = CD \qquad K_A = B$

$B \rightarrow U, Y$

$J_B = AC \qquad K_B = 1$

$C \rightarrow V, X$

$J_C = 1 \qquad K_C = B$

$D \rightarrow W$

$J_D = 1 \qquad K_D = 1$

FIGURE 8.36 Karnaugh maps for a four-FF counter to solve the die problem.

since two $J-K$ FFs come on each IC. Thus adding a fourth FF does not increase the number of ICs required for the FFs, but it does reduce the number of ICs required for the gate logic. The entire design (excluding clock and LED drivers) requires only three ICs. ∎

8.4   DETECTION OF CODE SEQUENCES

One type of problem often encountered in digital electronics is the design of a circuit that examines a serial stream of data looking for a particular code sequence. There are systematic techniques for solving this problem, but they are fairly lengthy and beyond the scope of this text. Instead of developing these techniques, we will outline a method for using the 7495 shift register to detect code sequences.

First, let us outline the problem in detail. Suppose a sequence of serial data is present at an input to our detection circuit. Suppose further that these data are synchronized to a clock so that the serial data are input to our circuit at the falling edge of each clock cycle. We will assume that the serial data are stable at this time and do not change until after the falling edge of the clock. To be specific, suppose we are looking for the code sequence consisting of 101. We would like the detection circuit to signal with a momentary high output anytime it detects this sequence. There is one choice we must now make. Is each sequence of 101 to be regarded as entirely distinct, or can one sequence form part of the next? That is, if a sequence such as 10101 is input to our circuit, will we regard this as two instances of the sequence 101 (with the middle 1 in 10101 common to both sequences), or will we require a completely distinct second sequence to occur before we recognize two sequences? Stated another way, will the circuit reset after it detects a sequence (the second possibility above) or not (the first possibility above)? The difference in output between these two choices is illustrated as follows:

Input sequence:	0 0 1 1 0 1 0 1 1 0 1 0 1 0 1 0
Output (no reset):	0 0 0 0 0 1 0 1 0 0 1 0 1 0 1 0
Output (with reset):	0 0 0 0 0 1 0 0 0 0 1 0 0 0 1 0

The difference between the two output conditions is apparent. With reset, a completely new sequence must be detected before a high output is produced, whereas with no reset the end of one sequence may simultaneously form the beginning of a new sequence. Using the 7495 shift register, we will implement a circuit that will detect a 101 sequence, first without reset and then with reset.

The basic sequence-detection circuit for the case of no reset is shown in Figure 8.37. You might wish to review the logic diagram of the 7495 shown in Figure 6.24 at this time. In Figure 8.37, as data are shifted into the register, whenever the first three FFs hold the sequence 101, the output of the AND gate will go high, signaling that the sequence has been detected. The operation of the circuit is quite straightforward. It is clear that any grouping of 101 will produce a high output as soon as this grouping reaches the ABC FFs.

A point to be considered is what happens when power is first applied to the circuit. The internal FFs of the 7495 will come on in a random state at power up. Thus the possibility exists that the circuit will initially produce a high output even

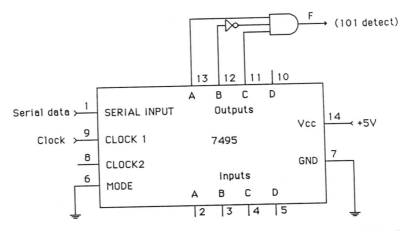

FIGURE 8.37 Detection of the serial sequence 101, with no reset following detection.

though no data have been received. If this possibility is unacceptable, some method of resetting the circuit at power-up must be used. Figure 7.16(a) shows one way to do this. The output of the 555 timer would be connected directly to the mode control of the 7495. In addition, the clock would be connected to both 7495 clock inputs. The 555 produces a temporary high output, which holds the 7495 in the parallel input mode. If the duration of this high from the 555 is adjusted to be a few clock cycles in duration and if the parallel inputs to the 7495 are all grounded, then the first clock pulse (in this case passing through the "parallel" clock input) will load zeros into all the 7495 FFs. To prevent F from going high before the first clock pulse is received, the 555 output should be inverted and ANDed with F to produce the actual circuit output. Thus, as long as the 555 output is high, the circuit output must remain low. Once the output of the 555 goes low, serial data will be shifted into the register.

If we wish to change the circuit so that reset occurs whenever a 101 sequence is detected, a couple of things need to be considered. First, we need to arrange for reset to occur. However, care must be taken that this happens without missing any incoming serial data. Thus the reset must occur as quickly as possible, leaving the circuit ready to accept more serial data. The circuit in Figure 8.38 is one way to achieve the desired objectives. Notice that the output of the code-detecting AND gate is fed back to the "parallel" clock $C2$ (pin 8) through an AND gate and also to the mode control (pin 6). As soon as F goes high, two things happen. First, the AND gate connected to pin 8 is enabled. The output of this gate will immediately go high, since the system clock has just gone low (remember, a serial bit of data has just been clocked in, causing F to go high). Second, the mode control will go high, setting the circuit into parallel mode. As soon as the system clock goes high (halfway through its cycle), this will disable the AND gate connected to pin 8, and its output will go low. This falling edge will clock the parallel data (all zeros) into the internal FFs and immediately turn off the high output F. When F goes low, this will place the register back into serial mode, ready for the next serial bit of data.

There is a flaw in both the power-up reset and the reset on pattern detection procedures. Consider what happens on reset (for either procedure). The shift register is reset to 0000. Suppose that the detection circuitry (essentially the AND gate) is set to detect the bit pattern 100. Since the shift register already holds the pattern 0000, if the first bit shifted in following reset is a 1, the shift register will now hold the pattern 1000. If the detection circuitry is looking at the left 3 bits, it immediately detects the desired pattern, even though only one new bit has been shifted into the register. One way to avoid this problem is to disable the detection circuitry until three clock pulses have been received following reset (note, we are *not* talking about disabling the shift register for three pulses, just the detection circuitry). One approach to designing the necessary circuitry would be to use a modulus-3 down counter that was loaded with the count 10 (decimal 2) by the code-detection pulse and that stopped counting when its count reached 00. An OR gate could be used on the output of the modulus-3 counter, and the high output of the OR gate could be used to disable the detection circuitry. Once the modulus-3 counter reached 00, the output of the OR gate would go low, enabling the detection circuitry.

The basic circuitry shown in Figures 8.37 and 8.38 can be used to detect any 2-, 3-, or 4-bit sequence by suitably adjusting the inputs to the AND gate producing F. For larger numbers of bits in the sequence, two or more 7495s could be connected

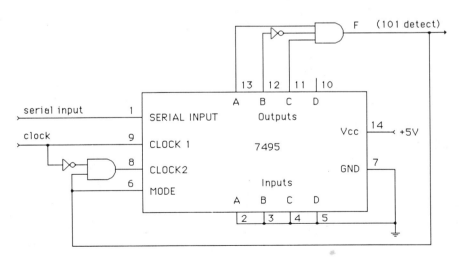

FIGURE 8.38 Detection of the serial sequence 101, with reset following detection.

together in series. We are not restricted to looking for only one bit pattern. More than one AND gate can be connected to the different register outputs to detect any bit patterns of interest.

PROBLEMS

1. From the Karnaugh maps in Figure 8.39, deduce the minimized Boolean function in each case. Try to take advantage of any don't-care states that are present. Remember, each 1 *must* be used, but only as often as necessary. That is, do not add any group that does not make use of an *unused* 1.

2. From the Karnaugh maps in Figure 8.40, deduce the minimized Boolean function in each case. Try to take advantage of any don't-care states that are present.

3. Use a Karnaugh map to find the minimized Boolean function for the truth table in Figure 8.41. Design an SSI gate circuit to implement this function. Show how the function could be implemented with an 8-input multiplexer.

4. Use a Karnaugh map to find the minimized Boolean function for the truth table in Figure 8.42. Design an SSI gate circuit to implement this function. Show how the function could be implemented with a 16-input multiplexer.

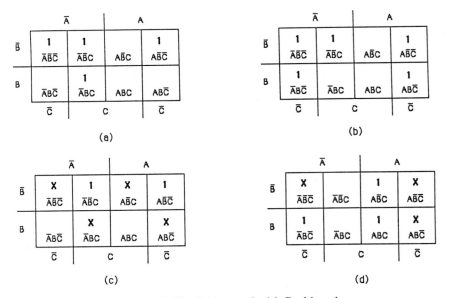

FIGURE 8.39 To be used with Problem 1.

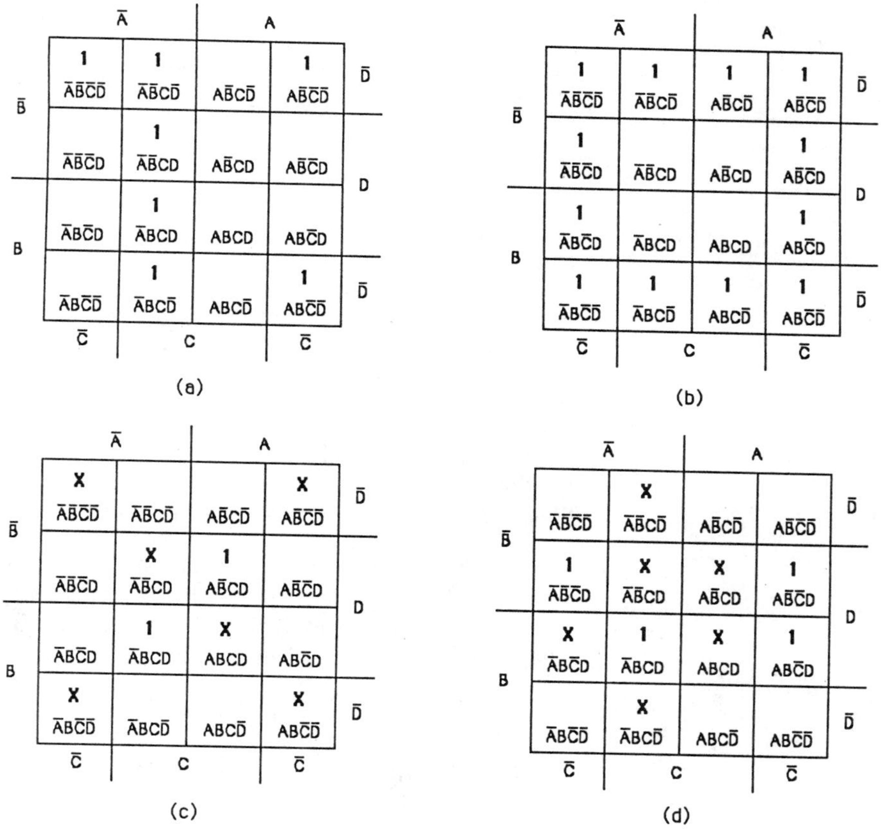

(a)

	Ā	Ā	A	A	
B̄	**1** ĀB̄C̄D̄	**1** ĀB̄C̄D	AB̄CD̄	**1** AB̄C̄D̄	D̄
B̄	ĀB̄C̄D	**1** ĀB̄CD	AB̄CD	AB̄C̄D	D
B	ĀBC̄D	**1** ĀBCD	ABCD	ABC̄D	D
B	ĀBC̄D̄	**1** ĀBCD̄	ABCD̄	**1** ABC̄D̄	D̄
	C̄	C		C̄	

(b)

	Ā	Ā	A	A	
B̄	**1** ĀB̄C̄D̄	**1** ĀB̄C̄D	**1** AB̄CD̄	**1** AB̄C̄D̄	D̄
B̄	**1** ĀB̄C̄D	**1** ĀB̄CD	AB̄CD	**1** AB̄C̄D	D
B	**1** ĀBC̄D	ĀBCD	ABCD	**1** ABC̄D	D
B	**1** ĀBC̄D̄	**1** ĀBCD̄	**1** ABCD̄	**1** ABC̄D̄	D̄
	C̄	C		C̄	

(c)

	Ā	Ā	A	A	
B̄	**X** ĀB̄C̄D̄	ĀB̄C̄D	AB̄CD̄	**X** AB̄C̄D̄	D̄
B̄	ĀB̄C̄D	**X** ĀB̄CD	**1** AB̄CD	AB̄C̄D	D
B	ĀBC̄D	**1** ĀBCD	**X** ABCD	ABC̄D	D
B	**X** ĀBC̄D̄	ĀBCD̄	ABCD̄	**X** ABC̄D̄	D̄
	C̄	C		C̄	

(d)

	Ā	Ā	A	A	
B̄	ĀB̄C̄D̄	**X** ĀB̄C̄D	AB̄CD̄	AB̄C̄D̄	D̄
B̄	**1** ĀB̄C̄D	**X** ĀB̄CD	**X** AB̄CD	**1** AB̄C̄D	D
B	**X** ĀBC̄D	**1** ĀBCD	**X** ABCD	**1** ABC̄D	D
B	ĀBC̄D̄	**X** ĀBCD̄	ABCD̄	ABC̄D̄	D̄
	C̄	C		C̄	

FIGURE 8.40   To be used with Problem 2.

Inputs			Output
A	B	C	F
0	0	0	1
0	0	1	X
0	1	0	0
0	1	1	0
1	0	0	1
1	0	1	X
1	1	0	0
1	1	1	0

FIGURE 8.41   To be used with Problem 3.

Inputs	Output
A B C D	F
0 0 0 0	0
0 0 0 1	1
0 0 1 0	X
0 0 1 1	1
0 1 0 0	0
0 1 0 1	0
0 1 1 0	1
0 1 1 1	X
1 0 0 0	0
1 0 0 1	0
1 0 1 0	1
1 0 1 1	1
1 1 0 0	0
1 1 0 1	0
1 1 1 0	X
1 1 1 1	1

FIGURE 8.42 To be used with Problem 4.

5. Develop a truth table for a circuit that examines four inputs and produces a high output if exactly two inputs are high. From the truth table produce a Karnaugh map, and from the map find the minimized Boolean function for the circuit. Then implement the circuit with NAND gates (and inverters).

6. Using the techniques you have learned so far, design a circuit that converts XS3 code into BCD code.

7. Using the Karnaugh map reduction techniques, reduce each of the following Boolean expressions to a minimum form:

(a) $F = \overline{A}\,\overline{B}\,\overline{C}D + \overline{A}CD + A\overline{B}CD + A\overline{C}D$

(b) $F = \overline{A}\,\overline{C} + CD + \overline{B}\,\overline{C} + AB\overline{C}$

(c) $F = \overline{A}\,\overline{B}\,\overline{C} + \overline{B}CD + \overline{C}\,\overline{D} + A\overline{C}D + \overline{A}BD + ABCD$

8. Suppose that you wish to produce the count sequence shown in Figure 8.43. There are a number of ways to do this, but you are to produce two designs. In the first design, use three $J-K$ FFs, and develop the input equations for each using the sequential Karnaugh mapping technique. Then produce a circuit diagram. In the second design you will take advantage of the fact that there are only four states in the sequence. Four states can be produced by only two FFs. Set up a standard four-state binary counter using two FFs. Then design

Count Sequence

```
0 0 0  ←──────┐
1 0 1     Recycle
0 1 0         │
1 1 1  ───────┘
```

FIGURE 8.43 To be used with Problem 7.

a combinational circuit that takes the two outputs from the counter (which have four possible states) and produces a three-output code in the sequence given in the table in Figure 8.43. (*Hint:* In this second design, use the four output states of the two-FF counter as combinational inputs and the count sequence in Figure 8.43 as three *separate* Boolean outputs.)

9. Redesign the die problem from the text using a counter that cycles through the first six binary states instead of the count sequence used in the text. You must first design the counter so that it recycles after reaching 101. Once you have designed the counter, design a combinational network that produces the correct output on the die.

10. Using the sequential Karnaugh mapping technique developed in the text, design a BCD counter. That is, the counter should cycle through the 10 BCD states in order.

11. Using the circuit in Figure 8.37 as an example, design a circuit that has separate detection circuitry for the following bit patterns: 100, 111, and 1100. Assume that the circuit does not reset following detection.

12. Using the circuit in Figure 8.38 as a guide, design a circuit that has separate detection circuitry for the bit patterns 100 and 001. Include circuitry that resets the circuit to 000 following detection of either bit pattern.

13. In Section 8.4, it was pointed out that sometimes it is necessary to pay special attention to what happens when power is first applied to the circuit. A scheme was discussed where a 555 timer was to be used to temporarily disable the sequence-detection circuitry for a brief period at power-up. Following this brief period, the shift register was to be reset to 0000. Using the discussion in the text as a guide, develop the complete circuit diagram for temporarily disabling the shift register at power-up and resetting the shift register to 0000 at the end of the disable period.

Chapter 9

DAC, ADC, AND MICROPROCESSOR INTERFACING

Learning Objectives

After completing this chapter you should know:

- The basic means by which a DAC operates.

- The operation of a typical DAC in IC form.

- Some common uses of DACs.

- The process of analog-to-digital conversion using a DAC.

- The operation of a typical ADC in IC form.

- The meaning and use of the basic control lines on a 6502 microprocessor.

- How to interface input and output ports to a 6502 microprocessor, and how these ports can be accessed from both machine language and BASIC.

- Why buffering of the microprocessor buses is necessary and how to do it.

9.1 INTRODUCTION

The subject of this text up to now has been digital electronics. However, in the real world most electrical quantities are created in an analog form. For example, a typical pressure transducer creates a voltage that is directly proportional to pressure. Since pressure varies in a continuous fashion, so does the voltage that represents it. The same holds true for the voltages produced by temperature, light intensity, force, motion, and other types of transducers. To treat these voltages with a digital system, it is first necessary to convert the analog voltages into a digital form. The type of device that does this is called an analog-to-digital converter (ADC).

Once the digital system has processed the analog information (in digital form), it is often necessary for the system to send out an analog voltage to control some sort of device. For example, suppose the job of the digital system is to monitor the speed of a motor and maintain this speed within certain narrow limits. The motor might have a tachometer attached to it that produces a voltage that is directly proportional to the speed of the motor. Using the ADC, the digital system can monitor the speed of the motor. To control the motor speed, the digital system must send an appropriate analog voltage to the motor (actually to the motor power controller). To produce the analog voltage, the digital system needs a device that converts digital information into an analog voltage. This type of device is called a digital-to-analog converter (DAC).

Both ADCs and DACs form the links between the digital and analog worlds. In this chapter we will investigate examples of both of these types of devices. ADCs and DACs can be used quite easily with simple digital systems. However, it has become very common to use these types of devices with simple microprocessor-based systems (which are, of course, digital systems). For this reason, we will look into the concept of interfacing simple digital devices to a microprocessor system. To keep things simple, we will limit the discussion to the widely used 6502 8-bit microprocessor. This microprocessor is typical of what is found in simple microcomputers and microprocessor-based controllers. We will see how to create both input and output ports for the 6502 and then see how ADCs and DACs are used as input and output devices, respectively.

9.2 THE TYPICAL DAC

One of the most common ways to convert digital information into an analog voltage is through the use of the $R-2R$ ladder. This is most easily explained by example. Consider the circuit in Figure 9.1. This circuit consists of a simple 7493A 4-bit counter. Clock signals are input into pin 14, and the counter simply counts up from 0 to 15 and then recycles. Actually, the counter really consists of separate 1- and

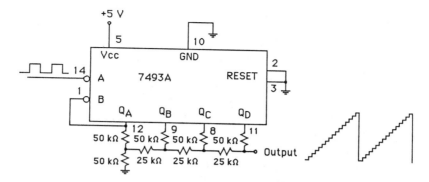

FIGURE 9.1 Digital-to-analog conversion using a counter and an R–$2R$ ladder.

3-bit counters in series. In the figure, the output of the 1-bit counter (pin 12) has been tied to the input of the 3-bit counter (pin 1), so the two counters have been tied together to form a single 4-bit counter. As the counter counts upward, the count appears at the outputs labeled Q_A to Q_D as combinations of 0- and 5-volt outputs. These output voltages are connected to a series of resistors that come in two sizes, one just twice the other (thus the name R–$2R$). Let us attempt to analyze how the circuit functions.

The superposition theorem tells us that, whenever a resistance network has a series of voltages connected to it, each voltage contributes *independently* to the voltage at any given point in the network (for example, at the output). Thus the outputs of the counter each contribute independently to the output of the network. If a given counter output is zero at any time, it contributes nothing to the output. But if a given counter output is at 5 volts, it will contribute a fixed fraction of this 5 volts (depending on where the output is in the network) to the network output. A careful analysis of the resistor network shows that *half* of the voltage at output Q_D appears at the network output. Further analysis shows that, for outputs Q_C, Q_B, and Q_A, the corresponding fractions that appear at the network output are $\frac{1}{4}$, $\frac{1}{8}$, and $\frac{1}{16}$, respectively. Thus the network output can be described by the equation

$$V = (5 \text{ volts}) \times \left(\frac{D_D}{2} + \frac{D_C}{4} + \frac{D_B}{8} + \frac{D_A}{16} \right)$$

where the D's are the digital output of the counter (1's and 0's).

If we remember that the digital output of the counter can be thought of as a binary number, where the most significant bit has a value of eight, the next a value of four, and so on, then this equation can be rewritten taking this into account as

$$V = 5 \text{ volts} \times \frac{4\text{-bit digital output}}{16} \tag{9.1}$$

For example, if the count has reached 1001 (which is equivalent to 9 decimal), then the output will be

$$V = 5 \text{ volts} \times \left(\frac{9}{16}\right) = 2.8125 \text{ volts}$$

Essentially, each count adds 5/16ths of a volt (0.3125 volt) to the output. The circuit will convert any digital number between 0000 and 1111 into a proportional analog voltage given by Eq. (9.1). The circuit in Figure 9.1 could easily be expanded by adding a second 4-bit counter and extending the pattern of the $R-2R$ ladder. This would produce an 8-bit DAC, where the output would be given by

$$V = 5 \text{ volts} \times \frac{8\text{-bit digital number}}{256} \tag{9.2}$$

In this case, each count would add 5/256ths of a volt (0.01953 volt) to the output. With 8 bits, the "fineness" of the control becomes much tighter. That is, it is possible to produce an analog output within 0.00977 volt (half of 0.01953) of any desired value between 0 and 4.98047 volts. You cannot quite get to 5 volts since the largest 8-bit binary number is 11111111, which is equivalent to 255. Thus the largest output voltage is one count away from 5 volts (that is, 5 volts × 255/256).

If you construct the circuit in Figure 9.1 and send in a fairly high frequency clock signal to pin 14 of the 7493A (say a few kilohertz), you can easily watch the output on an oscilloscope. You will see that the output is a series of steps, each increasing by the same fixed voltage (assuming your resistors are really in a 2/1 ratio). An 8-bit counter with an $R-2R$ ladder would produce a similar step output, but there would be many more steps and they would be much closer together. In a given application, the number of steps needed is determined by how closely the analog voltage must be controlled.

The circuit in Figure 9.1 generates an increasing series of analog voltages as the count of the counter increases. It would be a simple matter to replace the 7493A counter with a 7475 dual 2-bit latch. This is shown in Figure 9.2. The *input* to the latch would come from some other digital circuit, while the *output* of the latch would drive the resistor network. In this fashion, the latch could be used to output any analog voltage that conforms to Eq. (9.1). The latch accepts the *digital* information from some other digital circuit and, through the $R-2R$ ladder, outputs an analog representation of the digital information. The sequence of events necessary to use the latch is quite simple. The "other digital circuit" simply places the desired digital number on the inputs of the 7475, takes the enable line low temporarily to "latch in" the data, and then takes the enable line high so that the data are locked into the 7475. The output voltage will now remain fixed until a new digital number is sent to the latch.

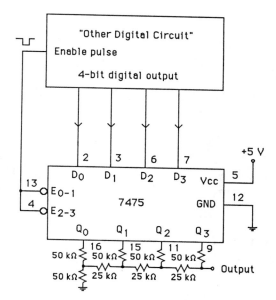

FIGURE 9.2 Digital-to-analog conversion using a 7475 latch.

DACs in IC Form

Although the circuits shown in Figures 9.1 and 9.2 will work fairly well, there are some drawbacks. For one thing, typical TTL outputs are not guaranteed to be 5 volts, so the voltages driving the $R-2R$ ladder will not necessarily be 5 volts (and might indeed vary from one IC to the next). In addition, it is cumbersome to fabricate a precision $R-2R$ network. A better solution is to include all the necessary electronics on a single integrated circuit. There are many examples of such a circuit, and we will now examine a fairly common one, the MC1408L.

The basic pinout diagram of the 1408 is shown in Figure 9.3(a). The 1408 is an 8-bit DAC that produces an output *current* (in conventional terms, this current actually flows *into* pin 4), which is proportional to the 8-bit digital input. To illustrate the use of the 1408, we will analyze it within a particular context, that of generating an analog voltage as shown in Figure 9.3(b). First, a reference current must be established flowing into pin 14. This is done by choosing values of V_{ref+}, V_{ref-}, and R_{14} such that the reference current is a few milliamperes (but not more than 5 milliamperes). Pin 14 is internally held at V_{ref-}, so the current through R_{14} is determined by the difference between V_{ref+} and V_{ref-}. This current is frequently chosen as 2 milliamperes. Since the value of V_{ref+} is often 5 volts and V_{ref-} is usually tied to ground, this would mean that R_{14} should be chosen as 2.5 kilohms. Notice that, in Figure 9.3(b), R_{14} is actually comprised of two smaller resistors with a capacitor to ground from their junction. This provides filtering of any noise that

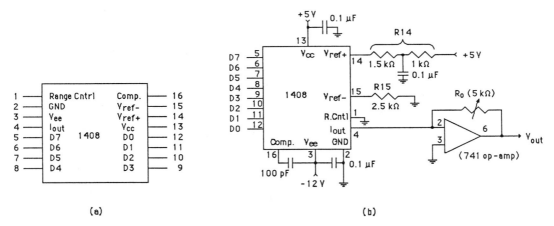

FIGURE 9.3 (a) The MC1408L DAC pin-outs. (b) DAC used in current-to-voltage conversion mode.

might be present on the 5-volt supply (unless you happen to have a precision 5-volt supply handy, in which case the filtering is not necessary). R_{15} is usually chosen equal to R_{14}, so it too would be chosen as 2.5 kilohms.

V_{cc} is typically 5 volts and is bypassed by a 0.1-microfarad capacitor to eliminate power supply fluctuations. V_{ee} is generally in the range of -10 to -15 volts [Figure 9.3(b) shows -12 volts] and is likewise bypassed by a 0.1-microfarad capacitor. A small capacitor (100 picofarads is suitable) must be connected between V_{ee} and pin 16 (compensation) to prevent oscillations. Pin 2 is supply ground, and pin 1 (range control) allows adjustments to the logic level that the 1408 will interpret as a logic high. For TTL level inputs, pin 1 should be grounded.

The digital input (D) is received on pins 5 to 12 (msb to lsb). As a result, an output current will be developed flowing into pin 4 and will have the following magnitude:

$$I_{out} = I_{ref} \times \frac{D}{256} = \frac{V_{ref+} - V_{ref-}}{R_{14}} \times \frac{D}{256} \tag{9.3}$$

With the values of V_{ref+}, V_{ref-}, and R_{14} shown in Figure 9.3(b), Eq. (9.3) becomes

$$I_{out} = 2 \text{ mA} \times \frac{D}{256} \tag{9.4}$$

This current must, of course, come from some other circuit element. Often it simply comes through a grounded resistor. In this case, the current flow through the resistor converts the output current into a small output voltage across the resistor. With pin 1 grounded, this output voltage (which shows up at pin 4) must not exceed -0.5 volt. This places a limitation on how large a resistor may be used. The circuit in Figure 9.3(b) uses another means to convert the output current into an output voltage.

In Figure 9.3(b), the output current flows directly from the inverting $(-)$ input of the op-amp into pin 4 of the 1408. Since an op-amp has a very high input impedance, essentially no current flows into or out of pin 2 of the op-amp. Thus the current flowing into the 1408 must come from the op-amp output (pin 6) *through* R_0. It is also true that, whenever an op-amp is connected in a negative feedback configuration (output tied back to inverting input, pin 2), the output of the op-amp will adjust itself so that the voltages on the two op-amp inputs stay equal. Since the inverting input (pin 3) is tied to ground, pin 2 will remain at 0 volts also. Thus the op-amp output will be determined by whatever voltage drop is produced by the output current flowing through R_0. Since the output current flows from the op-amp output through R_0 into pin 4 of the 1408, the output voltage will be positive and given by

$$V_{out} = I_{out} \times R_0 \tag{9.5}$$

Combining Eq. (9.4) with Eq. (9.5) gives

$$V_{out} = (2 \text{ mA} \times R_0) \times \frac{D}{256} \tag{9.6}$$

The only restriction on R_0 is that V_{out} should not exceed the output limitations of the op-amp. If we set R_0 equal to 2.5 kilohms (the same as R_{14}), then V_{out} will vary from 0 to nearly 5 volts as D varies from 0 to 255.

To easily maintain the digital inputs to the 1408, an 8-bit latch, such as the 74100 dual 4-bit latch, should be used. This will permit a digital system to send a digital number to the latch, which will maintain a constant output to the 1408.

A Programmable Gain Amplifier

Recall that the output current of the 1408 is actually a fraction of the reference current [Eq. (9.3)], which is in turn controlled by V_{ref+} (assuming V_{ref-} is at ground). With this in mind, the output voltage of Figure 9.3(b) becomes

$$V_{out} = \frac{R_0}{R_{14}} \times \frac{D}{256} \times V_{ref+} \tag{9.7}$$

Now imagine that instead of a fixed value for V_{ref+} we use a time-varying voltage of some sort, say a low-amplitude audio signal. Equation (9.7) shows that the output of the op-amp will faithfully follow this signal, but with a scaling factor (gain) that is proportional to D, the digital input to the 1408. By changing D, we change the scale factor (gain) in Eq. (9.7). Thus the 1408 can be used to create a variable-gain amplifier that can easily be set for 256 different gains simply by controlling the digital input to the 1408. It is not hard to imagine a simple system where the digital input to the 1408 comes from a set of dials or switches on the front panel of some sort of amplifier, say a stereo system.

9.3   ANALOG-TO-DIGITAL CONVERSION

In Section 9.2 we saw how to convert a digital number into an equivalent analog current or voltage. Quite frequently it is necessary to undertake the reverse of this process. There are many ways to achieve analog-to-digital conversion. One common method is to make use of a DAC and a voltage comparator. This idea is represented by the circuit in Figure 9.4. When the normally open momentary contact switch is closed, this grounds the active-low reset input of the 8-bit counter so the counter output goes to binary zero. The DAC then outputs 0 volts to the voltage comparator. This will be less than the analog input (V_{in}), so the comparator output will be high. This high will enable the NAND gate so that clock pulses can pass through to the counter input. When the reset switch is finally released, the counter begins to count up from zero. The count is output on a suitable display; but if the clock is of fairly high frequency, then the count will change so fast that only the final count will be seen. As the count proceeds upward, the output of the DAC ramps upward in unison. When the output of the DAC finally equals V_{in}, the output of the comparator will go low, disabling the NAND gate and preventing any further clock pulses from reaching the counter. The count then displayed represents the voltage of V_{in}. All the circuit elements in Figure 9.4 have been discussed previously

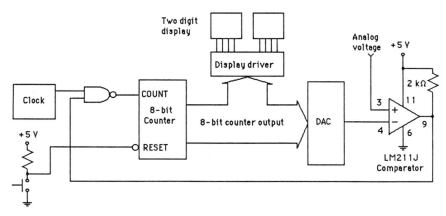

FIGURE 9.4   Analog-to-digital conversion using a DAC and a comparator.

with the exception of the LM211J voltage comparator. This comparator is typical of those that can be connected to a single-sided power supply (that is, operated between a positive voltage and ground). By using a 5-volt supply to power the LM211J, the output will be TTL compatible and thus suitable for direct connection to the NAND gate.

ADCs in IC Form

The approach taken in Figure 9.4 requires several circuit elements to implement and is also rather slow, especially for values of V_{in} that are close to 5 volts. Fortunately, there exist several ADCs that have been implemented as single ICs. Many, if not most, of these ICs are designed to be easily interfaced to a microprocessor system. Later in this chapter we look into microprocessor interfacing to some extent. For now, however, we will examine a typical ADC as a stand-alone IC.

Figure 9.5 shows the pin-outs of the ADC0809, which is an 8-bit ADC with eight separate analog inputs. As normally used, both V_{cc} and REF(+) are connected to 5 volts and both GND and REF(−) are connected to ground. The analog voltage to be converted to an 8-bit digital number is connected to one of the eight inputs IN0 to IN7. The analog voltage must be in the range from 0 to 5 volts. A number of inputs control the function of the circuit, specifically START, ALE, OUTPUT ENABLE, and ADD A to C. There is also a system clock, CLOCK, which is a TTL clock input in the frequency range of 10 kilohertz to 1.2 megahertz. To begin a conversion, a 3-bit address is placed on the ADD A to C inputs, denoting which analog input is to be used. While this address is on the ADD inputs, the START and ALE inputs are taken high and then low (a short positive start pulse of 1 microsecond or longer will suffice). Conversion then begins within the IC. After about 70 or so clock cycles, conversion will be complete. The IC signals that the conversion has been completed by taking the EOC output high (however, EOC is high when conversion starts and does not go low until after a few clock cycles).

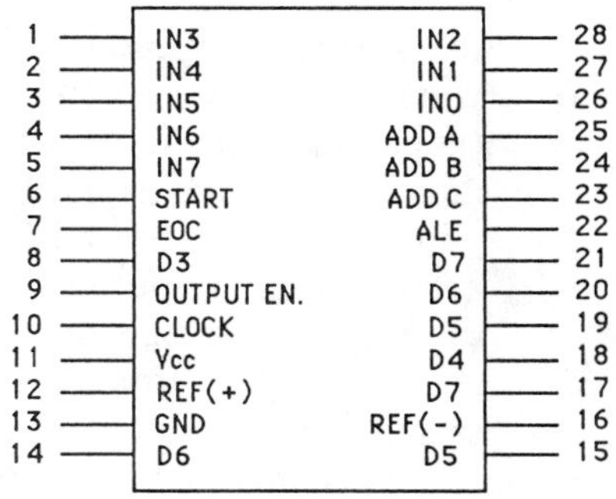

1 —	IN3	IN2	— 28
2 —	IN4	IN1	— 27
3 —	IN5	IN0	— 26
4 —	IN6	ADD A	— 25
5 —	IN7	ADD B	— 24
6 —	START	ADD C	— 23
7 —	EOC	ALE	— 22
8 —	D3	D7	— 21
9 —	OUTPUT EN.	D6	— 20
10 —	CLOCK	D5	— 19
11 —	Vcc	D4	— 18
12 —	REF(+)	D7	— 17
13 —	GND	REF(−)	— 16
14 —	D6	D5	— 15

FIGURE 9.5   The pin-outs of the ADC0809 analog-to-digital converter.

The digital result may now be read on the D7 to D0 outputs by taking OUTPUT ENABLE high. Until the OUTPUT ENABLE is taken high, the D7 to D0 outputs float in a tristate condition.

In a typical microprocessor system, the system clock (if it is 1 megahertz or less) can be used as the ADC clock as well. In addition, the microprocessor can send the necessary signals to determine which analog input is to be used, to start the conversion process, and to read the digital result. All these signals could also be designed into any digital system; but with the vastly reduced prices of 8-bit microprocessors, it usually proves cheaper to control a circuit such as the ADC0809 with a microprocessor.

Sample-and-hold Circuits

One problem with using an ADC such as the ADC0809 is that it operates on the basis of successive approximations. The analog voltage to be converted is assumed to lie in the range from 0 to 5 volts. Within the ADC, a series of comparison voltages is set up and compared with the analog voltage. First, the analog voltage is compared with the midpoint of the possible analog range, 2.5 volts (halfway between 0 and 5 volts). If the analog voltage is larger than 2.5 volts, then the analog voltage is compared with the midpoint of the remaining possible analog range, 3.75 volts (halfway between 2.5 and 5 volts). Now suppose the analog voltage is lower than 3.75 volts. Then the analog voltage is compared with the midpoint of the remaining analog range, 3.125 volts (halfway between 2.5 and 3.72 volts). This process of dividing the remaining analog range in half and then comparing the analog voltage to the halfway point continues for a total of eight such comparisons. After eight comparisons, the analog voltage will be known to lie between two voltages that are different by the smallest voltage resolution of the ADC.

For the ADC0809 running on a 1-megahertz clock, the time required to complete the comparisons is about 70 microseconds. The problem arises if the analog voltage to be measured *changes significantly* during this time. Ideally, the analog voltage should not change by more than the voltage resolution of the ADC during the time of conversion. Otherwise, the successive approximations of the ADC are attempting to home in on a moving target. One way of guaranteeing that the analog voltage does not change too fast is to use a low-pass filter on the analog voltage. A low-pass filter forces the analog voltage to change relatively slowly (depending on the components making up the filter). Of course, this approach modifies the analog voltage to be measured by removing any high-frequency components that are present, which may not be desirable.

A second approach employs a *sample-and-hold* circuit. This type of circuit samples the analog voltage for a brief time, perhaps 1 microsecond, and then *holds the value it finds* for a much longer period (perhaps several milliseconds if necessary). During the hold period, the ADC can carry out the successive approximations on a *fixed* analog voltage. A sample-and-hold circuit can be as simple as a capacitor (low-leakage type) and a switch (usually a CMOS switch such as the CD4016 bilateral switch). To sample the analog signal, the switch is closed, allowing the small capacitor to charge up to the analog voltage. During the hold period, the switch is opened, and the capacitor holds the voltage at its value when the switch was opened. If the capacitor is a low-leakage type, the voltage on the capacitor will drop very slowly during the ADC conversion period. Figure 9.6 (which will be discussed in more detail in the next section) shows a sample-and-hold circuit at the input of the ADC circuit.

Limitations on Sampling Rate

The purpose of an ADC is to sample an analog voltage. Sometimes this sampling is done infrequently, whenever information is needed. Often, however, the sampling is done at regular intervals of time, that is, at a definite frequency. The result of the conversion is a digital number that is usually processed or stored for future processing. The processing often involves attempting to reconstruct the original analog signal that was sampled. For example, an area of interest is voice synthesis. In attempting voice synthesis, it is first necessary to accurately measure the output of the human voice. This can be done by monitoring the voltage output of a microphone into which a volunteer speaks certain words or phrases. The human voice is made up of a range of frequencies that are present simultaneously during speech. The output of a microphone will therefore consist of the additive sum of several frequencies, all present together.

Voice analysis consists of measuring the output of a microphone using an ADC, storing the resulting information, and then later analyzing the information. There is a fundamental mathematical problem, however. Accurate analysis requires that the ADC sample the analog signal at *twice the frequency of the highest analog frequency present*. This is called the Nyquist theorem. For example, if the human voice contains frequencies as high as 5 kilohertz, then to accurately analyze a voice, the ADC would need to sample at a 10-kilohertz rate.

The Nyquist theorem must be taken into consideration whenever an ADC is used to sample a signal that contains relatively high frequencies. Generally, if there is any question that the signal might contain frequencies that exceed one-half the sampling rate, the analog signal is passed through a low-pass filter designed to eliminate any frequencies higher than one-half the sampling rate.

A Typical Application

ADCs and DACs are often used together to control complex processes. The simple example of motor control has already been mentioned. Figure 9.6 illustrates how such control might be achieved. An operator might input to a computer (by the

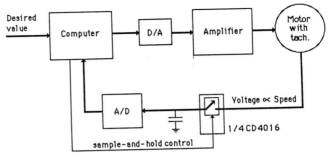

FIGURE 9.6   Simple motor control using a DAC and ADC with a computer.

keyboard or some other device) a desired speed for the motor. The computer uses an ADC to read the actual speed of the motor and compares this to the desired speed. If the speed is not correct, the computer uses the DAC to adjust the analog voltage input to the power amplifier that directly powers the motor. The computer then uses the ADC to again read the motor speed. The process essentially continues indefinitely, with the computer constantly making adjustments through the DAC to keep the actual motor speed as close as possible to the desired speed. The actual structure of the computer program that implements the control depends on the physical details of the motor and the load it must turn and also on how closely the speed must be controlled. Digital control theory is in fact a very large subject in itself.

9.4   MICROPROCESSOR INTERFACING

Up to this point in the text the digital ICs that we have discussed have used small-scale or at most medium-scale integration techniques. In the 1970s, integrated-circuit technology advanced to the point where much more complex functions could be implemented on a single integrated circuit. This has led to large-scale (LSI) and very large scale (VLSI) integration technologies. One of the most important developments has been the design of microprocessors. These circuits are widely used as the heart of microcomputer systems and of small intelligent instruments. It is not our goal in this text to describe the detailed function of microprocessors. However, since it is often necessary to interface a digital system to a microcomputer, it is useful to see how this can be done and why it is useful. You should treat this section on microprocessors as an opportunity to determine if you might have a more serious interest in the subject. If you find that you do have an interest, you should consider an introductory course on microprocessors. Such a course provides a very useful complement to a course on digital electronics.

The 6502 Microprocessor

One of the most widely used 8-bit microprocessors is the 6502. It is called an 8-bit microprocessor because it handles information 8 bits at a time (8 bits is called a *byte*). The 6502 forms the core of the Apple II line of computers as well as a number of others. It is also used extensively in industrial controllers. A number of other 8-bit microprocessors are available in addition to the 6502. Among these are the 6800 series, the 8085, the Z80, and the 8051. All these microprocessors have different instruction sets and differ somewhat in their detailed operation. However, all use a similar sequence to move data into and out of the microprocessor. Thus, the concepts developed in examining the 6502 will be useful in understanding the operation of other microprocessors. Therefore, let us begin by looking at the pin-outs of a 6502 shown in Figure 9.7 and the structure of a simple 6502-based system shown in Figure 9.8.

The 6502 can be thought of as the central controller or brain of the system. As a minimum, the system must have the 6502, a system clock that coordinates activities, permanent ROM memory, temporary RAM memory, and some input and output capabilities. These are shown in Figure 9.8. The 6502 communicates with other parts of the system by using three sets of wires called buses. First, there is the address bus (A0 to A15 in Figure 9.7). The 6502 uses the address bus to determine which device it is communicating with. There also is the data bus (D0 to D7 in Figure 9.7). Information flows into or out of the 6502 over this bus, 8 bits at a time. Finally, there is the control bus, which consists of various control signals that can be used by the 6502. Of these signals, the three that we will concern

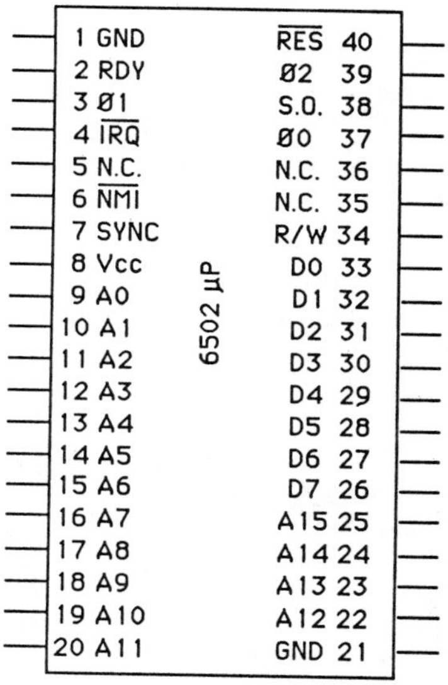

1 GND	$\overline{RES}$ 40
2 RDY	Ø2 39
3 Ø1	S.O. 38
4 $\overline{IRQ}$	Ø0 37
5 N.C.	N.C. 36
6 $\overline{NMI}$	N.C. 35
7 SYNC	R/W 34
8 Vcc	D0 33
9 A0	D1 32
10 A1	D2 31
11 A2	D3 30
12 A3	D4 29
13 A4	D5 28
14 A5	D6 27
15 A6	D7 26
16 A7	A15 25
17 A8	A14 24
18 A9	A13 23
19 A10	A12 22
20 A11	GND 21

6502 μP

FIGURE 9.7 Pin-outs of the 6502 microprocessor.

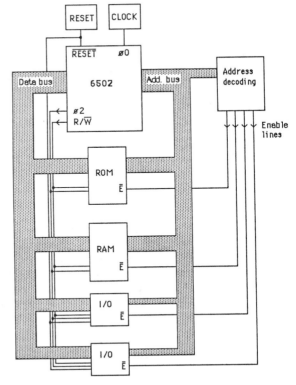

FIGURE 9.8 A typical 6502 system. Note that the φ2 and R/$\overline{W}$ lines might first pass through conditioning circuitry before going to some of the elements of the system (especially RAM and output ports).

ourselves with are the system clock (φ2), the read/write line (R/$\overline{W}$) and the RESET line.

The 6502 controls the system by executing very simple types of procedures. Even though the procedures are simple, very complex tasks can be undertaken because the 6502 can execute the procedures so rapidly. Essentially, the 6502 does only three basic types of operations. It reads in information from some place in the system, it sends out information to some place in the system, or it internally manipulates information that it has acquired. We are particularly interested in how the microprocessor reads in or sends out information. Once we understand how it does these two things, we will be in a position to interface digital circuitry to the 6502.

First, let us examine the system clock, φ2. This clock is a TTL-level square wave that operates at 1 megahertz (or 2 or 4 megahertz in newer versions). The clock signal is originally created with external circuitry and is input to the 6502 at pin 37 (φ0). This signal is shaped internally and sent out to the rest of the system from pin 39 of the 6502 (φ2). Figure 9.9 shows the clock signal. The clock signal is intimately involved in all data transfers into or out of the 6502. Next, let us consider the address bus. When the 6502 wants to communicate with a device, it

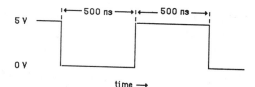

FIGURE 9.9   Basic system clock, φ2, of a 1-MHz 6502 system.

must have some means of uniquely specifying a particular device. It does this by means of the address bus, which consists of 16 output lines from the 6502. With 16 lines, 2^{16} possible combinations of 1's and 0's can be sent out over the address bus. This amounts to 64K different combinations or addresses (where 1K = 1024 in computer jargon). The 6502 can thus communicate directly with up to 64K different devices.

To communicate with a specific device, the 6502 places the address of the device on the address bus. It begins to do this while the system clock, φ2, is low. The address is stable on the address bus before φ2 goes high. The 6502 must also inform the system whether information is to be sent out from the 6502 (a *write* operation) or read into the 6502 (a *read* operation). The 6502 does this by taking the R/$\overline{W}$ line low prior to a write operation and high prior to a read operation. Like the address bus, the R/$\overline{W}$ line will be stable before φ2 goes high. The circuitry within the system must use the address bus, φ2, and the R/$\overline{W}$ line to respond properly to the 6502.

Let us examine in some detail exactly what takes place during a read operation and a write operation. Figure 9.10 summarizes the important signals that occur during a read operation. Notice that during the low phase of φ2 the address bus and R/$\overline{W}$ lines are in the process of changing. They take on their new values before φ2 goes high. The device being addressed then has the 500 nanoseconds of the high phase of φ2 to get its 8 bits of information on the data bus. This must be accomplished a few tens of nanoseconds before the end of the positive phase of φ2. The data on the data bus will be read into the 6502 at the falling edge of φ2.

FIGURE 9.10   The basic timing of a READ operation. The setup time, t_s, is the time required for the address bus and R/$\overline{W}$ line to become stable following the end of φ2. The time t_d is the time the data bus must be stable prior to the end of φ2 in order for the 6502 to success-fully read the data. The time t_h is the time the data bus must remain stable following the end of φ2 in order for the READ to be successful.

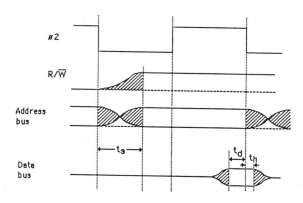

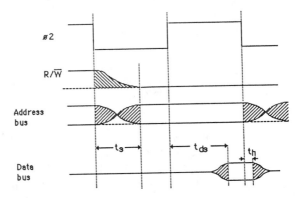

FIGURE 9.11 The basic timing for a WRITE operation. The times t_s and t_h have the same meaning as for a READ operation. The time t_{ds} is the length of time following the beginning of $\phi2$ needed for the 6502 to place stable data on the data bus.

During a write operation, shown in Figure 9.11, the address bus and $R/\overline{W}$ line again become stable prior to the end of the low phase of $\phi2$. However, during the high phase of $\phi2$, it is now the 6502 that places its data on the data bus. This is accomplished about 100 nanoseconds before the end of the high phase of $\phi2$. The device being written to then reads in the data at the falling edge of $\phi2$.

To prevent possible conflicts within the system that might occur when the address is in the process of changing (who knows what address is present at this time?), the devices within the system usually use the address bus, the $R/\overline{W}$ line, and $\phi2$ to determine if they are being addressed. Decoding circuitry must be provided for each device such that the device is enabled for a READ operation only when:

1. The correct address is present on the address bus.
2. The $R/\overline{W}$ line is high.
3. $\phi2$ is high.

A device is enabled for a WRITE operation only when:

1. The correct address is present on the address bus.
2. The $R/\overline{W}$ line is low.
3. $\phi2$ is high.

The circuitry that checks for these conditions is called address-decoding circuitry. Usually, the decoding is broken into two parts. One part, common to all devices, is a combination of conditions 2 and 3 (for both READ and WRITE operations). This decoding is shown in Figure 9.12. Notice that the output of the decoding circuitry goes low only when the $R/\overline{W}$ line is low AND $\phi2$ is high. The output of the circuit in Figure 9.12 is often called the RAM $R/\overline{W}$ signal. If this signal is used to enable devices, no device will ever be written to inadvertently

FIGURE 9.12 The forma-
tion of the RAM R/W signal.
Note that this signal can only
go low when φ2 is high AND
R/W is low. Thus RAM will
never be inadvertently
written to during a low phase
of φ2.

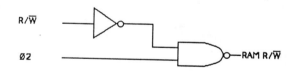

because of a spurious address during the low phase of φ2. The other decoding that is required is the decoding of the address bus. To decode a single unique address from the address bus requires that all 16 address lines be decoded. For example, two 8-input NAND gates (7430) could be used as shown in Figure 9.13 to respond with a low output only when the address shown is present on the address bus. Fortunately, major parts of the system, such as RAM and ROM memory, have several memory locations clustered on single ICs. Thus an 8K RAM or ROM chip has built-in address decoding for the 8K devices on the chip. This means that the low 13 address lines (sufficient to provide 8K unique addresses) are attached directly to the memory IC. Only the upper three address lines must be decoded by special decoding circuitry. For a very simple 6502 system, the address decoding scheme shown in Figure 9.14 could be sufficient. Here we see that the upper three address lines go to a 74138 one-of-eight decoder. The outputs of the 74138 break the 64K address space into eight equal pieces. For a simple system, an 8K RAM chip could be enabled by the 000 output of the 74138, while an 8K ROM chip could be enabled by the 111 output. This is a typical configuration, since the 6502 requires ROM to be at the high addresses and RAM to be at the lowest.

The other six outputs of the 74138 could be used to enable different input or output ports for the system. It is not necessary to decode all 16 address lines for a particular interface port as long as nothing else is connected to the system that will conflict with the addresses assigned to the port. For example, suppose that a single output port used the 001 output of the 74138 as an enable signal. Because

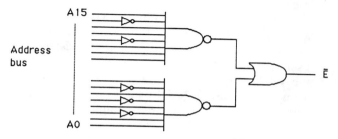

FIGURE 9.13 Two 8-input NAND gates used for complete address decoding of a single address. Note that the output of the OR gate will only go low when the address is 1011 0111 1010 1000, or B7A8 hex.

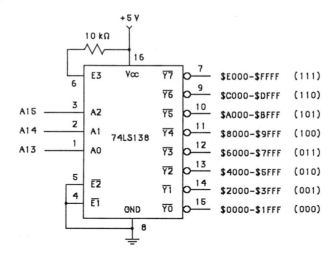

FIGURE 9.14 A 74LS138 one-of-eight decoder used to break the 64K address space of a 6502 into eight equal segments, each 8K in size. The status of the three address lines A15, A14, and A13 follows each address range in parentheses.

the lower 13 address lines are not decoded, *any* address that starts with 001 will enable the output port. That is, there are really 8K different addresses (all starting with 001) that will affect the output port in exactly the same way. But if no other device is enabled by any of these addresses, then it does not matter that 8K addresses are "wasted" on one device. On the other hand, if many devices must be interfaced or if a great deal of memory must be present, leaving only a few addresses for input or output, then more complete address decoding must be done. On commercial microcomputers it is often true that address decoding has already been done for user input/output ports. The Apple II family, for example, provides a series of expansion ports where the address decoding has already been done. It is then simply a matter of using the signals provided to enable interfaces that the user constructs.

To illustrate specific input and output techniques, let us interface a DAC and an ADC to a 6502 system. In fact, we will expand on the decoding scheme shown in Figure 9.14. Let us assume that we wished to add a 1408 DAC to the system using the 74138. We will use the 001 output of the 74138 to send information to the 1408. The 1408 requires that digital information be maintained at its input. Thus we will use a 74100 dual 4-bit latch as an intermediate output port to the 1408. The procedure is quite straightforward. Since the 6502 will be sending information to the 1408, a write operation is involved. Thus we want to enable the 74100 to accept data from the 6502 when the 001 output of the 74138 is low AND the output of the RAM R/$\overline{W}$ (from Figure 9.12) is low. The two enable inputs of the 74100 are active high. Thus, if we use a gate whose output is high only when both inputs are low (a NOR gate), we will be in business. Figure 9.15 shows the complete circuit. Information can be sent to the DAC by any 6502 instruction that writes information out. For example, the following two instruction sequences would succeed in sending the 8-bit number *N* to the device whose address was ADDRESS.

```
LDA #N              Load the 6502 accumulator
                    with a specific 8-bit number
                    N.
STA ADDRESS         Send the accumulator contents
                    to the specified address.
```

In BASIC, the same thing could be achieved by the single instruction

```
POKE ADDRESS, N     Send the value N to the
                    designated address.
```

The address would of course be any address starting with 001, that is, in hex notation, any address in the range from 2000 to 3FFF or, in BASIC, an address in the range from 8192 to 16383.

Now let us interface the ADC using the 010 output of the 74138. In this case we must prepare for both a read and a write operation. The write operation will be used to send the appropriate information to the ADC to select the desired

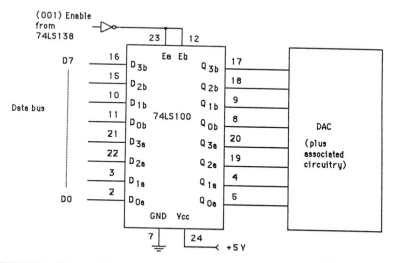

FIGURE 9.15 A 74LS100 dual 4-bit latch used as a single 8-bit latch (8-bit output port) to drive a DAC and associated circuitry (not shown). Data change when the (001) output of the 74LS138 goes low and RAM R/$\overline{W}$ is low, and are latched when RAM R/$\overline{W}$ goes high again at the end of the WRITE cycle. Since the 74LS100 has active-high enables, the NOR gate is necessary.

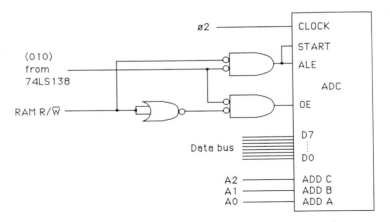

FIGURE 9.16   The circuitry necessary to interface an ADC to the micro-processor. Notice that the same address is used for writing to the ADC (to start conversion) and for reading from the ADC (to read the result). The RAM R/W line determines whether a write or a read operation is done.

analog input and start the conversion. The read operation will be used to read the digital result. Figure 9.16 shows the required circuitry. The low three address lines are connected to the ADD A to ADD C lines of the ADC. The system clock, $\phi2$, is adequate for the ADC clock. To start conversion, we need to place the desired 3-bit value on the low three address lines (to select the analog input) and then take both the START and ALE inputs high. The circuitry shown in Figure 9.16 will accomplish this in a single write operation, where the 6502 writes to an address whose high 3 bits are 010 (to take the START and ALE inputs high) and whose low 3 bits are the appropriate 3 bits to select the desired analog input. For example, to start conversion of the analog voltage on input 3, the 6502 would write to address 4003 hex. Thus the single 6502 command

```
STA $4003              (the $ symbol indicates that
                       hexadecimal notation is being
                       used)
```

will begin the conversion process for the analog voltage on input 3. In BASIC, this would be

```
POKE 16387,N           where N is any number up to
                       255.
```

To read the result of the conversion, a read operation must be performed. Instead of monitoring the EOC line to see when the conversion takes place, it is simpler to wait about 100 microseconds, since the conversion will be done without question in this period of time. This frees us from having to tie in the EOC output to our circuitry. Once 100 microseconds has passed, the digital result can be read by a read operation of any address starting with 010, such as 4000 hex. Thus the 6502 command

```
LDA $4000
```

would load the results of the conversion into the 6502 accumulator for later processing. The equivalent BASIC command would be

```
X = PEEK(16384)
```

which would set X equal to the result of the conversion.

A Simple General Input Port

It is often necessary to read the status of a group of digital devices into the microprocessor. For example, it could be necessary to read the status of a set of switches. This can be done very easily using an 8-bit tristate buffer such as the 74LS244. This buffer really has two separate 4-bit buffers, both with active-low enables. The use of the 74LS244 as an input port is shown in Figure 9.17. When the 74LS244 is not enabled, the outputs float in a tristate, or high-impedance state. Thus the outputs have no effect on the data bus unless the chip is enabled. Since the 74LS244 is connected as an input port to read the status of the switches, the chip should only be enabled during a read operation. If we use the 011 output of the 74138 to

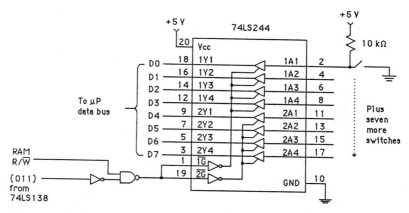

FIGURE 9.17 A general input port using the 74LS244 tristate buffer. The microprocessor data bus is not connected to the switches unless the buffer is enabled by the active-low signal (011) from the 74LS138. On the figure only one switch is shown, although you should visualize eight identical switches being present.

enable the 74LS244, then we want the 74LS244 to be enabled when the 011 output of the 74138 is low AND the RAM R/$\overline{\text{W}}$ line is high. The circuit in Figure 9.17 accomplishes this. During the read operation, the status of the switches will be read at the brief instant of time when the 74LS244 is enabled. Once the read operation has ended, the switches will have no effect on the output of the 74LS244. The single 6502 command

```
LDA $6000
```

will read the switch status into the 6502. In BASIC, the equivalent command would be

```
X = PEEK(24576)
```

A Caution about Buffering

The outputs on the 6502 such as the address bus, data bus, clock, and R/$\overline{\text{W}}$ line have very little drive capacity. In fact, these outputs are designed to drive only one LS TTL load. In a very simple 6502 system, this may pose no problem. The high address lines may only go to a single address-decoding chip, while the lower address lines may go to several RAM and ROM chips. The RAM and ROM chips use MOS technology so that they do not present a significant load. In all but the smallest systems, however, the address lines, $\phi2$, and the R/$\overline{\text{W}}$ line must drive several inputs. Thus these output lines should be buffered. Since these lines are all *output* lines, the simplest way to do this is to use noninverting buffers such as the 74LS241 8-bit tristate buffer. It is also possible to use back-to-back inverting buffers such as the 74LS04.

The data bus presents a more difficult problem. In this case we are dealing with a bidirectional bus. Thus during a write operation the buffered data bus must point away from the 6502, while during a read operation the buffered data bus must point toward the 6502. A bidirectional buffer of some type is needed. The 74LS245 octal bus transceiver is an example of this kind of circuit. Figure 9.18 shows how a 74LS245 could be used to buffer the data bus of a 6502. As shown in the figure, all data bus operations will pass through the 74LS245. The two alternate sets of tristate buffers are controlled by the RAM R/$\overline{\text{W}}$ line. When this line is high (in the read state), the set of buffers pointing *into* the 6502 is enabled; when the RAM R/$\overline{\text{W}}$ line is low (in the write state), the set of buffers pointing *out of* the 6502 is enabled.

Bus transceivers such as the 74LS245 are often used to buffer the data bus on individual expansion cards that are attached to microcomputers. In this case, the R/$\overline{\text{W}}$ line is combined with an address-decoding signal so that the transceivers

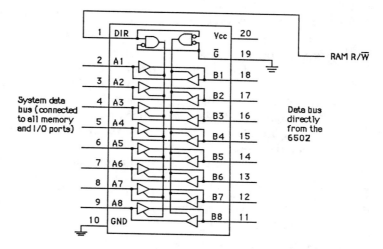

FIGURE 9.18 A 74LS245 bus transceiver used to buffer the data bus of a 6502.

normally point *away from* the microprocessor. Only when the $R/\overline{W}$ line is high AND the circuitry on the card is addressed will the transceivers pointing toward the microprocessor be enabled.

PROBLEMS

1. Imagine a 4-bit D/A converter that produces an analog output of 4.875 V when the digital input is 1111. What is the size of the voltage step produced when the digital output is increased by one? If the digital input is 0101, what is the analog output?

2. Suppose a D/A converter has a maximum analog output of 5V when its digital input is at its maximum count. It is desired that the voltage step produced when the digital count is increased by one be no more than 0.1 V. How many input bits must the converter have to satisfy this requirement?

3. A typical 8-bit A/D converter operates from a + 5V reference (that is, a + 5V analog input is converted into a digital output of 1111 1111). How much must the analog input increase to generate an increase of one in the digital output? If the analog input is 1.20 V, what is the digital output? Suppose that the analog signal is changing at the rate of 100 V/s. Also suppose that the ADC requires 100 μs to complete a conversion. Will the analog voltage change enough during

the conversion process to require the use of a sample-and-hold circuit? What if the analog voltage were changing at 400 V/s?

4. Suppose that you had a temperature probe that produced a voltage of 10 mV/kelvin (K). Thus at the ice point (0°C, 273K) the probe produces a voltage of 2.73 V. You wish to make temperature measurements of a bath that always remains between the ice point and the boiling point (373K). First, what range of output voltages will your probe produce over this temperature range? Second, if you are using a 8-bit D/A converter that produces an output of 1111 1111 when the input is +5 V, what is the size of the temperature step that corresponds to a single digital step? If you wanted this temperature step to be 0.5 K, how many bits would your A/D converter require? Can you think of any way to use an 8-bit A/D converter to obtain this accuracy for the temperature range to be measured (can the voltage from the temperature probe be adjusted somehow before it is input to the A/D converter)?

5. A particular ADC can sample at a rate of 12,000 samples per second. What is the highest frequency analog signal that this ADC should be applied to in order that the Nyquist theorem be satisfied?

Appendix

Laboratory Experiments

INTRODUCTION

This appendix contains a series of laboratory exercises that are designed to accompany and amplify the material in the text. The laboratories are mostly of the "theory proving" type in that they are designed to illustrate the operation of a particular type of circuit. Each laboratory contains specific directions and figures that outline the procedures for the laboratory. These are followed by a detailed discussion of the expected results that should have been obtained. In addition to the laboratory exercises, four projects have been included. These are relatively simple design exercises in which the student attempts to solve a simple digital problem. The problem is posed in sufficient detail to guide the student toward a possible solution, but no detailed circuit diagrams are provided, nor is any discussion of expected results given. The projects are optional in that they do not introduce any new concepts. However, they provide the student with an opportunity to be creative, rather than simply follow directions. Table A.1 contains a suggested timetable of when to do each experiment. It can be seen from this table that the experiments begin following Chapter 2.

To do the exercises in this appendix, a number of items of equipment are necessary. First is a voltmeter of some sort, preferably a DVM. Second, an assortment of parts must be assembled. A list is given in Table A.2. Third, and perhaps most important, is some type of digital trainer, such as the Heathkit digital design experimenter (or some equivalent piece of equipment). The digital trainer should have voltage supplies of $+5$ volts and ± 12 volts, a number of LED logic indicators, a number of digital data switches, a couple of debounced digital output switches, a digital clock with frequencies selectable at about 1 hertz and 1 kilohertz, and a small solderless breadboard for wiring simple digital circuits.

TABLE A.1 Suggested Schedule of Experiments

Laboratory	Following This Section	Laboratory	Following this Section
L1	2.2	L18	5.3
L2–L7	2.3	L19	5.4
L8	2.4	L20	6.2
L9	3.5	L21, L22	6.3
P1	3.5	L23	6.5
L10	4.2	P4	6.5
L11, L12	4.4	L24	7.3
P2	4.4	L25	7.5
L13	4.5	L26	8.2
L14–L16	4.6	L27	8.3
L17	5.2	L28	9.2

If no digital trainer is available, doing the experiments will require the fabrication of the essential functions of a digital trainer. A dc power supply should be available with outputs of $+5$ volts and ± 12 volts (these could come from separate supplies). A solderless breadboard should be available that is large enough to construct the experimental circuits and also leave room for supplementary circuits that replace some of the functions of a digital trainer. Circuits to provide these functions are shown in Figures A.1 to A.4.

Figure A.1 shows how to construct a simple digital output that can be switched from 0 to $+5$ volts by simply opening and closing the SPST switch. This switch is not debounced, so this type of digital output should not be used to drive any circuits that respond to digital pulses. Figure A.2 shows how to construct a debounced switch. The actual switch in this circuit should be a push-button SPDT switch with the normally closed contact connected as shown. This circuit provides two outputs that are always complementary, with the Q output normally high. Figure A.3 provides a simple logic indicator. Because an open TTL input acts like a high input,

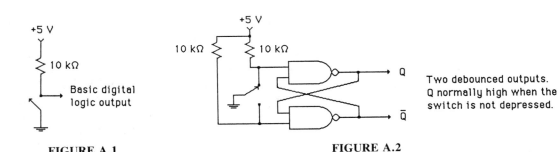

FIGURE A.1

FIGURE A.2

TABLE A.2   Summary Parts List

Quantity	Description	Quantity	Description
1	+5-V power supply (dc)	1	CD4001, CMOS quad 2-input NOR gate
1	+12-V power supply (dc)		
2	debounced switches	2	74LS00, quad 2-input NAND gate
4	logic indicators		
1	digital voltmeter	1	74LS02, quad 2-input NOR gate
1	TTL-compatible square-wave generator (≈ 1 kHz) (optional)	1	74LS03, quad 2-input NAND gate (O.C.)
1	dual-trace oscilloscope (optional)	1	74SL04, hex inverter
		1	74LS08, quad 2-input AND gate
3	2N2222A transistors (or equivalent)	1	74LS20, dual 4-input NAND gate
4	1N4148 diodes (or equivalent)	1	74LS32, quad 2-input OR gate
1	LED	1	74LS42, BCD to decimal decoder/driver
1	TIL-312 seven-segment LED		
2	220-μF capacitors (electrolytic)	1	74LS47, seven-segment decoder/driver
1	10-μF capacitor (electrolytic)	1	74LS74A, dual positive-edge-triggered D flip-flop
1	1-μF capacitor		
1	0.01-μF capacitor	1	74LS75, quad latch
1	130-Ω resistor (all resistors are 1/4 W)	2	74LS76A, dual negative-edge-triggered $J-K$ flip-flop with preset and clear
7	330-Ω resistors		
1	1-kΩ resistor	1	74LS83A, 4-bit adder with carry in and carry out
1	1.6-kΩ resistor		
1	2-kΩ resistor	1	74LS86, quad EXOR gate
1	4-kΩ resistor	1	74LS90, BCD counter
2	5-kΩ resistors	1	7495, 4-bit shift register
1	10-kΩ resistor	1	74LS123, dual retriggerable one-shot
2	15-kΩ resistors		
3	25-kΩ resistors	1	74LS138, 1-of-8 decoder
5	50-kΩ resistors	1	74LS151, 8-input multiplexer
1	56-kΩ resistor	1	74LS193, binary counter
1	100-kΩ resistor	1	74LS365, hex tristate buffer/driver
1	1-MΩ resistor		
1	555 timer (such as NE555)		

the LED will light when the input is left unconnected, in addition to when it is high. If this proves irritating, the LED can be temporarily disconnected from the 7404 when the indicator is not needed. Finally, Figure A.4 shows a simple clock circuit. The two values of capacitance shown in the figure (0.47 microfarad and 470 picofarads) give a slow (≈ 1 hertz) clock for easy visualization and a fast (≈ 1 kilohertz) clock for input into counter circuits when needed. Adjustments to the clock frequencies can be easily made by changing capacitors.

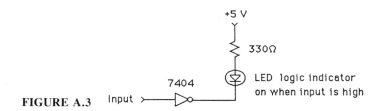

FIGURE A.3

The simple circuits shown in Figures A.1 to A.4 can be quickly fabricated as needed by the student in whatever quantity necessary. In fact, in most cases the logic output of Figure A.1 can be replaced by a simple wire that can be directly connected to either +5 volts or ground, and the logic indicator of Figure A.3 can be replaced by a DVM. If time permits or help is available, the circuits in Figure A.1 to A.4 could easily be fabricated on a single permanent circuit board and mounted on a suitable box. This would make the functions more readily available. The important point is that special trainers are nice but not absolutely necessary.

As an alternative to the experiments in this appendix, many equipment manufacturers provide lab manuals and parts kits that are designed to be used with their digital trainers. Most of these lab manuals provide a range of experiments that is similar to what is provided in this appendix. It is simply a matter of making a selection and choosing an order that emulates the experiments in this appendix.

Finally, it should be pointed out that time may not allow for all the experiments in this appendix to be performed. The list is of sufficient length and variety that, depending on the taste of the instructor, several experiments could be omitted.

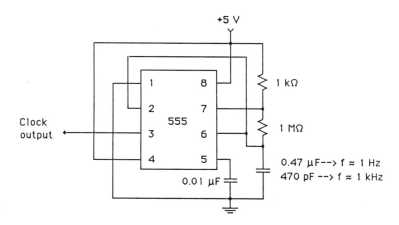

FIGURE A.4

LABORATORY 1   TRANSISTOR PROPERTIES

Objectives   To observe the operation of a transistor in the cutoff, saturation, and linear regions; to understand how a transistor can function as a logic inverter.

*Materials
Required*   +5-V power supply

1	DVM (digital voltmeter)
1	1-MΩ variable resistor
1	2-kΩ resistor ($\frac{1}{4}$ W)
1	10-kΩ resistor ($\frac{1}{4}$ W)
1	2N2222A transistor

Procedure   1. Construct the circuit shown in Figure L1.1. Begin with the 1-MΩ resistor disconnected from the +5-V supply. Observe the output voltage at the collector of the transistor (V_{out} as shown in Figure L1.1). Is there any base current flowing in this circuit? Is there any collector current flowing in this circuit? What is the state of the transistor (cutoff, saturated, in linear region)?

   2. Connect the 1-MΩ resistor to the +5-V supply, setting the resistor for *maximum* resistance. Measure the voltage across the 10-kΩ resistor. Use Ohm's law to compute the base current flowing in the circuit. Measure the voltage across the 2-kΩ load resistor and use Ohm's law to compute the collector current flowing in the circuit. Now reduce the resistance of the 1-MΩ variable resistor such that the voltage across the 10-kΩ resistor approximately doubles (that is, the base current doubles). Again measure the voltage across the 2-kΩ resistor. What has happened to the collector current? Continue to reduce the setting of the 1-MΩ resistor such the base current doubles each time. Measure the collector current after each change in setting (by measuring the voltage across the 2-kΩ resistor). How does the collector current depend on the base current? As you continue to increase the base current, does the collector current continue to

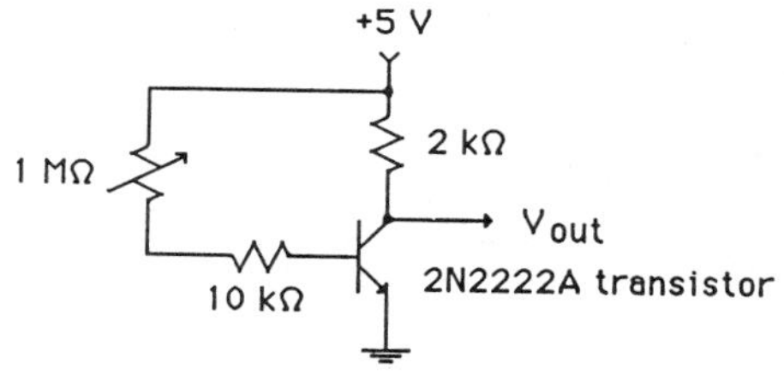

FIGURE L1.1

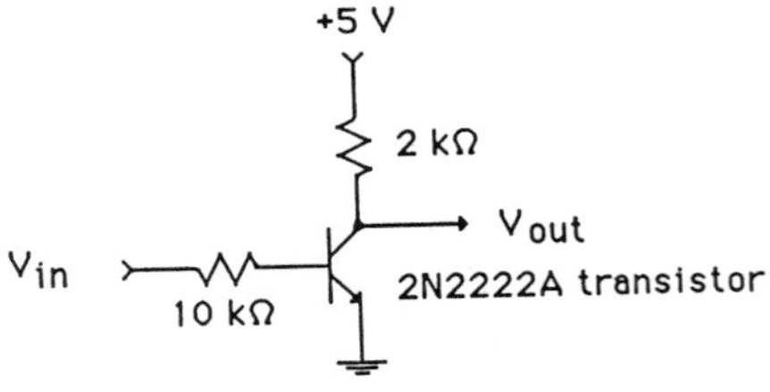

FIGURE L1.2

increase in more or less direct proportion, or does there come a time when further increases in the base current seem to produce very little change in the collector current? As long as the collector current increases in proportion to the base current, what is the state of the transistor? What state is the transistor in when the base current increases to the point where the collector current no longer increases when the base current is increased?

3. Remove the 1-MΩ resistor from your circuit. The circuit should now be as shown in Figure L1.2. The free end of the 10-kΩ resistor will now be considered the circuit input, and the collector voltage will be considered the circuit output. Consider *two* possible input voltages, 0 and +5 V. Input 0 V to the circuit by grounding the circuit input. What is the circuit output? Now input +5 V to the circuit by tying the input directly to +5 V. What is the circuit output? Make up a small truth table for the circuit showing the input voltages in one column and the resulting output voltages to the adjacent column. What logical function does this circuit perform, assuming you assign logical 1 to +5 V and logical zero to 0 V?

Discussion   1. In part 1 you observed that, with the 1-MΩ resistor disconnected, no base current could flow. Since no base current could flow, then no collector current could flow either. The transistor is in the cutoff state. The 2-kΩ resistor and the transistor form a voltage divider, and with the transistor cut off, essentially all the supply voltage appears across the transistor (with no collector current, there is no voltage drop across the 2-kΩ resistor). Thus the output voltage is +5 V.

2. When the 1-MΩ resistor is connected at its maximum resistance setting, a small base current can flow. Since the base–emitter voltage is about 0.7 V, this leaves about 4.3 V across the 1-MΩ (and 10-kΩ) resistor. The current flow should therefore be about 4.3 µA. This 4.3-µA base current should produce a collector current that is beta times larger (review the meaning of beta from Chapter 2). Since a 2N2222A transistor typically has a beta of around 50 to 200, the collector current produced should be in the neighborhood of 0.4 mA (for a beta of 100). This current would produce a voltage drop of 0.8 V across the 2-kΩ resistor (the actual figures that you measure may vary from this by several tenths of a volt). As you reduce the setting of the 1-MΩ resistor, the base current increases and so does the collector current (although the increase in collector current may

not exactly follow the base current since beta is not really constant). As long as the collector current increases more or less in proportion to increases in base current, the transistor is in the linear region. Finally, a point is reached where further increases in base current produce no further increase in collector current. At this point the voltage drop across the transistor (collector–emitter voltage) will only be a few tenths of a volt. The transistor is now said to be saturated. In the circuit of Figure L1.1, the setting of the 1-MΩ resistor will be about 200 kΩ when saturation sets in (although this figure could vary by several tens of kilohms depending on the particular transistor). Any lower setting of the 1-MΩ resistor will increase the base current, but not change the collector current.

3. When you removed the 1-MΩ resistor to produce the circuit of Figure L1.2, you set up a situation where a +5-V input produced so much base current that the transistor went well into saturation. Likewise, when you grounded the input, no base current could flow and the transistor was cut off. Thus, assuming inputs are restricted to +5 and 0 V, the transistor could only be in one of two well-defined states, cutoff or saturated. Your truth table showed you that when the input was high the output was low, and vice versa. Considering a high voltage (+5 V) as logic 1 and a low voltage as logic 0, the transistor circuit clearly performs the logical inversion or NOT function.

LABORATORY 2   DISCRETE TTL NAND GATE

Objectives   To analyze the function of a TTL NAND gate constructed from discrete components.

Materials Required

+5-V power supply

1	DVM
1	56-kΩ resistor ($\frac{1}{4}$ W)
1	4-kΩ resistor ($\frac{1}{4}$ W)
1	1.6-kΩ resistor ($\frac{1}{4}$ W)
1	1-kΩ resistor ($\frac{1}{4}$ W)
1	130-Ω resistor ($\frac{1}{4}$ W)
3	2N2222A transistors
4	1N4148 diodes

Note: Since transistors come in various types of packages, make sure your instructor explains to you which connections of your 2N2222A transistors are the emitter, base, and collector.

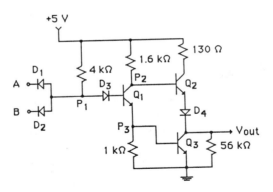

FIGURE L2.1

Procedure   1. Construct the circuit shown in Figure L2.1. This circuit is essentially identical to Figure 2.10 with Q_1 replaced by the three diodes D_1, D_2, and D_3. The 56-kΩ resistor on the output is simply meant to load the output of the circuit somewhat, since normally the output of a logic gate goes to other circuits. Begin with both inputs A and B grounded. Turn on the power to the circuit and measure the output voltage. Also measure the voltage at points P_1, P_2, and P_3 in the circuit. Make a table with the headings A, B, P_1, P_2, P_3, and V_{out}. In this table, enter the readings that you have just made (where the entries for A and B are the voltages at inputs A and B, respectively: 0 V in this case).

2. Repeat the measurements in step 1 for each of the following combinations of input voltages: $A = +5$ V, $B = 0$ V; $A = 0$ V, $B = +5$ V; $A = +5$ V, $B = +5$ V. If we make the logic assignments that any voltage higher than $+2$ V is logic 1 and any voltage lower than $+0.8$ V is logic 0, construct a truth table for the circuit (that is, A and B are the logical inputs and V_{out} is the logical output). What logical function does the circuit perform?

3. Referring to your table of measured voltages, for each line on the table determine the state of each transistor in the circuit (cutoff or saturated). Can you understand *why* each transistor is in the observed state?

4. Disconnect both inputs A and B. How do the various voltages in the circuit compare to the case where inputs A and B were both $+5$ V? Now ground input A. How do the various voltages in the circuit compare to the case where input A was grounded and input B was $+5$ V? What do you conclude about the effect of leaving an input unconnected?

5. The circuit that you have constructed will be used with only minor modification in Laboratory 3. Therefore, if possible, leave the circuit assembled in preparation for Laboratory 3.

Discussion   1. In part 1 you grounded both inputs to the circuit. This should cause both the input diodes, D_1 and D_2, to become forward biased, holding point P_1 at about 0.7 V. With point P_1 at such a low voltage, neither D_3 nor Q_1 can be biased on (this would require at least enough voltage to forward bias three diodes — remember the base–emitter junctions of Q_1 and Q_2 — a voltage of nearly 2 V). With Q_1 cut off, there is no current to provide base current for Q_3. Thus point P_3 should be at ground and Q_3 should be cut off. Also, with Q_1 cut off, base current can flow through the 1.6-kΩ resistor into the base of Q_2, thus turning it on (saturating it). With Q_2 saturated and Q_3 cut off, only the base current for Q_2 will flow through the 1.6-kΩ resistor. This produces a neglible voltage drop, so P_2 should be at essentially +5 V, and V_{out} should be two diode drops lower at about 4 V (with the low level of output current drawn by the 56-kΩ resistor, a diode drop is more like 0.5 V).

2. In part 2 you tried other combinations of voltages on inputs A and B. You should have found that as long as either A or B were held at 0 V the results were virtually the same as in part 1. However, when both A and B were made +5 V, the circuit output changed to 0 V. With both A and B at 5 V, diodes D_1 and D_2 were reverse biased. This allowed D_3 to become forward biased and provide base current to Q_1. Q_1 then saturated and allowed base current to reach Q_3, in turn saturating Q_3. With Q_3 saturated, the output voltage should have been nearly 0 V. P_3 should have been one diode drop above ground at about 0.7 V, while P_2 should have been nearly equal to P_3 since Q_1 was saturated. Finally, P_1 should have been about two diode drops above P_3 at about 2 V. The truth table you constructed should have indicated that the circuit was providing the NAND function.

3. In part 4 you should have found that an open input behaves in virtually the same manner as an input tied to +5 V. In either case, the input diode becomes reverse biased.

LABORATORY 3   THE OPEN-COLLECTOR OUTPUT

Objectives   To understand the operation of the typical open-collector output.

*Materials
Required*   +5-V power supply

+12-V power supply

1   DVM

1   56-kΩ resistor ($\frac{1}{4}$ W)

1	10-kΩ resistor ($\frac{1}{4}$ W)
1	4-kΩ resistor ($\frac{1}{4}$ W)
1	1.6-kΩ resistor ($\frac{1}{4}$ W)
1	1-kΩ resistor ($\frac{1}{4}$ W)
2	2N2222A transistors
3	1N4148 diodes

Procedure 1. Construct the circuit shown in Figure L3.1 (note, if you have not yet dismantled the circuit from the previous lab, simply modify it to conform with Figure L3.1). Again, as in Laboratory 2, the 56-kΩ resistor is present simply as a load resistor and is not inherently part of the basic circuit. To start, leave Vcc unconnected. You will be using input voltages (at A and B) of 0 and +5 V in various combinations. Turn on the power to the circuit and examine the output voltage obtained for each of the four possible combinations of input voltages.

2. Now connect +5 V as your Vcc voltage. Repeat step 1, recording your results in a truth table for convenient reference. How did the results change from step 1? How do the results compare from those obtained in step 2 of Laboratory 2?

3. Remove the +5 V from the Vcc connection (*not* from the rest of the circuit) and replace it with +12 V. Repeat step 1, again recording your results in a truth table. How did the results change from step 2?

Discussion 1. In part 1 you found that with no supply voltage (Vcc) connected to the output transistor the output of the circuit remained at 0 V. Transistor Q_3 will be turned on *only* when inputs A and B are both high (+5 V). Otherwise, Q_3 will remain cut off. However, with no voltage supply connected to the collector, the output remains at zero regardless of the state of the transistor.

2. In part 2 you found that when +5 V was used as a supply voltage for Q_3 the results were very similar to those found in part 2 of Laboratory 2. When Q_3

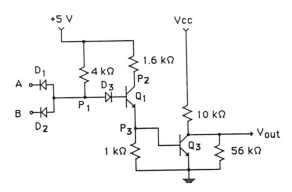

FIGURE L3.1

was in saturation, the collector was held at essentially 0 V. This occurred only when inputs A and B were both high. When either input A or input B was low, Q_3 was cut off and the output voltage was then determined by the voltage divider consisting of the 10- and 56-kΩ resistors. Since the 56-kΩ resistor has a much larger resistance, the output rose to nearly 5 V when Q_3 was cut off (the output would rise to almost exactly 5 V if the 56-kΩ resistor were removed).

3. When 12 V was used as the supply voltage for Q_3, you found that the output was about 10 V when either A or B was held low. This output would be nearly +12 V if the 56-kΩ resistor were removed. Thus the open-collector type of output can be used to convert a TTL-level logic input (0 and +5 V) into a very different voltage level for the logical high state.

LABORATORY 4   PROPERTIES OF 74LSXX TTL LOGIC ICS

Objectives   To become familiar with the function and properties of some simple 74LSXX integrated circuits.

Materials Required   +5-Volt power supply

1	DVM or logic indicator
1	74LS00 quad 2-input NAND gate (a standard 7400 will also suffice)
1	74LS04 hex inverter (a standard 7404 will also suffice)
1	2-KΩ resistor ($\frac{1}{4}$ W)

Procedure   1. Refer to Figure L4.1 for the pin-outs of the 74LS04. Select the 74LS04 and attach +5 V to the *Vcc* connection (pin 14) and ground to the GND connection (pin 7). Now focus your attention on the first inverter, the one with pin 1 as the input and pin 2 as the output. First, apply +5 V at the input and observe the output. Is it high or low (high being anything over +2 V, low being anything less than 0.8 V; if you are using a logic indicator, a high output will light the LED and a low will not)? Now apply 0 V to the input and observe the output. Construct a logic truth table for the inverter. Does it conform to Figure 1.2? Remove any input from the inverter (that is, leave pin 1 unconnected). What output do you observe? How do you conclude that an open TTL input behaves?

2. Construct the circuit shown in Figure L4.2. Apply a high input at pin 1, the input to the first inverter. Check the output of each inverter in the chain. Now

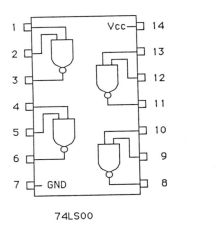
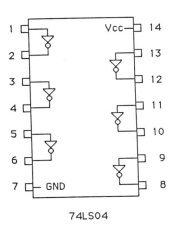

FIGURE L4.1 Pin-outs of the TTL circuits 74LS00 (quad 2-input NAND gate) and 74LS04 (hex inverter).

apply a low input to the first inverter and again check the output of each inverter in the chain. Are the results as you expected them to be?

3. Refer to Figure L4.1 for the pin-outs of the 74LS00. Connect power to the circuit (remember, for TTL logic Vcc will always be $+5$ V). Focus your attention on the first NAND gate (the one with inputs at pins 1 and 2 and output at pin 3). Four possible input combinations can be applied to this NAND gate (refer to Figure 1.10 to refresh your memory). Construct a truth table for the NAND gate by applying each of the four possible input combinations and recording the outputs that result. Do your results conform to those in Figure 1.10? Now leave one input unconnected and observe the output with the other input high and then low. How does the unconnected input behave? Does this conform with the behavior you found in part 1 for the 74LS04?

4. Construct the circuit in Figure L4.3(a). With the two NAND inputs tied together, the circuit effectively has a single input (which is labeled A in the figure). Apply a low input at A and observe the output. Now apply a high input at A and observe the result. Construct a logic truth table for the circuit. What logic function does the circuit perform? Now construct the circuit in Figure L4.3(b). Repeat the preceding steps for this new circuit. How do your results compare with the results you obtained with the circuit in Figure L4.3(a)?

5. Construct the circuit in Figure L4.4. Think of the circuit as one complete unit with two inputs (A and B) and one output (C). What logic function do you

FIGURE L4.2

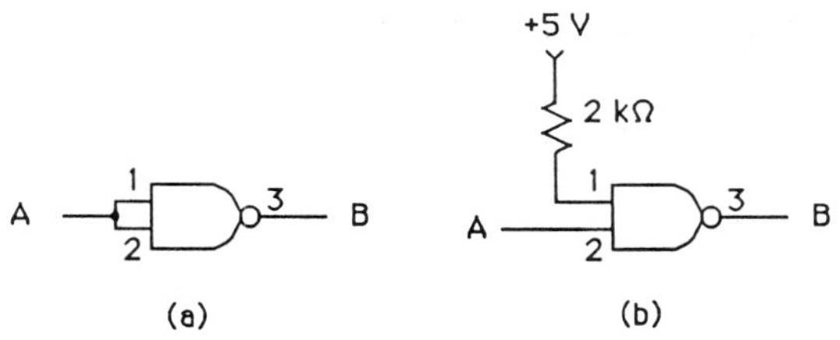

(a)   (b)

FIGURE L4.3

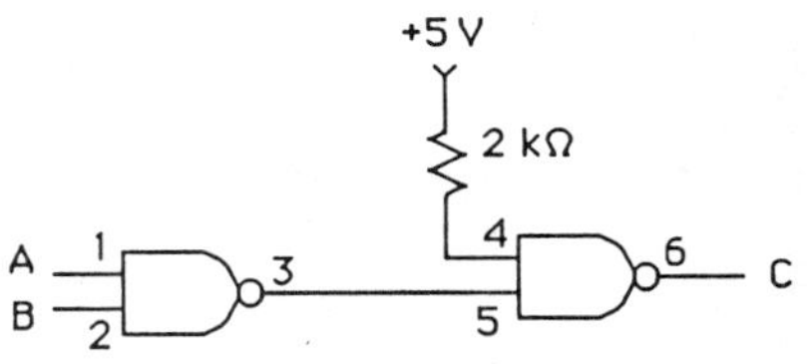

FIGURE L4.4

expect this circuit to perform? Apply the four possible input combinations to the circuit and record the outputs for each combination. Place your results in a truth table. From your analysis of the truth table, what logic function do you conclude is being performed? Did you anticipate this result?

Discussion

1. In part 1 you observed that an inverter produces an output that is logically the inverse of the input. You also saw that an unconnected input behaves as a high input.

2. In part 2 you observed that each inverter in the chain simply inverts the input to it. Thus, as you progressed through the chain, you found that each inverter had the opposite output as the previous inverter. From this it is clear that two inverters in series produce an output that is the same as the *original* input, as do any *even* number of inverters. Any *odd* number of inverters acts logically like a single inverter.

3. In part 3 you confirmed that a NAND gate does indeed have the truth table given in Figure 1.10. You also observed that when you left one input unconnected the circuit behaved as if that input were high. This confirms the behavior that you observed for the inverter.

4. In part 4 you constructed two circuits that used a single NAND gate to emulate the behavior of an inverter. In each case, the output (B) was the inverse of the input (A). Either of these two approaches can be used to make an inverter out of an unused NAND gate that might be available.

5. In part 5 you found that the circuit in Figure L4.4 behaved as an AND gate. This makes sense since the inverter (constructed from the second NAND gate) simply inverts the output of the first NAND gate, in effect cancels the inverter indicated by the small circle at the output of the first NAND gate.

LABORATORY 5 OPEN-COLLECTOR OUTPUTS AND THE WIRED-OR CONNECTION

Objectives To observe the functioning of a typical TTL open-collector output, and to understand how a number of open-collector outputs can be connected directly together to form an effective OR function.

Materials Required +5-V power supply

1 DVM or logic indicator

1 5-kΩ resistor ($\frac{1}{4}$ W)

1 56-kΩ resistor ($\frac{1}{4}$ W)

1 74LS03 quad 2-input NAND gate with open-collector output

Procedure 1. Refer to Figure L5.1 for the pin-outs of the 74LS03. Construct the circuit shown in Figure L5.2(a); don't forget to apply power to the circuit through pins 14 and 7. The 56-kΩ resistor is meant to act as a load on the output. Apply the standard four combinations of digital inputs to inputs *A* and *B*, and observe the resulting output in each case. Record your results in a truth table.

2. Add the 5-kΩ resistor as shown Figure L5.2(b). Repeat step 1. Again record your results in a truth table.

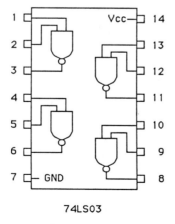

FIGURE L5.1

74LS03

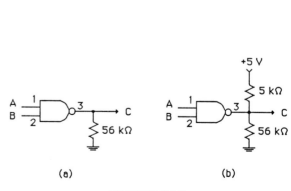

(a)   (b)

FIGURE L5.2

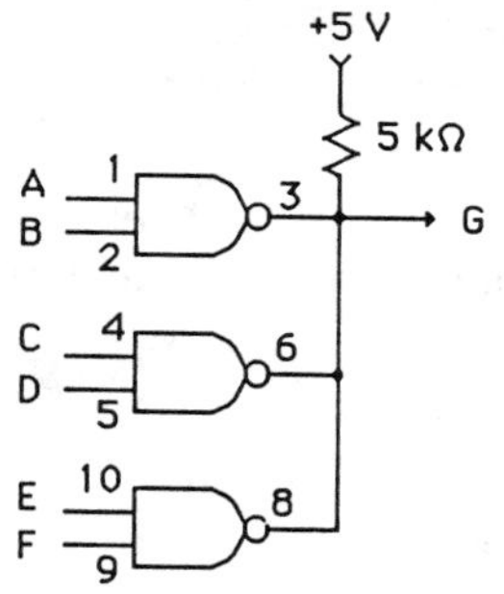

FIGURE L5.3

3. Construct the circuit shown in Figure L5.3. Since we know that the output of a NAND gate only goes low when *both* inputs are high, we will investigate two sets of circumstances. First, set inputs A, C, and E low and inputs B, D, and F high. The outputs of each NAND gate should be high (the output transistor of each gate should be cut off). Verify that G is high with this set of inputs. Now take input A high. What happens to the circuit output? Take input C high also. Does this change the output? To get a low output at G, does it matter which NAND gate has both inputs high? Does it matter how many NAND gates have both inputs high?

Discussion   1. In part 1 you observed that the output was always low regardless of the state of the inputs. This occurred because the output transistor in an open-collector output has no connection to $+5$ V unless you supply such a connection. The *state* of the transistor does indeed change from cutoff to saturated when both inputs are taken high. But without an external pull-up resistor, the output voltage will remain at 0 V. The output of an open-collector IC should never be input directly to another TTL IC without the use of a pull-up resistor.

2. In part 2 you attached a 5-kΩ pull-up resistor to the open collector output and observed that the NAND gate functioned in the expected manner. In this case, whenever the output transistor is cut off, the output voltage is pulled up to $+5$ V through the 5-kΩ resistor. When the output transistor is saturated, the output is grounded through the transistor.

3. In part 3 you connected several open-collector outputs directly together. As long as all the output transistors are in the cutoff state, the circuit output (G) is pulled up through the 5-kΩ resistor to $+5$ V. However, if *any one* (or more) of the output transistors becomes saturated, the circuit output will be grounded through the saturated transistor. That is, if the output of *any* of the NAND gates goes low, the circuit output goes low. Unlike normal totem-pole TTL outputs, open collector outputs may be directly connected together.

LABORATORY 6   TRISTATE OUTPUTS

Objectives   To understand the function of a typical TTL tristate output.

Materials
Required   + 5-V power supply

1   DVM

1   330-Ω resistor ($\frac{1}{4}$ W)

1   LED

1   74LS365 hex tristate buffer/driver

Procedure   1. Place the 74LS365 in your breadboard and apply power (the pin-outs for the 74LS365 are shown in Figure L6.1). Leave pins 1 and 15 unconnected for the moment. Measure the voltage at pin 3, the output of the first buffer on the IC. Apply + 5 V to the input of this buffer (pin 2) and observe the output. Apply 0 V to the buffer input and observe the output. How do all these measurements compare? Repeat these steps with pins 1 and 15 connected to + 5 V. Are your results any different? Try grounding pin 1 and repeating the steps. Do your results change?

2. Now ground both pins 1 and 15 on the 74LS365 (these are inputs to the onboard 2-input NOR gate, which enables the tristate buffers). Repeat step 1. How do your results compare with those obtained in step 1?

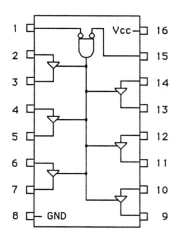

FIGURE L6.1   74LS365

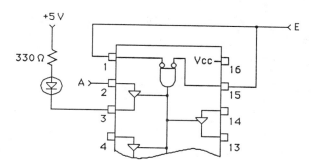

FIGURE L6.2

3. Construct the circuit shown in Figure L6.2. Allow input E to remain unconnected for the moment. Apply $+5$ V to input A and observe the state of the LED. Also, measure the voltage at pin 3. Now ground input A and again observe the state of the LED and measure the voltage at pin 3. Are the results any different with input A grounded?

4. Ground the enable input E, and repeat step 3. How do your results compare with those obtained in step 3?

Discussion

1. In step 1 you found that as long as you did not ground both enable inputs (pins 1 and 15) the output of the tristate buffer remained unchanged regardless of the input. This output was probably a few tenths of a volt.

2. In step 2 you found that with both enable inputs grounded, the buffer output followed the buffer input. This is because when both enable inputs are taken low the onboard NOR gate output goes high, enabling all six tristate buffers. These are noninverting buffers, so the output of each buffer follows the input (that is, high input → high output).

3. In step 3 you found that with E held high the LED did not light and the voltage at pin 3 was nearly 5 V. Input A had no effect. With the tristate buffer disabled, the output is effectively electrically disconnected from the buffer. The output thus floats up to nearly 5 V because of the connection through the LED and resistor to $+5$ V. No current can flow through the LED since this current has no place to go (the buffer output being electrically disconnected).

4. In step 4 you found that when E was taken low the tristate buffer was enabled. Now, with $+5$ V applied to input A, the buffer output was also $+5$ V. Thus no current flowed through the LED–resistor combination because both ends of the combination were at the same potential. When A was taken low, the buffer output became nearly 0 V. This placed a 5-V potential across the LED–resistor pair, causing current to flow and thus lighting the LED.

LABORATORY 7   CMOS INTEGRATED CIRCUITS

Objectives   To observe the function of a typical CMOS integrated circuit and become familiar with using such circuits.

Materials Required

+5-V power supply

+12-V power supply

1   DVM

1   CD4001 quad 2-input NOR gate

Procedure

1. Before handling the CD4001, remember that you are dealing with a CMOS IC. An input on a CMOS integrated circuit will be the gate of a MOS transistor. This gate is not designed to permit any actual current flow; that is, the gate is insulated from the body of its transistor. The gate essentially forms one side of a very small capacitor. For this reason, even a *very small* static charge deposited on the gate results in a large voltage across this small capacitor, easily enough to break down the metal oxide insulating layer between the gate and the body of the transistor. It is thus imperative that a CMOS IC not be handled carelessly. The CMOS IC will be packaged in conducting foam (usually) or inside a conducting plastic container. It should be removed from its package and placed directly into the breadboard. No voltage should be applied to the circuit until all connections have been made. When the laboratory has been completed, the CMOS IC should be immediately placed back in its package.

2. The pin-outs of the CD4001 quad 2-input NOR gate are shown in Figure L7.1. Place the CD4001 in your breadboard. With your +5-V power supply turned off, connect the power supply output to the V_{DD} input of the IC (pin 14), and connect the supply ground to the V_{SS} input (pin 7). If you are using a digital trainer, connect the output of two digital outputs to pin 1 and pin 2 of the CD4001 (the inputs to the first NOR gate). Connect your DVM to the output of the first NOR gate (pin 3). If you do not have a digital trainer, you should construct two digital outputs such as the one shown in Figure A.1. These outputs can be switched back and forth from +5 V to 0 V without danger to the CD4001.

3. Turn on your +5-Volt power supply. Input each of the four basic digital input combinations (such as shown in Figure 1.11) to the NOR gate and measure the output that results in each case. Record your results in a truth table. Do they conform to what is shown in Figure 1.11?

4. Set one input to the NOR gate at 0 V and the other at +5 V. The output of the NOR gate should be 0 V. Now disconnect the +5-V input; that is, simply

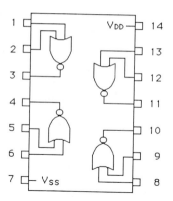

FIGURE L7.1 CD4001

remove the connecting wire entirely from the NOR gate, leaving a totally unconnected input. What happens to the output? Leave the input disconnected for a few minutes and observe the output. Does it stay constant?

5. You are now going to repeat step 3 using the +12-V power supply to replace the +5-V supply. *You must also replace the +5-V logic outputs with +12-V outputs.* That is, when you wish to input a logic high to the NOR gate, *you must use +12 V.* Rebuild two basic logic outputs as shown in Figure A.1 using +12 V to replace +5 V. Repeat step 4 with +12 V operating the CD4001 and +12 V used as logic high on the inputs to the NOR gate. How do your results compare?

Discussion 1. In step 3 you found that the NOR gate acted as you would expect from examining Figure 1.11. You may have noticed that the output voltages were almost exactly +5 and 0 V, whereas for TTL outputs the voltages are not always so close to the supply voltages.

2. In step 4 you observed how an unconnected input behaves. You may have observed that the output did not immediately change when the input was disconnected, but did change after a short time. The output may even have fluctuated for a period of time before settling down. The point is that unconnected CMOS inputs behave in an unpredictable fashion. All unused CMOS inputs should always be tied to another input or to +5 V or ground as appropriate.

3. In step 5 you saw that changing the supply voltage to +12 V resulted in a new logic high voltage of +12 V rather than +5 V. This shift could not be done with a TTL IC. TTL integrated circuits must always be operated from a +5-V supply.

LABORATORY 8   TTL-TO-CMOS INTERFACING

Objectives   To understand how +5-V TTL circuitry can be interfaced to CMOS circuitry operated at non-TTL voltages.

Materials Required
+5-V power supply

+12-V power supply

1   DVM

1   CD4001 quad 2-input NOR gate

1   74LS03 quad 2-input NAND gate with open-collector outputs

2   5-kΩ resistors ($\frac{1}{4}$ W)

Procedure   1. Refer to Figure L8.1. In this figure, the two NAND gates are from the 74LS03, and the NOR gate is from the CD4001. The 74LS03 operates on a +5-Volt supply, while the CD4001 operates on anywhere from +3 to +15 V depending on design requirements. In this lab you will operate the CD4001 using a +12-V supply. Thus you should connect pin 14 (V_{DD}) of the CD4001 to your +12-V supply and pin 7 (V_{ss}) to ground. Construct the circuit shown in Figure L8.1 (remember to handle the CMOS CD4001 with care).

2. Inputs $A \rightarrow D$ in Figure L8.1 are TTL-level inputs. Inputs E and F to the NOR gate must be at a level appropriate for the CD4001 operating at +12 V. Thus inputs E and F must use 0 and +12 V as logical 0 and 1, respectively. Construct a truth table such as shown in Figure L8.2. Input all 16 possible combinations of A, B, C, and D, making your entries in terms of voltages. Record the voltages at E, F, and G (as shown in the first line of Figure L8.2).

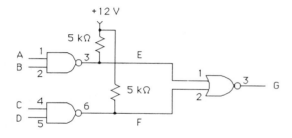

FIGURE L8.1

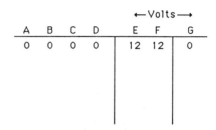

FIGURE L8.2

Discussion   1. You found that although inputs $A \rightarrow D$ were at TTL levels the high-level outputs E and F were shifted to $+12$ V (with the low level remaining zero). This new high level is appropriate for operating the CD4001. You also found that output G responded correctly to the E and F inputs. The use of open-collector TTL outputs allows the high-level voltage output to be shifted as necessary to make it compatible with CMOS inputs. It should be noted that the *same* voltage level must be connected to the pull-up resistors on the open-collector outputs as is used to power the CMOS circuit.

LABORATORY 9   BOOLEAN ALGEBRA

Objectives   To observe how a Boolean function can be implemented directly with SSI gates, implemented using only NAND gates, and implemented in a reduced form.

Materials Required $+5$-V power supply

1	DVM or logic indicator
1	74LS00 quad 2-input NAND gate
1	74LS04 hex inverter
1	74LS08 quad 2-input AND gate
1	74LS32 quad 2-input OR gate

Procedure   1. Consider the Boolean expression

$$F = \overline{B} + C(\overline{A} + AB)$$

Construct a truth table with the following headings:

$$\underline{A\ \ B\ \ C\ \ \overline{A}\ \ \overline{B}\ \ (\overline{A} + AB)\ \ C(\overline{A} + AB)\ \ F = \overline{B} + C(\overline{A} + AB)}$$

For the eight unique input combinations of A, B, and C, fill in each column of the table.

2. Design a circuit using 2-input OR gates, 2-input AND gates, and inverters that directly implements the preceding expression for F. The pin-outs of the 74LS04 hex inverter are shown in Figure L4.1, while the pin-outs of the 74LS08 (quad 2-input AND gate) and 74LS32 (quad 2-input OR gate) are shown in Figure L9.1. Input all eight combinations of A, B, and C from your truth table and

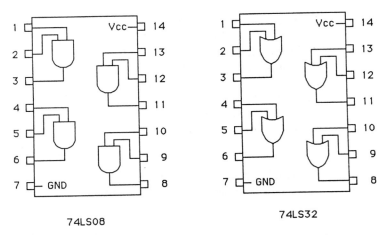

74LS08

74LS32

FIGURE L9.1

record the output of the circuit. How does the output of your circuit compare with what your truth table predicted?

3. Redesign your circuit so that it consists entirely of 2-input NAND gates. Review Section 3.4 if you are unsure of how to carry out the redesign. Now implement your design using a 74LS00 (the pin-outs are shown in Figure L4.1). Again input the eight combinations of *A*, *B*, and *C* from your truth table and record the output of the circuit. Are the results the same as in part 2?

4. Apply the rules of Boolean reduction to the function *F*, reducing it to a minimum form. Implement your result using only 2-input NAND gates. Again input the eight combinations of *A*, *B*, and *C*. from your truth table and record the output of the circuit. Are the results the same as in parts 2 and 3?

Discussion   1. In part 1 your truth table should be the same as that shown in Figure L9.2. If you examine the truth table closely, you will see that *F* is high anytime either *C* is high OR *B* is low. This should tell you that *F* can be reduced to a very simple expression.

A	B	C	$\bar{A}$	$\bar{B}$	$(\bar{A} + AB)$	$C(\bar{A} + AB)$	$F = \bar{B} + C(\bar{A} + AB)$
0	0	0	1	1	1	0	1
0	0	1	1	1	1	1	1
0	1	0	1	0	1	0	0
0	1	1	1	0	1	1	1
1	0	0	0	1	0	0	1
1	0	1	0	1	0	0	1
1	1	0	0	0	1	0	0
1	1	1	0	0	1	1	1

FIGURE L9.2

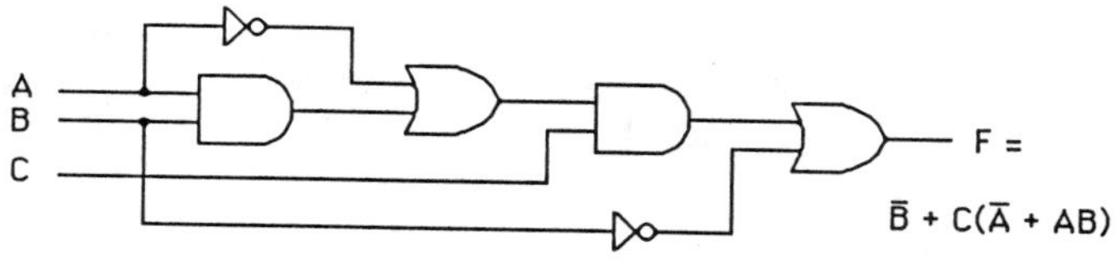

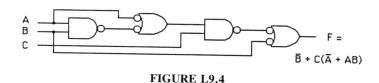

FIGURE L9.3

FIGURE L9.4

2. In part 2 you designed a direct implementation of F using inverters and 2-input AND and OR gates. Your design should be identical to Figure L9.3. This circuit will generate the same output as your truth table (assuming your table matches Figure L9.2). The circuit in Figure L9.3 requires three separate integrated circuits to implement.

3. In part 3 you converted your design to a circuit that consisted only of 2-input NAND gates. You should have obtained the circuit in Figure L9.4. This circuit will also generate the same output as predicted by the truth table in Figure L9.2.

4. In part 4 you used Boolean algebra to reduce F to its minimum form. The reduction should have proceeded as follows:

$$F = \overline{B} + C(\overline{A} + AB)$$
$$= \overline{B} + C(\overline{A} + B)$$
$$= \overline{B} + CB + C\overline{A}$$
$$= \overline{B} + C + C\overline{A}$$
$$= \overline{B} + C$$

This reduction could have been easily deduced from the truth table. A look at Figure L9.2 shows that F is high whenever *either* B is low *or* C is high. This is exactly what the reduction says. Figure L9.5(a) shows the implementation of the reduced form of F. Figure L9.5 (b) uses a NAND gate to replace the inverter. Notice that input A is not needed at all in the reduced implementation of F.

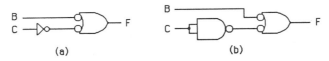

$$F = \bar{B} + C$$

FIGURE L9.5

■ PROJECT 1   A SEAT-BELT ALARM

Minimum Materials Required

+5-V power supply
1   74LS02 quad 2-input NOR gate (for pinouts, see Laboratory 17)
1   74LS04 hex inverter
1   LED
1   330-Ω resistor ($\frac{1}{4}$ W)

The purpose of this project is to design a seat-belt alarm system that uses an LED as an indicator that a seat belt is unfastened. The design specifications involve examining five sensors or inputs and causing the LED to light when various combinations of these inputs assume certain states. The first sensor is a switch that indicates whether a key is in the ignition. This switch will produce a low when a key is *in* the ignition. The driver's seat also has a switch that closes and produces a low when the driver's seat is *occupied*. The driver's seat belt has a switch that produces a low when the driver's seat belt is *fastened*. The passenger's seat also has a switch that produces a low when the seat is *occupied*, and the passenger's seat belt has a switch that produces a low when the belt is *fastened*.

The seat-belt alarm (the LED) should indicate when an alarm condition exists. An alarm condition exists if a key is in the ignition AND either (1) the driver's seat is occupied but the seat belt is unfastened, OR (2) the passenger seat is occupied but the seat belt is unfastened. Your task is to design a digital circuit that lights the LED whenever any of these alarm conditions exist. Viewed as a black box, your circuit has five inputs (to sense the state of the various switches) and one output (to light the LED). Since you will be using LS TTL logic, an open input will act like an open switch (a logic high), while a grounded input will produce a logic low. You can test your design, therefore, by simply grounding or not grounding the various inputs.

Before you test your design, be sure you have your instructor review it for any serious problems that might damage your equipment. Once your design has been approved, implement the design and test its performance.   ■

LABORATORY 10   DECODERS

Objectives   To construct and understand the function of simple decoders composed of SSI gates; to understand the function and use of MSI decoders.

Materials Required   +5-V power supply

1   DVM or logic indicator

1   74LS00 quad 2-input NAND gate

1   74LS04 hex inverter

1   74LS20 dual 4-input NAND gate

1   74LS138 1-of-8 decoder

Procedure   1. Select a 74LS04 and a 74LS20 (the pin-outs are shown in Figure L10.1) and construct the circuit shown in Figure L10.2. Make a table with the five headings *A*, *B*, *C*, *D*, and *F*. Under the headings *A* to *D*, list the 16 possible input combinations from 0000 to 1111. Now, in turn, input each of these combinations into your circuit and record the output *F*. How many combinations are there that produce a low output? Replace the inverter on the input of the NAND gate with a combination that will cause a low output for *F* when the input combination is $ABCD = 0101$. Did your design work properly? In the circuits in this section, the decoder output, *F*, is active low. That is, *F* goes low *only*

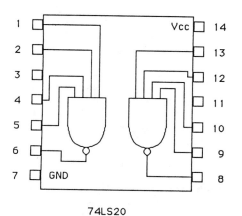

74LS20

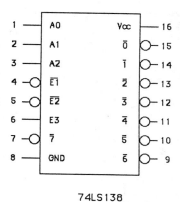

74LS138

FIGURE L10.1

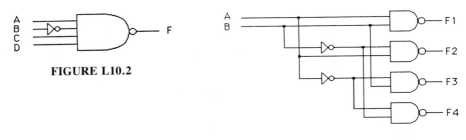

FIGURE L10.2

FIGURE L10.3

when the appropriate code is present on the inputs. How would you change the circuits to make F an active-high output?

2. Construct the circuit shown in Figure L10.3. Make a table with the headings B, A, $F1$, $F2$, $F3$, and $F4$. Under the headings B and A, list the four possible input combinations 00 to 11. Input each of these combinations into your circuit and record the results. What is the connection between the input combination and the state of the outputs $F1$ to $F4$?

3. Select the 74LS138 (pin-outs are shown in Figure L10.1). Construct the circuit shown in Figure L10.4. Make a table with the headings $A2$, $A1$, $A0$, and 0 to 7. Under the headings $A2$ to $A0$, list all the possible input combinations to the system. Apply each of these inputs to the system and record the status of outputs 0 to 7. Is there any obvious connection between the input combinations and the status of the outputs? Now disconnect the wire connecting pin 4 ($E1$) to ground.

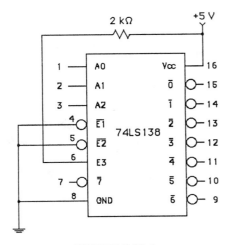

FIGURE L10.4

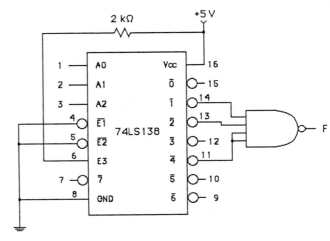

FIGURE L10.5

Check the status of the outputs. Change the input combination A2 to A0. Does this have any effect on the outputs as long as pin 4 is disconnected?

4. Construct the circuit shown in Figure L10.5. Make a table with the headings A2, A1, A0, and F. Under the headings A2 to A0, list all the possible input combinations to the system. Apply each of these inputs and record the status of F in each case. Your circuit has generated a truth table for a system with three inputs and one output. Notice that F will be high anytime *any* of the inputs to the NAND gate is low. By selecting the appropriate outputs from the 74LS138 to connect to the NAND gate, any truth table (with up to four 1's under the F column) can be generated. Change the connections to the NAND gate so that the following truth table is generated.

A2	A1	A0	F
0	0	0	1
0	0	1	0
0	1	0	0
0	1	1	1
1	0	0	1
1	0	1	0
1	1	0	0
1	1	1	1

Discussion   1. In part 1 you observed that the simple decoder circuit responded with a low output only when the input (ABCD) was 1011. The NAND gate itself required all its inputs to be high to produce a low output. Because of the inverter on input B, input B had to be low for the output of the inverter (and thus the input to the NAND gate) to be high. To change the code that produced a low output of F, it was necessary for you to move and/or add inverters on the inputs of the NAND gate. To produce a circuit that responded with a low output to the input combination 0101, you should have constructed the circuit shown in Figure L10.6. Finally, to change the outputs of the decoders in Figures L10.2 and L10.6 from active low to active high, you simply add an inverter to the output of the NAND gate. An alternative approach would replace the NAND gate with a 4-input AND gate.

2. In part 2 you constructed a 1-of-4 decoder. The two input lines B and A could take an four unique states, 00, 01, 10, and 11. Each of these input combinations caused a different output to be enabled (active low). The principle used in Figure L10.3 can easily be expanded to more than two inputs by using NAND

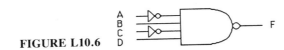

FIGURE L10.6

gates with more than two inputs. Thus it is a straightforward matter to construct a 1-of-8 or a 1-of-16 decoder.

3. In part 3 you investigated the basic properties of the 74LS138 1-of-8 decoder. You found that it worked very much like the 1-of-4 decoder you constructed in part 2, except now there were three inputs with eight possible combinations. Each combination caused one of the outputs to become active (low). You also found that if pin 4 was disconnected from ground (thus allowing enable E1 to float high) *all* the outputs became inactive (high). The 74LS138 has three enable inputs, and *all three* must be active for the chip to be enabled. When the chip is disabled, the code inputs A2 to A0 have no effect.

4. In part 4 you saw how the 74LS138 coupled with a 4-input NAND gate could be used to generate a truth table (that is, implement a Boolean function). In the case of the circuit in Figure L10.5, the truth table was

A2	A1	A0	F
0	0	0	0
0	0	1	1
0	1	0	1
0	1	1	0
1	0	0	1
1	0	1	0
1	1	0	0
1	1	1	0

The basic design principle is that wherever a 1 occurs in the truth table the corresponding output from the 74LS138 should be connected to the NAND gate. You were then asked to design a circuit that would implement the truth table given in procedure part 4. This required changing the connections from the 74LS138 to the 4-input NAND gate. Your design should have been as shown in Figure L10.7. If you wished to implement a truth table with more than four 1's under the F column, this would simply require a NAND gate with more than four inputs. Obviously, an 8-input NAND gate could handle any 3-input truth table. Likewise, additional NAND gates could be used so that the *same* 3-to-8 line decoder could generate any number of Boolean functions (that is, any number of truth tables).

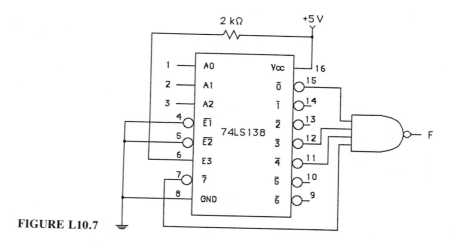

FIGURE L10.7

LABORATORY 11   THE 74LS42 BCD-TO-DECIMAL DECODER

Objectives   To understand the function and uses of a 74LS42 binary coded decimal-to-decimal decoder.

Materials Required   +5-V power supply

1   DVM or logic indicator
1   74LS42 BCD-to-decimal decoder

Procedure   1. Select a 74LS42 (pin-outs are shown in Figure L11.1) for insertion into your breadboard. Construct a truth table with the five headings *D, C, B, A,* and *F.* Under headings *D* to *A* list the 16 possible input combinations from 0000 to 1111. Apply power to your circuit and input, in turn, each of these combinations into the 74LS42. Record which output, if any, is active for each input combination under column *F* in your table. What happens for inputs 1010 to 1111?

2. Imagine that you had an ample supply of 74LS42s, but that you needed to implement a 3-line to 8-line decoder. Design a circuit using a 74LS42 that will accomplish this result.

Discussion   1. The 74LS42 has active-low outputs. When the *DCBA* input is in the range from 0000 to 1001 (0 to 9 decimal), one output goes low, the one numbered the same

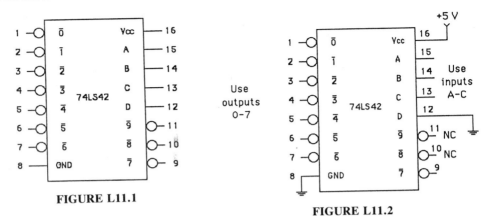

FIGURE L11.1

FIGURE L11.2

as the decimal *value* of the input. For example, if the input is 0111 (decimal 7), then output 7 will go low. For inputs in the range from 1010 to 1111, no output goes low. The 74LS42 only responds to decimal inputs (that is, binary inputs in the range from 0000 to 1001).

2. To use the 74LS42 as a 3-line to 8-line decoder, it is necessary to use only the low three inputs (and thus the low eight outputs). Simply tie input *D* to ground and use the remaining inputs (*A* to *C*) as the three inputs. Use outputs 0 to 7, and ignore outputs 8 and 9 (See Figure L11.2).

LABORATORY 12   THE 74LS47 SEVEN-SEGMENT DECODER DRIVER

Objectives   To investigate the properties of the 74LS47 seven-segment decoder driver and understand its use in controlling seven-segment displays.

Materials Required   +5-V power supply

1	DVM
1	74LS47 seven-segment decoder driver
1	TIL-312 seven-segment display LED
1	74LS193 binary counter
7	identical resistors (150 to 330 Ω, 1/4 W)
1	1-kΩ resistor (1/4 W)
1	debounced logic switch
4	logic indicators

Note: If you do not have a trainer with a debounced logic switch and logic indicators, these will need to be constructed as discussed in the introduction to this appendix.

Procedure

1. You will be investigating the properties of the 74LS47 seven-segment decoder/driver used in conjunction with a TIL-312 seven-segment LED. The pin-outs for these ICs are shown in Figures L12.1 and L12.2. The basic function of the 74LS47 is to take a BCD input (on inputs *DCBA*) and produce the appropriate active-low outputs to drive a common-anode, 7-segment LED so that the BCD number on the 74LS47 inputs is displayed on the LED. Non-BCD inputs produce a variety of nonalphanumeric characters. The 74LS47 outputs are of the open-collector variety, and therefore limiting resistors must be used so that current through a given output transistor does not exceed 24 mA.

2. Begin by constructing the circuit in Figure L12.3. A number of points should be made about this circuit. The area boxed in by the dotted line contains four logic indicators. If you have these on your trainer, simply connect directly to them. Otherwise, they must be fabricated as shown on the figure. Likewise, the signal shown coming from a "debounced switch" assumes you have one on your trainer. If not, you will need to construct one as shown in Figure A.2 at the beginning of this appendix. Finally, the 74LS193 in the figure is a binary counter used in this lab simply as a means of producing a 4-bit binary input to the 74LS47. The detailed operation of the 74LS193 is not of concern at the moment other than to note that, when the debounced switch is toggled, the counter will add 1 to its count (or roll over to 0000 if the count is 1111).

3. Apply power to your circuit. You should find that some binary count is present at the output of the counter (as indicated by the four logic indicators) and that some figure is present on the 7-segment LED. Make a table with column headings *Binary Count* and *Displayed Figure*. Under the binary count column, list the

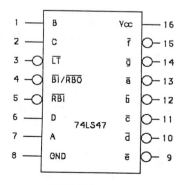

FIGURE L12.1

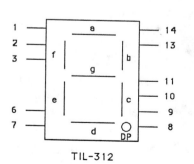

TIL-312

FIGURE L12.2

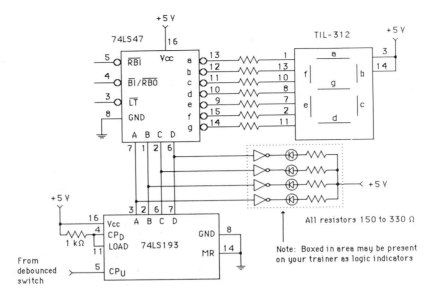

FIGURE L12.3

16 possible inputs 0 to 15 (corresponding to the binary inputs 0000 to 1111). Repeatedly toggle the debounced switch until all 16 inputs have been obtained and make a record of the figure displayed by the 7-segment LED for each input.

4. We will now investigate the $\overline{LT}$ input to the 74LS47. Ground this input. What happens? Try toggling the debounced switch. Does this have any effect? What do you conclude is the function of the $\overline{LT}$ input?

5. Unground the $\overline{LT}$ input, and ground the $\overline{RBI}$ input. Make sure the binary count is not 0000, and measure the voltage at the $\overline{BI/RBO}$ connection and record this value. Continue to toggle the debounced switch. As long as the count is not 0000, what voltage do you measure on the $\overline{BI/RBO}$ pin? Continue to toggle the switch until a count of 0000 is reached. What is displayed on the 7-segment LED? What voltage is present on the $\overline{BI/RBO}$? What do you conclude is the function of the $\overline{RBI}$ input? Can you imagine of what use the $\overline{BI/RBO}$ output might be?

6. Unground the $\overline{RBI}$ input. Now ground the $\overline{BI/RBO}$ pin. What is displayed on the 7-segment LED? Try toggling the debounced switch. Does this have any effect? What do you conclude is the function of the $\overline{BI/RBO}$ pin when used as an *input*?

7. Try this little design problem. Suppose that you had a display comprised of eight 7-segment LEDs. The first three are to be used to deal with integers in the range from 000 to 999. The fourth is to permanently display a decimal point.

The last four are to deal with decimal numbers in the range from .0000 to .9999. It is desired that the display have the following properties.

a. Any leading zeroes on the first *two* displays should be blanked. That is, if the first numeral is zero, it should be blanked, and if the first two numerals are *both* zero, they should both be blanked.

b. Any trailing zeros on the last *three* displays should be blanked. That is, if the last numeral is zero, it should be blanked; if the last two numerals are zero, they should both be blanked; if the last three numerals are zero, they should all be blanked.

Assume that each display is driven by its own 74LS47, and do not be concerned at this time with where each 74LS47 gets its binary input. Your task is to properly connect the $\overline{\text{RBI}}$ and the $\overline{\text{BI}/\text{RBO}}$ pins of the various 74LS47s so that the desired results are achieved.

Note: Since much of the circuit shown in Figure L12.3 is required in the next laboratory, you may wish to delay dismantling this circuit if you plan to move immediately on to Laboratory 13. The same comment applies to the debounced switch and the four logic indicators (if it was necessary for you to construct these).

Discussion

1. In procedure part 3 you investigated what was displayed for each binary input 0000 to 1111. You found that for inputs 0 to 9 the LED simply displayed the appropriate numeral. For inputs 10 to 13, a set of strange U-shaped characters were displayed in various orientations. For input 14 an inverted F was displayed, and for input 15 nothing was displayed. Clearly, the 74LS47 is primarily meant to display decimal numbers.

2. In part 4 you grounded the $\overline{\text{LT}}$ input. This caused all segments of the LED to be illuminated, including the decimal point. The binary inputs to the 74LS47 had no effect. The purpose of the $\overline{\text{LT}}$ input is to allow all LED segments to be tested, thus the name Lamp Test.

3. In part 5 you grounded the $\overline{\text{RBI}}$ input. As long as the binary input was not 0000, you found that the display continued to work as in part 3, and the $\overline{\text{BI}/\text{RBO}}$ pin was high. However, when the binary input reached 0000, the display went blank and the $\overline{\text{BI}/\text{RBO}}$ pin went low. The purpose of the $\overline{\text{RBI}}$ input is to allow leading zeros to be suppressed. The $\overline{\text{RBO}}$ output goes low when a zero is suppressed. This $\overline{\text{RBO}}$ output can be connected to an adjacent $\overline{\text{RBI}}$ input to allow multiple leading zeros to be suppressed.

4. In part 6 you grounded the $\overline{\text{BI}/\text{RBO}}$ pin. This caused the output to go blank regardless of the binary input. When used as an input, the $\overline{\text{BI}}$ input simply suppresses any display.

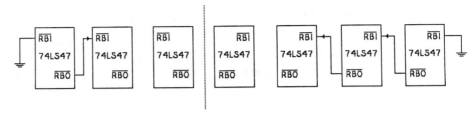

FIGURE L12.4

5. Finally, in part 7 you were asked to complete a design for a seven-digit display according to certain rules. To suppress the first leading zero of the display, simply ground the $\overline{RBI}$ input of the first 74LS47. To allow for two leading zeros to be suppressed, connect the $\overline{RBO}$ output of the first 74LS47 to the $\overline{RBI}$ input of the second 74LS47. To suppress a trailing zero on the last display, simply ground the $\overline{RBI}$ input of the last 74LS47. To allow for multiple trailing zeros to be suppressed, simply connect the $\overline{RBO}$ of the last 74LS47 to the $\overline{RBI}$ input of the next to last, and so on. This scheme is shown in Figure L14.2. Notice that the two 74LS47s immediately on either side of the decimal point (located by the dotted line) are allowed to display whatever their binary inputs are. Thus a complete decimal input of 000.0000 would be displayed as 0.0, and 030.0500 would be displayed as 30.05.

■ PROJECT 2   COPIER ERROR INDICATOR

Minimum Materials Required

+5-V power supply
1   74LS00 quad 2-input NAND gate
1   74LS47 seven-segment decoder/driver
1   TIL-312 seven-segment LED
7   330-Ω resistors ($\frac{1}{4}$ W)

Many newer copying machines use 7-segment displays to indicate various error conditions, rather than having individual trouble lights for each condition. Imagine a copier that might suffer from any one of three errors: (1) out of paper, (2) out of toner, and (3) no auditron present (an auditron is a device that must be inserted into a slot by the user to keep track of the number of copies made). Assume that any of the three error conditions is indicated by a low on a particular output line. Also, to simplify the design, assume that *only one error condition will occur at any given time*. Design a circuit that will produce the following results. (1) If no error is present, the 7-segment LED is blank. (2) If the copier is out of paper, the LED indicates a 3. (3) If the copier is out of toner, the LED indicates a 2. (4) If no auditron is present, the LED indicates a 1.

Before testing your design, have your instructor check it. Once your instructor has approved your design, implement your design and test it. ∎

LABORATORY 13   THE 74LS151 EIGHT-INPUT MULTIPLEXER

Objectives   To understand the basic function of the 74LS151 eight-input multiplexer, and observe some of its uses.

Materials Required   +5-V power supply

1   74LS151 eight-input multiplexer
1   74LS193 binary counter
1   1-kΩ resistor ($\frac{1}{4}$ W)
1   debounced logic switch
4   logic indicators

Procedure   1. In this lab you will investigate the 74LS151 eight-input multiplexer (the pin-outs are shown in Figure L13.1). Construct the circuit shown in Figure L13.2. In this circuit the 74LS193 is used merely to generate a 3-bit binary code for input into the multiplexer.

2. Toggle the debounced switch until the binary count indicated by the logic indicators (CBA) is 000. Observe the state of the Y output (as shown by the logic indicator).

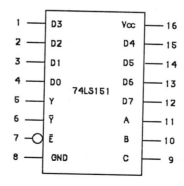

FIGURE L13.1

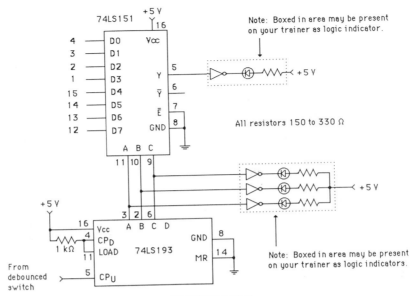

FIGURE L13.2

Now ground input $D0$ (pin 4 on the 74LS151) and observe the Y output. Remove the ground from the $D0$ input and in turn ground each of the other inputs ($D1$ to $D7$). Does this produce any effect? With all grounds removed from the inputs, unground the enable input $\bar{E}$. What happens to the Y output?

3. Reconnect enable $\bar{E}$ to ground. Toggle the debounced switch once (so the binary count is 001). By grounding each input in turn, determine which input affects the Y output when the binary count is 001. Continue to toggle the debounced switch and, following each toggle, determine which input is in control of the output. Do your results conform with what you predicted?

4. Modify your circuit as shown in Figure L13.3. Construct a table with columns C, B, A, and F. Under columns C to A, enter the eight possible binary inputs 000 to 111. Toggle the debounced switch until the count is 000. Note the output of your circuit and enter this in column F of your table. Continue to toggle the debounced switch and enter the circuit output into your table for each binary count 000 to 111. You have generated a truth table for a specific Boolean function. Do you see how any 3-input Boolean function could be generated using a 74LS151 eight-input multiplexer?

Discussion 1. In part 2 you found that, with the binary input to the multiplexer at 000, whatever data were on input $D0$ appeared at output Y (it should be noted that output $\bar{Y}$

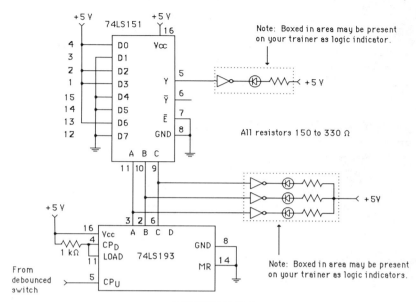

FIGURE L13.3

is always the inverse of output *Y*). With *D*0 open, the input behaved like a logic high, so output *Y* was high. When you grounded input *D*0, output *Y* went low. You also found that, with the binary input at 000, no other input (*D*1 to *D*7) had any effect on *Y*. With input *D*0 ungrounded (that is, high) and the binary input still at 000, you ungrounded the enable input *E*. This caused output *Y* to go low. Nothing you could do on *any* of the inputs *D*0 to *D*7 now had any effect on the output.

2. In part 3 you found that, as the binary input was incremented, the next higher input was placed in control of the output. That is, when the binary input was 001, the output was equal to input *D*1, and so on. Thus the binary code on inputs *C* to *A* determines which input is connected to output *Y*.

3. In part 4 you found that, as you toggled through the eight possible binary inputs, the outputs that were generated were as shown in Figure L13.4. This truth table

C	B	A	F
0	0	0	1
0	0	1	0
0	1	0	1
0	1	1	1
1	0	0	0
1	0	1	0
1	1	0	1
1	1	1	0

FIGURE L13.4

represents a 3-input (or three-variable) Boolean function. Ones in the table represent those inputs to the 74LS151 that are tied to +5 V, while zeros represent those inputs that are tied to ground. Clearly, any three-variable table can be generated simply by choosing the proper combination of inputs tied to +5 V or to ground.

LABORATORY 14   EXOR AND EXNOR GATES

Objectives   To observe the functioning of EXOR and EXNOR gates, and investigate a simple application of EXOR gates.

*Materials
Required*   +5-V power supply

1   DVM or logic indicator

1   74LS86 quad EXOR gate

1   74LS04 hex inverter

1   74LS20 dual 4-input NAND gate

Procedure   1. The pin-out of the quad EXOR gate is shown in Figure L14.1. Select one of the four gates on the IC and experimentally determine the truth table for the gate. That is, apply each of the four possible binary input combinations to the gate and record the resulting output.

2. An exclusive-NOR gate can be fabricated from an EXOR gate by simply inverting the output of the EXOR gate. Construct an EXNOR gate by adding an inverter to the output of one of your EXOR gates (pin-outs for the 74LS04 are shown

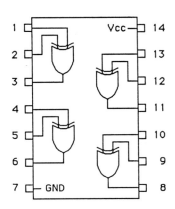

FIGURE L14.1

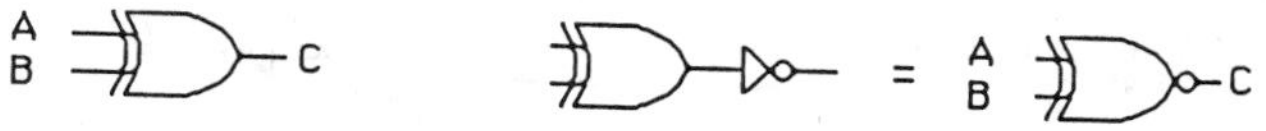

A	B	C
0	0	0
0	1	1
1	0	1
1	1	0

(a)

A	B	C
0	0	1
0	1	0
1	0	0
1	1	1

(b)

FIGURE L14.2

in Figure L4.1) and determine the truth table for this gate (remember, the output from the "gate" is the output from the inverter). Can you see why the EXNOR gate is often referred to as an equality detector?

3. There are four EXOR gates on the 74LS86, six inverters on a 74LS04, and two 4-input NAND gates on a 74LS20 (pin-outs shown in Figure L10.1). Use these three ICs to design a circuit that compares two 4-bit binary numbers and produces a high output only when the two numbers are equal. The numbers should be compared bit by bit (that is, bit 0 of one number is to be compared to bit 0 of the second number, bit 1 to bit 1, and so on). After you have finished your design, construct it and verify that it works properly.

Discussion   1. In part 1 you should have obtained the truth table shown in Figure L14.2(a). Note that the output of the EXOR gate is high only when both inputs are *different*. An EXOR gate could also be thought of as an active-low equality detector since its output only goes low when the two inputs are equal.

2. In part 2 you constructed the circuit shown in Figure L14.2(b) and obtained the truth table also shown in that figure. You found that the output from the EXNOR gate is high only when both inputs are *equal*. For this reason, the EXNOR gate is also called an *equality detector*.

3. In part 3 you designed a 4-bit equality detector. Your design should have looked like the circuit shown in Figure L14.3. If each EXOR plus inverter is visualized as an EXNOR gate, then only when the outputs of all four EXNOR gates are high will the NAND output go low and the output of the final inverter go high. That is, binary number *A* must be identical to binary number *B*, bit by bit, in order for output *F* to be high. Your design should have been verified by inputting various 4-bit numbers into the two sets of inputs and observing that only when the numbers were equal was output *F* high. Notice that the combination of four inverters on the input of the NAND gate plus the inverter on the output of the NAND gate is entirely equivalent to a single 4-input OR gate. Thus the design

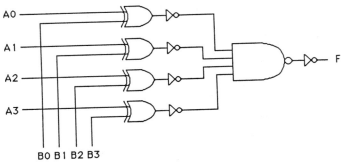

FIGURE L14.3

in Figure L14.3 could be achieved more efficiently by replacing the 74LS04 and 74LS20 by a single 4-input OR gate or by a combination of 2-input OR gates (three of them) from a 74LS32.

LABORATORY 15   PARITY

Objectives   To understand how EXOR gates can be used to detect the parity of a digital input, and to generate a parity bit for use in data transmission.

Materials Required   +5-V power supply

1   DVM or logic indicator

1   74LS86 quad EXOR gate

Procedure   1. Construct the circuit shown in Figure L15.1 (the pin-outs of the 74LS86 are shown in Figure L14.1). Construct a table with the headings $A3$, $A2$, $A1$, $A0$, and F. Under the headings $A3$ to $A0$ record the 16 possible binary input combinations 0000 to 1111. Input each of these combinations in turn to the circuit and record the state of F. From your table, can you deduce what simple set of circumstances is required for F to be high? Refer to Discussion item 1 at this time before proceeding to the next section of the Procedure.

2. Using two 74LS86s and a 74LS04 (hex inverter), design an 8-bit even-parity detector (that is, a circuit whose output is high when the parity of the 8-bit input is even).

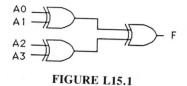

FIGURE L15.1

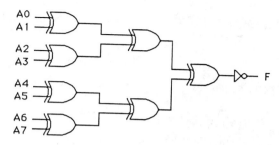

FIGURE L15.2

Discussion 1. The circuit that you constructed in procedure part 1 was a 4-bit odd-parity detector. Recall that the parity of a binary number refers to whether the total number of 1's in the number is even or odd. If the number of 1's is even, the number is said to have even parity; if the number of 1's is odd, the number has odd parity. If you examine your truth table, you will find that *F* is high *only* when the number of 1's contained in the input is odd. Your circuit could be changed to an even-parity detector simply by adding an inverter on the output. Detecting parity is also the first step toward *generating* a parity bit to be used in data transmission. When data are transmitted in serial fashion, each bit of the data "word" is transmitted individually, one at a time. At the receiving end it is possible that one of the transmitted bits could be received incorrectly (a 1 received as a 0, and vice versa). To help recognize this situation, the parity of the transmitted word is often also transmitted as a final bit of data. As a practical example, suppose an 8-bit data word is to be transmitted. Then a ninth bit is added to the transmission so that the parity of *all 9 bits* is always the same (either even or odd, as agreed upon in advance). For example, if 8 bits are to be transmitted and a parity bit is to be added to give the whole 9 bits odd parity, then first an 8-bit even-parity detector examines the 8 data bits. The output of this detector is then added to the transmission as the ninth (parity) bit.

2. In part 2 you designed an 8-bit even-parity detector. Your design should be as shown in Figure L15.2. One way to think of the circuit in Figure L15.2 is to view the final EXOR gate and inverter as an equality detector (an EXNOR gate). Thus the output of the circuit will only go high when the two inputs to the equality detector are identical. Now the equality detector gets its two inputs from two 4-bit odd-parity detectors. For the final equality detector to have a high output, *both 4-bit odd-parity detectors must produce the same output.* That is, the upper 4 bits and the lower 4 bits *must have the same parity.* But if this is true, then the 8 bits taken as a whole must have even parity. The entire circuit could be changed to an 8-bit odd-parity detector simply by removing the inverter.

LABORATORY 16 HALF-ADDERS AND FULL ADDERS

Objectives   To understand the construction and function of half-adders and full adders.

*Materials
Required*   +5-V power supply

  3 logic indicators

  1 74LS86 quad EXOR gate

  1 74LS00 quad NAND gate

  1 74LS04 hex inverter

Procedure   1. Construct the circuit shown in Figure L16.1(a). Make a table with the four headings A, B, C, and S. Under the headings A and B, record the four possible input combinations 00 to 11. Input each of these combinations into the circuit and record both outputs C and S for each case. You should find that your table is identical to that shown in Figure 4.28(a). You have constructed a half-adder circuit. Make sure you understand the difference between the C output and the S output. Proceed to the next part of the procedure, *but do not dismantle your half-adder.*

  2. Construct the circuit shown in Figure L16.2(a) *without using any of the gates from your half-adder circuit.* That is, construct the circuit in Figure L16.2 (a) entirely separate from your half-adder. You will use both circuits together later. Notice that one of the NAND gates in Figure L16.2 (a) is shown as a low-level input OR gate. Construct a table with five headings, A, B, C_{in}, S, and C_{out}. Under the three headings A, B, and C_{in}, record the eight possible input combinations 000 to 111. Input each of these combinations into the circuit and record both outputs S and C_{out} for each combination. You should find that your table is identical to that shown in Figure 4.29(a) of the text. You have constructed a full-adder circuit. Make sure you understand the significance of each output

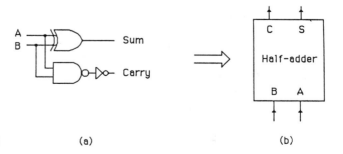

FIGURE L16.1   (a)   (b)

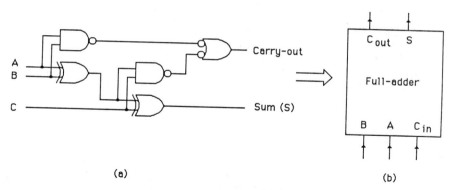

(a)

(b)

FIGURE L16.2

and each input. Before proceeding to the next step, read Discussion parts 1 and 2. *Do not dismantle your full-adder circuit.*

3. Construct the adder circuit shown in Figure L16.3. This circuit is designed to add two 2-bit numbers (*A* and *B*) and produce a 3-bit sum. The circuit can be constructed by simply tying together your half-adder and full-adder circuits. Copy the table shown in Figure L16.4. This table contains double entries for the two 2-bit numbers to be added. One entry shows the individual bits (in binary form) of the numbers. The second entry shows the decimal value of each 2-bit number. Since we know the circuit is designed to simply add the two numbers being input, the sum is easy to predict, especially in decimal form. This predicted sum is also listed under column *S* in the table. Once you have constructed the circuit, input each combination listed in the table and record the output bits *S*2, *S*1, and *S*0. Does the binary sum conform to the predicted value in column *S* for each input combination?

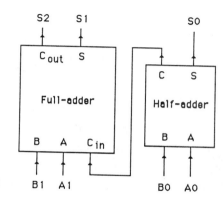

FIGURE L16.3

A1	A0	A	B1	B0	B	S2	S1	S0	S
0	0	0	0	0	0				0
0	1	1	0	0	0				1
1	0	2	0	0	0				2
1	1	3	0	0	0				3
0	0	0	0	1	1				1
0	1	1	0	1	1				2
1	0	2	0	1	1				3
1	1	3	0	1	1				4
0	0	0	1	0	2				2
0	1	1	1	0	2				3
1	0	2	1	0	2				4
1	1	3	1	0	2				5
0	0	0	1	1	3				3
0	1	1	1	1	3				4
1	0	2	1	1	3				5
1	1	3	1	1	3				6

FIGURE L16.4

Discussion

1. In part 1 you constructed a half-adder and verified its operation. Once designed and constructed, the half-adder can be considered a black box [such as shown in Figure L16.1(b)] whose properties are described by your truth table. It then becomes easier to simply use the black box figure in future designs involving half-adders, at least in the preliminary design phase.

2. In part 2 you constructed a full-adder and verified its operation. Like the half-adder, once the full-adder has been designed, it can be considered a black box with properties as defined by its truth table [see Figure L16.2(b)]. In designing larger addition circuits, the black box form is easier to manipulate.

3. In part 3 you constructed a circuit to add two 2-bit numbers and produce a 3-bit result. Notice that the design simply showed the necessary half-adder and full adder as black boxes. Since you already had constructed these circuits, it was only necessary to tie them together in the manner shown in Figure L16.4. It was really of no concern at this point how the half-adder and full adder functioned internally.

■ PROJECT 3   A DEMAND INDICATOR

Minimum Materials Required

+ 5-V power supply

2	74LS00 quad 2-input NAND gates
1	74LS04 hex inverter
1	74LS47 seven-segment decoder/driver
1	74LS83A 4-bit adder
1	74LS86 quad EXOR gate

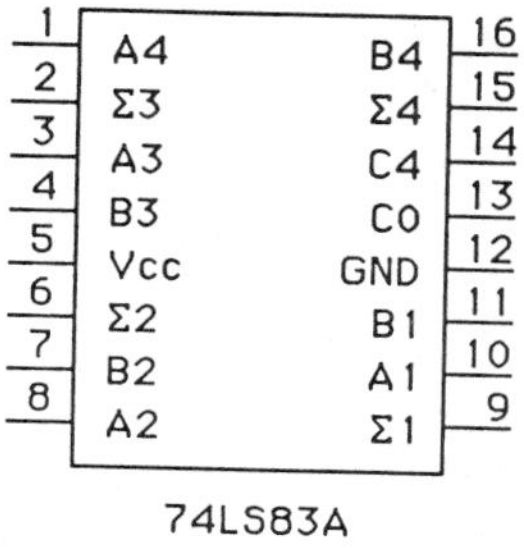

```
  1 ┌──────────┐ 16
────┤ A4    B4 ├────
  2 │          │ 15
────┤ Σ3    Σ4 ├────
  3 │          │ 14
────┤ A3    C4 ├────
  4 │          │ 13
────┤ B3    C0 ├────
  5 │          │ 12
────┤ Vcc  GND ├────
  6 │          │ 11
────┤ Σ2    B1 ├────
  7 │          │ 10
────┤ B2    A1 ├────
  8 │          │ 9
────┤ A2    Σ1 ├────
    └──────────┘
```

FIGURE P3.1   74LS83A

1   TIL-312 seven-segment LED

7   330-Ω resistors

Imagine that there are six locations within an establishment where people might walk up to a service window and request some type of service. Suppose that in the floor at each service window is a sensor that senses the presence of a person and signals this presence with a logical low. Now imagine that somewhere within the establishment there is a manager's office, and that in this office there is a single 7-segment LED that is supposed to indicate the number of windows where a person is requesting service. You are to design the circuitry that will indicate the number of windows where a person is currently requesting service.

Viewed as a black box, your circuit has six inputs and drives a single 7-segment LED as an output. The circuit is supposed to count the number of inputs that are low and indicate this number on the LED (including zero, when appropriate). To help you with your design, consider using the 74LS83A, which is a MSI circuit that adds two 4-bit binary numbers and produces a 4-bit sum plus a carry. The pin-outs for this circuit are shown in Figure P3.1. The 74LS83A is very simple to use. To add two binary numbers (up to 4 bits each), simply tie the C0 line (carry-in) low. Input the two 4-bit numbers on the inputs $A4$ to $A1$ and $B4$ to $B1$, respectively. Read the 4-bit sum on outputs $\Sigma 4$ to $\Sigma 1$ and any carry-out on $C4$. You might consider dividing the six inputs to your circuit into two groups of three. Recall that a full adder can add three single-bit numbers to produce a 2-bit output. The output of two full adders could then be fed to the 74LS38A adder.

Have your instructor check your circuit before you build and test it. ■

LABORATORY 17   R—S FLIP-FLOPS

Objectives   To construct and understand the function of $R—S$ flip-flops constructed from NAND gates and NOR gates.

Materials
Required   +5-V power supply
    2   logic indicators
    1   74LS00 quad 2-input NAND gate
    1   74LS02 quad 2-input NOR gate

Procedure 1. Construct the circuit shown in Figure L17.1 (do not construct the logic indicators if they are available on your trainer). The pin-outs of the 74LS00 quad 2-input NAND gate are shown in Figure L4.1. Construct a table with the headings S, R, Q, and $\overline{Q}$. Under the S and R headings, record the four possible input combinations 00 to 11. Input each of these combinations and record the state of Q and $\overline{Q}$. The input state 1,1 requires some care. Set the inputs at 0,1 for S, R, respectively. Change S to 1 and observe the outputs. Now set the inputs at 1, 0. Change R to 1 and observe the outputs. Is there a *unique* output state associated with the input state 1, 1? Under normal logic situations, the Q and $\overline{Q}$ outputs of an R–S flip-flop should always be in opposite states. If this is to be true, which input state must be avoided? If we call the flip-flop *set* when the output state is 1, 0 for Q, $\overline{Q}$, respectively, which input controls the setting of the FF, and what logic level must be used on this input to set the FF? What about resetting the FF (producing the output state 0, 1)?

     2. In this section you will work with the 74LS02 quad 2-input NOR gate. The pin-outs for this IC are shown in Figure L17.2. Construct the circuit shown in Figure L17.3. Notice that, compared to Figure L17.1, the labeling of the inputs has been reversed with respect to the outputs. As in part 1, determine the truth table for this circuit. This time pay particular attention to the input combination 0, 0. That is, set the inputs at 0, 1 for R, S, respectively, and then ground S.

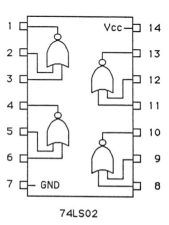

74LS02

FIGURE L17.2

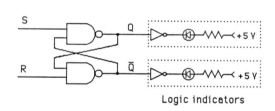

Logic indicators

FIGURE L17.1

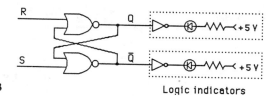

FIGURE L17.3   Logic indicators

Observe the state of the outputs. Now set the inputs at 1, 0 and then ground *R*. Now what are the outputs? Does the input state 0, 0 produce a unique output state? Like the FF from part 1, under normal logic conditions the *Q* and $\overline{Q}$ outputs should always be in opposite states. If this is to be true, which input combination must be avoided? If we call the FF *set* when the output is 1, 0 for *Q*, $\overline{Q}$, respectively, which input controls the setting of the FF, and what logic level must be used on this input to set the FF? What about resetting the FF?

Discussion   1. In part 1 you constructed an *R–S* FF using two 2-input NAND gates. This type of construction produces an *R–S* FF with active-low inputs. That is, a low on the *S* input will set the FF, and likewise a low on the *R* input will reset the FF. For the simple circuit that you constructed, an ambiguous logic state occurs if both inputs become active (low) at the same time. In this case, both outputs go high. This is *electrically* well defined, but logically ambiguous since the FF is, in a sense, both set and reset at the same time. You also found that the input state 1, 1 does not, *by itself,* produce a unique output. Instead, the input state 1, 1 leaves the state of the FF unchanged from whatever the state was prior to the time the inputs became 1, 1. You observed this when you changed the inputs from 1, 0 to 1, 1 and then from 0, 1 to 1, 1.

2. In Part 2 you constructed an *R–S* FF using two 2-input NOR gates. This type of construction produces an *R–S* FF with active-high inputs. That is, a high on the *S* input will set the FF, and a high on the *R* input will reset the FF. For your circuit, an ambiguous state occurs if both inputs become active (high) at the same time. In this case, both outputs go low, which is logically ambiguous. You also found that the state 0, 0 does not, by itself, produce a unique output, but rather leaves the state of the FF unchanged from its previous state.

3. In comparing the flip-flops constructed from NAND and NOR gates, you can see that they behave in a very similar fashion, but that there are significant differences. The NAND gate FF has active-low inputs, while the NOR gate FF has active-high inputs. The ambiguous and the no-change input conditions are exactly reversed. And, finally, the *S* and *R* inputs are physically connected to opposite gates with respect to the *Q* and $\overline{Q}$ outputs. That is, in the NAND gate FF the *S* input is connected to the same NAND gate that produces the *Q* output, while in the NOR gate FF the *S* input is connected to the NAND gate that produces the $\overline{Q}$ output.

LABORATORY 18   *D*-TYPE FLIP-FLOPS AND DATA LATCHES

Objectives   To become familiar with the *D*-type flip-flop and with its use as a data latch.

Materials
Required   + 5-V power supply

1   74LS00 quad 2-input NAND gate
1   74LS74A dual positive edge-triggered *D* flip-flops
1   74LS75 quad latch
1   debounced switch
4   logic indicators

Procedure   1. Construct the circuit shown in Figure L18.1 (the pin-outs of the 74LS00 are shown in Figure L4.1). Notice that this circuit contains an *R*–*S* FF as its output stage. Two of your logic indicators will be monitoring the *R* and *S* inputs to this FF. The *D* input to the circuit is called the data input, and the *T* input is called the clock, enable, or gate input. Begin with the *T* input grounded, and turn on the circuit. Observe the state of *R* and *S* and the state of the outputs. Try changing the state of the *D* input back and forth between 1 and 0 (+5 V and ground). Does this have any effect on the state of the outputs? Now take the *T* input high. Change the state of the *D* input back and forth between 1 and 0. What happens to the outputs? What happens to the points labeled *R* and *S*? With *D* set high, return *T* to 0 (ground *T*). What is the state of the outputs?

2. Select one of the flip-flops on the 74LS74A (pin-outs shown in Figure L18.2) and construct the circuit shown in Figure L18.3. In this circuit the CL and PR inputs are left open (unconnected) for the moment. In this state they behave as if they were at logic 1 (high). Since, by the small circles on the figure, the

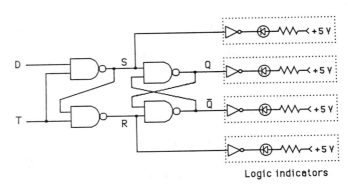

FIGURE L18.1

Logic indicators

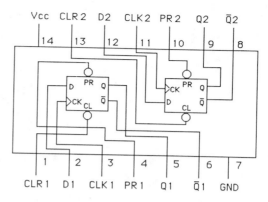

FIGURE L18.2

74LS74A

CL and PR inputs are obviously active low, when these inputs are left open they are in the inactive state. A debounced switch (normally high) should be attached to the CK (clock) input. Turn on your circuit and observe the state of the outputs. *Alternately* ground the CL input and the PR input a few times and observe what effect this has on the outputs. What do you conclude that the CL and PR inputs do? With the CL and PR inputs left open for the remainder of this section, try changing the data at the *D* input back and forth between 0 and 1. Does this have any effect on the outputs? (*Note:* The CK input should be high at this time.) Now use the debounced switch to make the CK input low and *hold it in this state* while you toggle the data back and forth between 0 and 1 a few times. Do the outputs change? With the data input *in the opposite state as the Q output,* release the debounced switch on the CK input. What happens to the outputs (in particular, what state does *Q* take on)? Without changing the *D* input, toggle the debounced switch a few times. What happens? Now again make the state of the data input opposite to that of the *Q* output. Depress the debounced switch and keep it depressed (so that the state of the CK input changes from high to low, and remains low). What happens to the *Q* output?

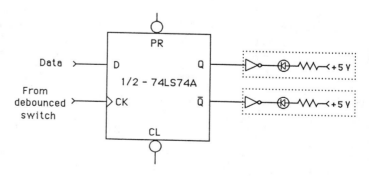

FIGURE L18.3

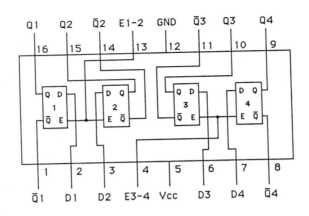

FIGURE L18.4   74LS75

Now release the debounced switch (so that the CK input changes from low to high). What happens to the Q output? From your observations, what are the conditions that allow Q to change?

3. In this section you will investigate the operation of the 74LS75 quad latch. The pin-outs for this IC are shown in Figure L18.4. Using these pin-outs, design a circuit in the following way. Outputs Q1 to Q4 should be attached to logic indicators. Inputs E1-2 and E3-4 should be grounded. Inputs D1 to D4 should be connected to some random assortment of 1's and 0's. The $\overline{Q}$ outputs will not be used and so may be left unconnected. Connect power lines to your circuit and turn on the power. Observe the state of the four outputs and compare this with the state of the four inputs. Are they the same? Try changing the inputs and observe what happens to the outputs. Now remove the grounds from the E1-2 and E3-4 inputs. How are the inputs and outputs related now? Change the inputs and observe the outputs. What happens? Set the inputs at some desired configuration (your choice) and observe the outputs. Now ground the E1-2 and E3-4 inputs once again. Are the outputs preserved?

Discussion   1. In part 1 you constructed a simple clocked D-type flip-flop. This circuit contains a simple R–S FF as its output stage, but it has some additional circuitry on the input end. The first two NAND gates hold the R and S inputs high *as long as the T input is low*. Thus, with the T input low, the R–S FF is held in its no-change state, and the D input has no effect. When the T input is taken high, then the NAND gate with the D input essentially functions as an inverter, and the lower NAND gate also inverts the output of the D NAND gate. This guarantees that the R and S inputs will be in opposite states. If D is high, then S is low and the FF will be set. If D is low, then R will be low and the FF is reset. In either case, Q simply follows D as long as the T input remains high.

When T is returned low, the data are latched (Q is frozen in whatever state D was in just prior to T going low).

2. In part 2 you investigated a more sophisticated D-type flip-flop. Each FF on the 74LS74A has a *preset* (PR) and *clear* (CL) input that can be used to set (PR) or reset (CL) the FF regardless of the state of D or of the clock (CK). You observed this behavior when you momentarily grounded the PR and the CL inputs. Assuming that the PR and CL inputs are held high, then the circuit is controlled by the D and CK inputs. However, you found that data could only be changed *on the rising edge of the CK input.* That is, the state of D had no effect as long as the CK input was held fixed (either high or low). The Q output was found to take on the state of the D input *at the instant the CK input changed from low to high.* The flip-flop is said to be positive-edge-triggered. The allows data to be captured at a precise instant of time. At no other time will the state of the D input have any effect.

3. In part 3 you investigated a simple quad latch. This IC essentially contains four clocked D-type flip-flops of the kind you constructed in part 1. When the E inputs are high, the Q outputs follow the D inputs. When the E inputs are taken low, the outputs remain as they were at the time of the high to transition on the E inputs. This type of circuit is often used with microcomputers as a simple output latch. The data of interest are placed on the D inputs by the microcomputer data bus. A short positive pulse is sent (by an appropriate address on the address bus) to the E inputs. The data are thus captured by the latch. Because of the detailed timing of signals within the microcomputer, the short positive "latching pulse" may actually begin before the data to be output are stable on the data bus, but the pulse ends *while the data are stable on the data bus.* The latch then continues to present its data to the outside world indefinitely (or until new data are sent to the latch).

LABORATORY 19   J−K FLIP-FLOPS

Objectives   To observe and become familiar with the properties of the typical $J–K$ flip-flop.

Materials Required   +5-V power supply

1	74LS76A dual negative-edge-triggered $J–K$ flip-flop with preset, clear
2	logic indicators
1	debounced switch

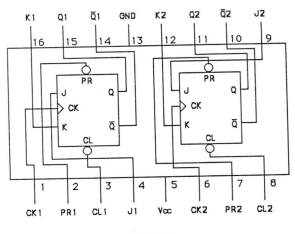

FIGURE L19.1   74LS76A

Procedure 1. You will be working with the 74LS76A dual *J*–*K* flip-flop. The pin-outs for this IC are shown in Figure L19.1. Select one of the two *J*–*K* flip-flops to work with and construct the circuit shown in Figure L19.2 (the logic indicators are shown in the dotted boxes). Begin by leaving the *J*, *K*, PR, and CL inputs unconnected as shown. In this configuration, these inputs act as a TTL logic high. Apply power to your circuit and observe the outputs. If the FF is set (*Q* high), then apply a momentary ground to the CL input and observe the result. Then apply a momentary ground to the PR input and observe the result. Repeat this procedure a few times to be sure you understand what is happening. If the FF is originally reset when you apply power (*Q* low), then initially apply a low to the PR input and then follow the procedure.

2. In the remaining parts of the procedure, leave the PR and CL inputs unconnected. At this point the *J* and *K* inputs should be unconnected (behave as logic high inputs). Toggle the debounced switch several times *slowly* (that is, hold the

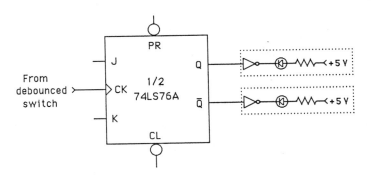

FIGURE L19.2

switch in each position for a short period of time) and observe what happens to the outputs and *exactly when it happens*. What do you conclude is the effect of a clock pulse when J and K are both high?

3. Leave the J input unconnected (high) and ground the K input. Use the CL input to reset the FF. Now toggle the CK input once and observe the result. Toggle it a few more times. What happens? What do you conclude is the effect of a clock pulse when the J, K inputs are 1, 0?

4. Ground the J input and leave the K input unconnected (high). Use the PR input to set the FF. Toggle the CK input once and observe the result. Toggle the CK input a few more times. What happens? What do you conclude is the effect of a clock pulse when the J, K inputs are 0, 1?

5. Ground both the J and K inputs. Use the PR input to set the FF. Toggle the CK input several times. What happens? Use the CL input to reset the FF. Again toggle the CK input several times. What happens? What do you conclude is the effect of a clock pulse when the J, K inputs are 0, 0?

6. Summarize the results of your observations by constructing a truth table for the effect of the four possible states of the J, K inputs.

Discussion   1. In part 1 you found that the PR and CL inputs were asynchronous preset (set) and clear (reset) inputs, respectively. That is, the FF can be set *at any time* by taking the PR input low and can be reset at any time by taking the CL input low. If you take *both* PR and CL low at the same time, both Q and $\overline{Q}$ outputs go high, a logically ambiguous state.

2. In part 2 you found that when the J, K inputs are 1, 1 the FF will *reverse its output state* each time the CK input goes from *high to low* (on the negative edge of the CK pulse). The 74LS76A is a negative-edge-triggered device.

3. In part 3 you found that when the J, K inputs are 1, 0 the FF will be set on a clock pulse. If the FF is already set, it will remain so.

4. In part 4 you found that when the J, K inputs are 0, 1 the FF will be reset on a clock pulse. If the FF is already reset, it will remain so.

5. In part 5 you found that when the J, K inputs are 0, 0 a clock pulse has no effect on the FF. That is, the FF remains in whatever state it was in prior to the clock pulse.

6. In part 6 you should have constructed a truth table similar to Figure 5.8.

LABORATORY 20 BINARY RIPPLE COUNTERS

Objectives To understand the construction and function of simple binary up and down ripple counters.

Materials Required
+5-V power supply
2 74LS76A dual $J-K$ flip-flops
4 logic indicators
1 debounced switch

Procedure
1. Construct the circuit shown in Figure L20.1. You should probably begin by redrawing this circuit in your lab book and then carefully assigning pin numbers to the various connections that need to be made (the pin-outs for the 74LS76A dual $J-K$ FF are shown in Figure L19.1). Notice that the J, K inputs are left unconnected and are thus high. Apply power to your circuit and observe the resulting output. It will probably be easier to write down the state of the outputs in the order D, C, B, A since this is the natural binary sequence.

2. Temporarily ground each CL input so that the circuit output becomes 0000 (that is, all outputs are low). Now begin to toggle the debounced switch (the leftmost CK input) and observe the state of the outputs following each toggle of the CK input. Note the results in a table. Do you observe a familiar sequence? Does

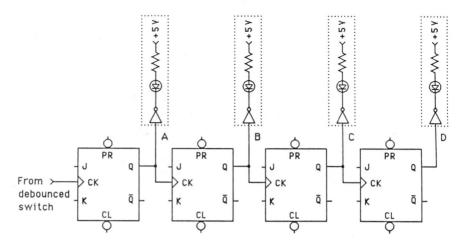

FIGURE L20.1

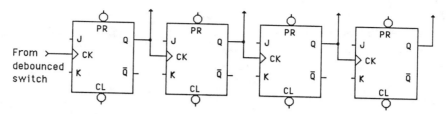

FIGURE L20.2

the sequence repeat? What do you conclude is the function of the circuit?

3. Rewire the connections of the circuit to conform to Figure L20.2. In this circuit the $\overline{Q}$ outputs are being used as the CK input for the next FF. However, the Q outputs are still considered the circuit output. Repeat part 2 of the procedure for the new set of connections. Now what function does the circuit implement?

Discussion   1. In part 2 you observed that the circuit of Figure L20.1 is a simple binary up-counter. If the circuit output is considered as binary number *DCBA*, then each clock pulse at the CK input causes the count to advance by 1. When the count 1111 is reached, the next pulse roles the count over to 0000. This type of counter is called a ripple counter because each FF cannot change its state until the previous FF changes state. Thus the count ripples through the counter, although this happens so fast that you cannot see it. However, a very fast circuit tied to the *DCBA* outputs would actually discern some invalid intermediate states that occur between the time the earlier flip-flops change but before the later ones have changed.

2. In part 3 you found that the circuit of Figure L20.2 is a binary down-counter. That is, each clock pulse causes the binary count to decrease by 1. When the count 0000 is reached, the next clock pulse causes the output to role over to 1111. Like the circuit of Figure L20.1, the down-counter circuit is a ripple counter, and the same comments apply.

LABORATORY 21   THE 74LS90 BCD COUNTER

Objectives   To investigate the operation and use of the 74LS90 BCD counter.

Materials
Required   +5-V power supply
    1   74LS90 BCD counter

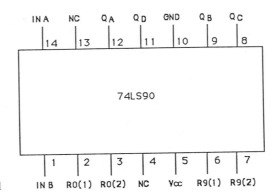

FIGURE L21.1

4   logic indicators

1   debounced switch

Procedure   1. In this laboratory you will investigate the 74LS90 BCD counter (pin-outs shown in Figure L21.1). Begin by constructing the circuit shown in Figure L21.2 (don't forget the *Vcc* and GND connections not explicitly shown in the figure). Notice that one of the "reset to 0" inputs, *R*0(1), is grounded, and likewise one of the "reset to 9" inputs, *R*9(1), is also grounded. The table at the right of Figure L21.2 shows that in this configuration neither reset input is active. Apply power to the circuit and observe the state of the outputs. Alternately disconnect the ground first to *R*0(1) and then to *R*9(1) a few times. Observe the outputs. Do the outputs behave according to the table in Figure L21.2?

2. Reset the count to 0000. Now toggle the debounced switch a number of times and observe the outputs. Reading the outputs as *DCBA*, what count sequence do you observe? What happens when the count is at 1001 and another pulse is applied to the *A* input? How many count states are there?

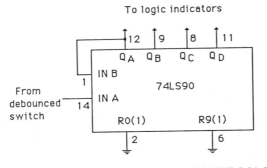

Reset to 0	Reset to 9
Both R0s must be high AND at least one R9 low	Both R9s must be high

FIGURE L21.2

3. Break the connection between pin 1 (IN B) and pin 12 (Q_A). Now toggle the debounced switch several times. Which outputs change?

4. Connect the debounced switch to IN B (pin 1) rather than IN A (pin 14); make sure there is still no connection from pin 1 to pin 12. Toggle the switch several times and observe the count on outputs Q_D, Q_C, Q_B as a 3-bit binary number DCB. What count sequence do you observe? Input B is therefore connected to what type of counter?

Discussion   1. In part 1 you saw that it was possible to reset the count to zero by simply disconnecting the $R0(1)$ input. When you did this, *both R0* inputs became high, and one $R9$ input was low. According to the table at the right of Figure L21.2, this is the condition necessary to cause a reset to zero. You also saw that it was possible to reset the count to 9(1001) by disconnecting the $R9(1)$ input, with the $R0(1)$ input again grounded. When you did this, both $R9$ inputs became high, which, according to the table in Figure L21.2, is sufficient to cause a reset to 9.

2. In part 2 you observed that with output $A(Q_A)$ connected to input B, the circuit functioned as a BCD counter. That is, the count proceeded from 0000 to 1001 before rolling over to 0000.

3. In part 3 you saw that with the connection between Q_A and input B broken, only Q_A toggled back and forth as you sent pulses into input A. That is, your input pulses were only reaching a 1-bit counter. Outputs Q_D, Q_C, and Q_B remained unchanged.

4. When you sent pulses into input B, you found that outputs Q_D, Q_C, and Q_B formed a 3-bit modulus-5 counter. The count sequence proceeded from 000 to 100 and then rolled over to 000 again. Thus the 74LS90 contains a modulus-2 counter and a modulus-5 counter, which can be connected together to form a modulus-10 counter. Several 74LS90s can be connected together by connecting the Q_D output of one to the A input of the next. In this manner, counters of any power of 10 can be created easily.

5. It should also be noted that it is perfectly possible to send the original input pulses into input B and connect the Q_D output to input A. In this manner, a square wave of frequency X on input B becomes a square wave of frequency $X/10$ at output Q_A. The configuration in Figure L21.2 would also take an incoming square wave at input A and produce a division by 10 at output Q_D, but the output would not have a 50% duty cycle. It is often useful to preserve the 50% duty cycle during the division process. If you have a square-wave generator handy with a frequency of about 1 kHz and an oscilloscope available, you may wish to directly observe the difference in the output signal produced by the two alternate ways of connecting the 74LS90 mentioned here.

LABORATORY 22   THE 74LS193 BINARY COUNTER

Objectives   To investigate the operating modes of the 74LS193 binary counter.

Materials
Required   +5-V power supply

1	74LS193 4-bit binary counter
1	74LS00 quad 2-input NAND gate
4	logic indicators
2	debounced switches

Procedure   1. The detailed logic diagram and pin-outs for the 74LS193 are shown in Figure 6.8. You will need to refer to this diagram as you undertake your laboratory activities. First, you will investigate the *load* and *clear* features of the IC. Begin by constructing the circuit shown in Figure L22.1. Set the debounced switch connected to the LOAD input in the high output state. Set the debounced switch connected to the CLEAR input in the low output state. In this way neither the LOAD nor CLEAR inputs will be active unless you toggle the appropriate switch. Turn on the power to your circuit. You will probably find that the outputs indicate a count that is neither 0000 nor 1111 (although it is possible that one of these counts might occur). Toggle the switch connected to the CLEAR input (that is, toggle the switch low and then high). What happens to the outputs?

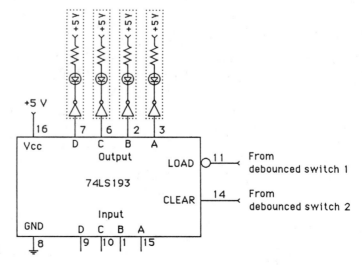

FIGURE L22.1

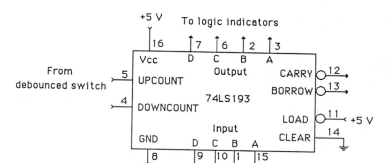

FIGURE L22.2

Now toggle the switch connected to the LOAD input. What happens to the outputs? Ground the *C* and *A inputs*. Now toggle the LOAD input. What happens to the outputs?

2. Rewire your circuit to conform to that shown in Figure L22.2. Set the debounced switch connected to the UPCOUNT input to the logic high output state. Apply power to your circuit and note the state of the outputs. Now toggle your debounced switch from high to low. Does anything happen at the outputs? Now toggle the switch from low to high. Does anything happen now at the outputs? Continue to toggle the switch several times from high to low and then back to high. How do the outputs change, and *when* do they change? Toggle the switch until your outputs reach 1110. Now disconnect the logic indicator from output *D* and connect the indicator to the CARRY output. Record the state of the CARRY output and the *DCBA* outputs for each setting of the debounced switch in the following sequence. Starting with the switch high (as it should now be if you have just reached a count of 1110), toggle the switch to (1) low, (2) high, (3) low, (4) high. It is particularly important to note under exactly which condition the CARRY output is low.

3. Restore the logic indicator from the CARRY output to the *D* output, and move the output of the debounced switch from the UPCOUNT input to the DOWNCOUNT input. Toggle the switch several times and note how the outputs change and exactly when they change. Now toggle the switch until the outputs are 0001. Remove the logic indicator from output *D* and connect it to the BORROW output. Again starting with the switch output high, record the state of the BORROW output and *DCBA* outputs as you toggle the switch (1) low, (2) high, (3) low, (4) high. As before, pay very close attention to exactly which condition is necessary to make the BORROW output go low.

4. Modify your circuit so that it conforms to that shown in Figure L22.3. Apply power to your circuit and momentarily ground the LOAD input (that is, take a grounded wire and tap it against the LOAD input). Do the outputs conform to the state of the inputs? Now toggle the debounced switch attached to the

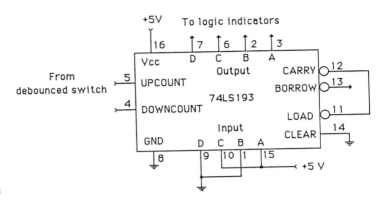

FIGURE L22.3

UPCOUNT input until the count reaches 1110. Notice that, as you toggled the switch, each count output remained unchanged for a full clock cycle (high and low on UPCOUNT). Remove the logic indicator from the D output and connect it to the CARRY output. With the count at 1110, record the state of the CARRY output and $DCBA$ outputs as you toggle the switch (1) low, (2) high, (3) low, (4) high. How long do counts 1111 and 0101 persist in terms of a full clock cycle? Continue to toggle the switch several times and record the output sequence. How many distinct output states does the counter have? It is therefore a modulus-? counter.

5. Modify your circuit to conform to that shown in Figure L22.4. Both debounced switches should initially be set in the high output state. Apply power to your circuit and toggle the debounced switch connected to the LOAD input low and then high. What happens to the outputs? Now begin to toggle the debounced switch connected to the DOWNCOUNT input. Observe and record the state of the outputs as you complete each toggle of the switch. What happens when the output reaches 0000? Does additional toggling of the switch have any effect after this?

Discussion   1. In part 1 you observed that the outputs of the 74LS193 can be cleared to 0000 by activating (taking high) the CLEAR input. This can be done at any time, regardless of the current count on the outputs. Likewise, the output can be preset to whatever value is on the inputs $DCBA$ by making the LOAD input active (low). This also can be done at any time.

2. In part 2 you observed that with the LOAD and CLEAR inputs held inactive, the 74LS193 counted up when clock pulses were input to the UPCOUNT input, and counted down when clock pulses were input to the DOWNCOUNT input. The count always changed *at the rising edge of the clock*. When the count reached 1111, you observed that the CARRY output became active (low) during the *half-cycle* when the count was at 1111 AND the UPCOUNT input was low.

That is, when the count first reached 1111 (following the rising edge of the UPCOUNT input), the CARRY output did not yet become active. Only when the UPCOUNT input went low again with the count still at 1111 did the CARRY output go active.

3. In part 3 you investigated the down-count operation of the circuit. When clock pulses were input into the DOWNCOUNT input, the circuit counted down, changing state at each rising edge of the clock. When the count reached 0000, you observed that the BORROW output became active (low) during the half-cycle when the count was at 0000 AND the DOWNCOUNT input was low. This is similar to the operation found in part 2 for the up-counting operation.

4. In part 4 you constructed a circuit that was a modulus-11 up-counter. As clock pulses were input into the UPCOUNT input, the circuit counted upward until the count reached 1111. At this point the following occurred. The count reached 1111 on the rising edge of the UPCOUNT input. During the positive half-cycle of the UPCOUNT input, the count held at 1111. When the UPCOUNT input went low, the CARRY output also went low and immediately caused the count 0101 (as dictated by the $DCBA$ inputs) to be loaded. This new count allowed the CARRY output to return high again and persisted while the UPCOUNT remained low (a half-cycle). On the rising edge of the UPCOUNT input, the count advanced to 0110. Thus the two consecutive counts 1111 and 0101 persisted for only a half clock cycle each, while all other counts persisted for a full clock cycle.

5. In part 5 you constructed a down-counter that stopped counting when the count reached 0000. The key to this was NAND gate 1 in Figure L22.4. The passage of clock pulses from debounced switch 1 is controlled by the lower input to NAND gate 1. If this lower input is zero, then the output of NAND gate 1 is held fixed at logic 1. The lower input to NAND gate 1 in turn comes from the

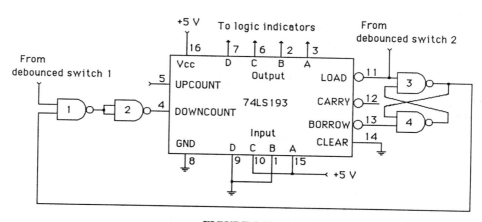

FIGURE L22.4

output of the $R-S$ flip-flop comprised of NAND gates 3 and 4. Operation begins when the $R-S$ flip-flop is set by a low from debounced switch 2. This low also activates the LOAD input and causes the count on the $DCBA$ inputs to be loaded into the counter. With the $R-S$ flip-flop set, counts can pass through NAND gate 1. In effect, NAND gate 1 now acts as an inverter for pulses from debounced switch 1. NAND gate 2 is connected as a second inverter, so clock pulses reach the DOWNCOUNT input having been inverted twice (and thus in their original logic state). Down-counting proceeds normally until the count reaches 0000. When the count reaches 0000 AND the DOWNCOUNT input goes low, the BORROW output goes low for a half-cycle, thus resetting the flip-flop. This essentially closes NAND gate 1 so that no further clock pulses can pass.

LABORATORY 23   THE 7495 SHIFT REGISTER

Objectives   To investigate the various operating modes of the 7495 4-bit shift register.

Materials Required   + -V power supply

1	7495 4-bit shift register
4	logic indicators
1	debounced switch

Procedure   1. The complete logic diagram and pin-outs for the 7495 are shown in Figure 6.24. Construct the circuit shown in Figure L23.1. Now connect your debounced switch to CLOCK 2 (pin 8) with the switch output set in the low logic state. Finally, ground the inputs $DCBA$. Apply power to your circuit and observe the state of the outputs. Now toggle your debounced switch from low to high. Do the outputs change? Now toggle the switch from high to low. Do the outputs change now? To what? Connect inputs D and B to $+5$ V. Toggle the switch again, first to high and then to low. How do the outputs behave? When do changes occur (at the rising or falling edge of clock input)? Now ground the MODE input, and again connect inputs D and B to ground. Try toggling the debounced switch a few times. Does anything happen? Disconnect the MODE input and move the debounced switch output from CLOCK 2 to CLOCK 1. Toggle the switch a few times. Does anything happen? Summarize your observations by stating which clock input controls parallel loading of data from the inputs to the outputs, and what the state of the MODE input must be for this to occur.

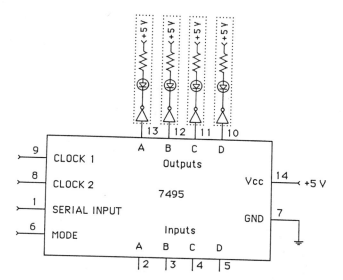

FIGURE L23.1

2. Connect your debounced switch to CLOCK 1 and ground the MODE input. Ground the SERIAL INPUT. Toggle the debounced switch through one cycle and observe what happens to the outputs and what type of clock transition causes the change to occur. Disconnect the SERIAL INPUT from ground and again toggle CLOCK 1 through one cycle. What is happening to the outputs? Again ground the SERIAL INPUT, and again toggle the CLOCK 1 input. Are the results as expected? Now disconnect the MODE input from ground. Try toggling the CLOCK 1 input. Does anything happen? Again ground the MODE control, but move the output of your debounced switch from the CLOCK 1 to the CLOCK 2 input. Now toggle the switch. Does anything happen? Summarize your findings by stating which clock controls the right-shifting of data from the SERIAL INPUT through the *ABCD* outputs of the circuit, and what the state of the MODE input must be for this to occur.

3. The 7495 is basically a shift-right register that can also be loaded in parallel. However, the register can also be made to shift left by using a few external connections. Modify your circuit as shown in Figure L23.2. In this circuit, the MODE input will be used to control whether the circuit shifts data right or left. Recall that CLOCK 1 is normally associated with shifting data to the right, coming from the SERIAL INPUT. This occurs when the MODE input is held in the low state. The circuit in Figure L23.2 should continue to operate in this fashion. Set the MODE input low and verify that the circuit does indeed still function as a right-shift register under these conditions. The new behavior occurs when the MODE input is set high. This normally causes data to be loaded from the parallel *ABCD* inputs to the *ABCD* outputs. Because of the external wiring

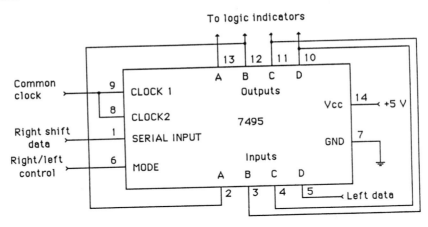

FIGURE L23.2

in Figure L23.2, the D output data are fed back to the C parallel input, the C output to the B input, and the B output to the A input. The D input data are an external input. Now every time the clock is toggled, the D output data move through the C input to the C output (that is, to the left). Likewise, the C output data move through the B input to the B output (again to the left), and similarly the B output ends up at the A output. New data (for the D output) come from the outside via the D input. Connect the SERIAL INPUT to ground, and the D input to $+5$ V (or leave it unconnected). With the MODE input at ground, toggle the debounced switch four times. How do the outputs behave? Now set the MODE input to high, and again toggle the switch four times. Now have do the outputs behave?

Discussion 1. In part 1 you investigated the operation of the 7495 in the parallel load mode. You found that data on the parallel inputs were loaded on the falling edge of CLOCK 2, provided the MODE input was held high (thus when the MODE input is set high, it is said to be in the parallel mode). With the MODE input taken low (the serial mode), pulses on CLOCK 2 had no effect. Likewise, with the MODE input held high (the parallel mode) pulses on CLOCK 1 had no effect.

2. In part 2 you investigated the operation of the 7495 in the serial mode. You found that data on the SERIAL INPUT were shifted right into the $ABCD$ outputs on the falling edge of CLOCK 1, provided the MODE input was held low (in the serial mode). With the MODE input taken high, pulses on CLOCK 1 had no effect. Likewise, with the MODE input held low, pulses on CLOCK 2 had no effect. One common type of operation for the 7495 is to set the circuit to parallel mode, clock in 4 bits of data at one time, switch to the serial mode, and clock out the 4 bits one at a time through the D output. Conversely, it is

possible to first set the circuit to serial mode, clock 4 bits of data in through the SERIAL INPUT (1 bit at a time), and then read out the 4 bits of data in parallel directly from the *ABCD* outputs.

3. In part 2 you saw that with suitable external connections the 7495 could be configured as either a shift-right or a shift-left register. The MODE input then controlled the direction of the shift. Information to be shifted right came in via the SERIAL INPUT, while information to be shifted left came in via the *D* input.

■ PROJECT 4   DIGITALLY CONTROLLED DUTY CYCLE

Minimum Materials Required

+5-V power supply
1   7495 4-bit shift register

It is sometimes useful to be able to precisely control the duty cycle of a signal. Assume that you have a 1-kHz square wave available (the duty cycle of which is of no concern). You wish to input this signal into a digital circuit and have this circuit produce a 250-Hz signal that has a duty cycle of 25%, 50%, or 75% depending on the settings of a series of switches. A user of your circuit would proceed as follows: (1) Set four switches to determine the duty cycle. (2) Press a single button to cause the duty cycle to change as per the switch settings.

Consider using the 7495 4-bit shift register as a recirculating shift register. Recall that the 7495 has two modes of operation, a serial shift mode and a parallel load mode. Instead of switches at the inputs to your system, simply tie the inputs low or leave them unconnected. You could use one of your debounced switches as the button to cause the duty cycle in change. Have your instructor check your design before you build and test it.   ■

LABORATORY 24   SIMPLE IC OSCILLATORS

Objectives   To construct and analyze the operation of simple IC astable and monostable oscillators.

Materials Required

+5-V power supply
1   74LS04 hex inverter
1   74SL123 dual retriggerable one-shot

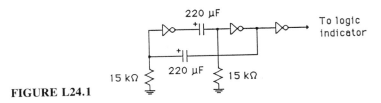

FIGURE L24.1

2   220-μF electrolytic capacitors
2   15-kΩ resistors (1/4 W)
1   100-kΩ resistor (1/4 W)
2   logic indicators
1   debounced switch

Procedure   1. Construct the circuit shown in Figure L24.1. This circuit is a simple astable oscillator. Apply power to your circuit and observe the results. Measure the length of one full cycle of oscillation. If you have extra resistors available, try replacing the two 15-kΩ resistors with much smaller values (say a few hundred ohms). What happens? You might also try replacing the resistors with much larger values (say several tens of kilohms). What happens? If you have an oscilloscope available, replace the two 220-μF capacitors with capacitors of approximately 0.1 μF. This will produce a much higher oscillation frequency, which can be easily observed on your scope. In particular, observe the waveforms at the input to the first inverter and then the input to the second inverter (if you have a dual-trace scope, you can observe both waveforms simultaneously). You should also observe the shape of the waveform at the output of the third inverter. Make a careful drawing of all these waveform shapes.

2. Construct the circuit shown in Figure L24.2. The basic circuit consists of a 74LS123 dual retriggerable one-shot with external capacitors and resistors to control timing. Set your debounced switch so that its output is low. Observe the state of both logic indicators. Now toggle the debounced switch high. Does anything happen? Now toggle the switch low. Observe what happens. Repeat the process a few times, and estimate the length of time that each output (Q_1 and Q_2) is high. What type of transition on input $A1$ is necessary to trigger the first monostable oscillator? From the logic diagram of the 74LS123, what type of transition *at the output of the AND gate* triggers the monostable? Then what type of transition at input $B1$ would trigger the monostable (assuming that $A1$ is low and CLR is high)? You might wish to ground $A1$ and move your debounced switch to input $B1$ in order to test your answer to the previous question.

Discussion   1. In part 1 you observed the operation of a simple astable oscillator constructed from a single IC (a 74LS04) and a few external components. This oscillator

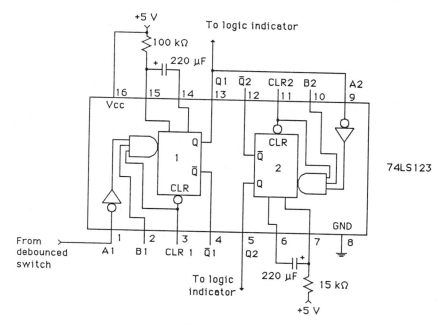

FIGURE L24.2

produces a square wave with a nearly 50% duty cycle. The time period for the oscillator should be roughly RC (where R is in ohms and C is in farads). With your values, $RC = (15 \times 10^3 \ \Omega)(220 \times 10^{-6} \ F) = 3.3$ s. The actual value you measured should be within a factor of 2 of this value. The frequency of oscillation can easily be changed by changing the value of the capacitors. However, the value of the resistors should not be changed much. If the value of the resistors is lowered significantly, the current output of the inverters will not be sufficient to produce a high enough voltage drop across either resistor to generate a logical high. If the value of the resistors is raised too much, the input at each inverter will remain high (remember, an open TTL input acts high, and a large enough resistor will approximate an open input). If you used an oscilloscope to observe the waveforms at the inputs of the first two inverters, you observed that these waveforms exhibited a high phase and a low phase. During the high phase, the observed voltage is produced by current charging the appropriate capacitor and flowing through the resistor in series with the capacitor. As the capacitor is charged, the current is reduced and thus the voltage drops. When the voltage drops to the critical point where the inverter input sees it as a logical low, the inverter will change its output from low to high. This will cause all the other inverters in the circuit to reverse their state also. The waveforms at the inputs to the two inverters are 180 degrees out of phase. This can be observed directly on a dual-trace scope. On a single-trace scope, this can also be observed by using the output of the third inverter as a trigger signal for the scope. In this way, any signal observed on the scope will show its absolute time relationship to the same fixed reference.

2. In part 2 you investigated the operation of the 74LS123 dual retriggerable one-shot. The circuit you constructed consisted of two one-shot (monostable) multivibrators, where the output of the first was used as a trigger for the second. The external resistor and capacitor determined the length of the pulse produced by the one-shot. Your debounced switch was used to trigger the first one-shot. The trigger occurred on the falling edge of the $A1$ input. Once triggered, the output of the one-shot went high for a time given approximately by $0.4\ RC$. At the end of this time the output returned to zero and remained in this state until triggered again. With the values shown in Figure L24.2, the output pulse from the first one-shot should have been approximately 8.8s. The falling edge of the pulse from the first one-shot was used to trigger the second one-shot. With the values of R and C shown in Figure L24.2, the length of the pulse from the second one-shot should have been about 1.3 s. If you investigated the use of input $B1$ as a trigger input, you found that the one-shot was triggered on the rising edge of this input. The different trigger inputs are offered as a matter of convenience to individuals using the 74LS123 in a circuit design.

LABORATORY 25   THE 555 TIMER AND ITS USES

Objectives   To use the 555 timer to construct an astable multivibrator and a monostable multivibrator.

Materials Required   +5-V power supply

1	555 timer
1	10-kΩ resistor (1/4 W)
1	1-MΩ resistor (1/4 W)
1	0.01-μF capacitor
1	1-μF capacitor
1	10-μF capacitor (electrolytic)
1	digital voltmeter (DVM)

Procedure   1. Construct the circuit shown in Figure L25.1. You may wish to refer to Figure 7.14 in your text to review the internal structure of the 555 timer. Apply power to your circuit and observe the output as indicated by your logic indicator. Time

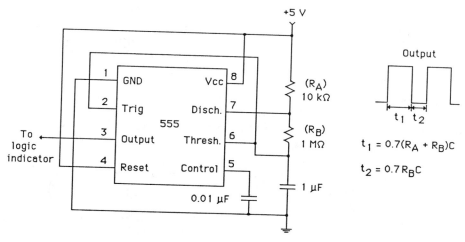

FIGURE L25.1

the period of the oscillations. If you have an oscilloscope, you may wish to replace the 1-μF capacitor with a much smaller value (say around 0.001 μF) and observe the output waveform on your scope. You may also wish to replace the 1-MΩ resistor with a much smaller value (say between 2 and 5 kΩ) and observe the change in the duty cycle of your signal.

2. Construct the circuit shown in Figure L25.2. Attach your DVM across the 10-μF capacitor and apply power to the circuit. Nothing should happen at this point. Now briefly ground the trigger input (that is, briefly tap a ground wire against the trigger input, pin 2). Note what happens to the output, and observe the reading on your voltmeter. According to Figure 7.15 in your text, how large should the voltage across the 10-μF capacitor become before it causes comparator 1 to reset the internal flip-flop and therefore discharge the capacitor, ending

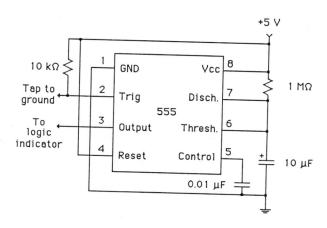

FIGURE L25.2

the output pulse? What is the reading on your DVM when the output pulse actually ends? If you have time, you may wish to experiment with other values of capacitance and resistance to obtain output pulses of different durations. However, do not use a resistor smaller than about 1 kΩ.

Discussion
1. In part 1 you observed that the 555 timer formed the core of a simple astable oscillator. With the components used in your circuit, the period of the output signal should have been slightly under 1 s. Also, because R_B was so much larger than R_A, the output waveform had a nearly 50% duty cycle. If you had an oscilloscope and tried the additional activities suggested, you found that replacing the 1-μF capacitor with a much smaller value decreased the period of the output signal in direct proportion to the reduction of the capacitance. If you replaced the 1-MΩ resistor by a much smaller value, you noticed that the duty cycle of the output signal was no longer near 50%, but in fact increased significantly. The duty cycle can be predicted from the formulas for t_1 and t_2 given in Figure L25.1.

2. In part 2 you observed the operation of a monostable multivibrator constructed from a 555 timer. In this circuit, no output is generated until the circuit is triggered. You found that briefly grounding the trigger input provided the necessary start pulse. Once triggered, the output of the circuit went high and the voltage across the 10-μF capacitor slowly increased from zero. According to Figure 7.15, the voltage across the capacitor needs to reach two-thirds of Vcc in order to cause comparator 1 to reset the circuit. Since Vcc was +5 V in your circuit, the voltage across the capacitor needed to reach about +3.34 V in order to end the output pulse. You should have observed that the output pulse ended when the DVM read approximately +3.34 V. If you experimented with various combinations of resistance and capacitance values, you saw how easy it was to obtain pulses with durations ranging over a wide range of time periods. The 555 timer is a very versatile circuit that is easy to use for both astable and monostable applications. There are a number of other uses as well, such as frequency division, pulse width modulation, and linear ramp generation. You should consult a manufacturer's data book for many additional suggestions.

LABORATORY 26   DIGITAL DESIGN USING COMBINATIONAL KARNAUGH MAPS

Objectives   To complete a digital design task from concept to truth table to circuit implementation with the aid of combinational Karnaugh mapping.

Materials
Required   1   +5-V power supply

1   74SL00 quad 2-input NAND gate

1   74LS02 quad 2-input NOR gate

2   logic indicators

Procedure   1. You are to assume that someone has designed a 4-bit counter that counts through a sequence that consists *only of numbers divisible by 3 or by 4*. That is, the counter counts through the decimal sequence 3, 4, 6, 8, 9, 12, and 15. It cycles endlessly through this sequence. Your task is to design two separate circuits, one that responds with a high output anytime a number divisible by 3 is input to it and one that responds with a high output anytime a number divisible by 4 is input to it. Figure L26.1 shows the structure of the circuitry. As a first task, develop a truth table with the headings A, B, C, D, $F1$, and $F2$. The headings A to D are for the 4-bit counter output (A the most significant bit, D the least), $F1$ is the output of the modulus-3 detector (detects numbers divisible by 3), and $F2$ is for the output of the modulus-4 detector. List the 16 possible combinations of counts from 0000 through 1111 under the A to D headings. Under the headings for $F1$ and $F2$, place 1's whenever the corresponding output should be high, 0's whenever the corresponding output should be low, and X's for those counts that are never produced by the counter (these are don't-care states).

2. Construct two 4-variable Karnaugh maps (such as shown in Figure 8.8). Use the $F1$ column of your truth table to fill in one of the Karnaugh maps and the $F2$ column to fill in the second Karnaugh map. For each map, construct appropriate groupings of the 1's (and X's where this would prove useful) and determine the minimum Boolean function for $F1$ and $F2$.

3. The Boolean functions that you obtained in part 2 can be implemented in various ways. Devise a way to implement the $F1$ and $F2$ Boolean functions using a 74LS00 quad 2-input NAND gate and a 74LS02 quad 2-input NOR gate. The pin-outs for the 74LS00 are shown in Figure L4.1, and the pin-outs for the 74LS02 are shown in Figure L26.2. Test your circuit implementation by inputing the counts that your counter would produce and observing the $F1$ and $F2$ outputs

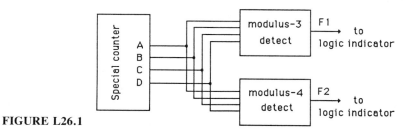

FIGURE L26.1

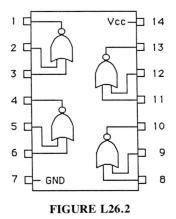

FIGURE L26.2

A	B	C	D	F1	F2
0	0	0	0	X	X
0	0	0	1	X	X
0	0	1	0	X	X
0	0	1	1	1	0
0	1	0	0	0	1
0	1	0	1	X	X
0	1	1	0	1	0
0	1	1	1	X	X
1	0	0	0	0	1
1	0	0	1	1	0
1	0	1	0	X	X
1	0	1	1	X	X
1	1	0	0	1	1
1	1	0	1	X	X
1	1	1	0	X	X
1	1	1	1	1	0

FIGURE L26.3

that result for each count. It is irrelevant how your circuits behave for the don't-care count states since these will not be generated by the counter.

Discussion 1. In part 1 you constructed a truth table to represent the Boolean functions $F1$ and $F2$ to detect numbers divisible by 3 and 4, respectively. Since the counter that was to generate the numbers in the first place only generated the sequence 3, 4, 6, 8, 9, 12, and 15, all other numbers could be considered don't-care states and labeled with X's in the truth table. Your truth table should have been as shown in Figure L26.3. Notice that one number (1100 = 12) is divisible by both 3 and 4 and thus has a 1 under both the $F1$ and $F2$ columns.

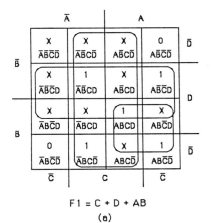

$F1 = C + D + AB$

(a)

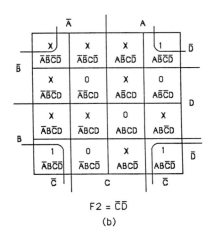

$F2 = \bar{C}\bar{D}$

(b)

FIGURE L26.4

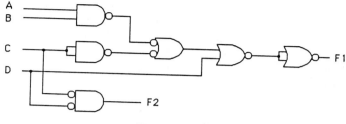

FIGURE L26.5

2. In part 2 you used the truth table from part 1 to fill in two Karnaugh maps and determine the minimum Boolean function for $F1$ and $F2$. Your maps should have been as shown in Figure L26.4. The minimum Boolean function for each map is also shown in Figure L26.4.

3. In part 3 you were to design a circuit that implemented the $F1$ and $F2$ functions by using a 74LS00 quad-2 input NAND gate and a 74LS02 quad 2-input NOR gate. An example of such a circuit is shown in Figure L26.5. Your circuit may have varied slightly from this depending on how you implemented inversions at various places in the circuit. Notice that in Figure L26.5 two different symbols have been used for both NAND gates and NOR gates.

LABORATORY 27   DIGITAL DESIGN USING SEQUENTIAL KARNAUGH MAPPING

Objectives To design and implement a custom digital counter with the aid of sequential Karnaugh mapping.

Materials Required

+5-V power supply

2	74LS76A dual $J-K$ flip-flops
1	74LS08 quad 2-input AND gate
4	logic indicators
1	debounced switch

Procedure 1. Your task in this laboratory is to design the special counter discussed in laboratory 26. Recall that this counter cycled through the counts 3, 4, 6, 8, 9, 12, and 15. You will use four $J-K$ flip-flops (plus whatever additional logical gates are required) to implement your design. The design procedure is that used in Section 8.3. Begin by drawing a four-variable Karnaugh map such as found in Figure 8.8. Consider the output of the four flip-flops as $ABCD$, where A is the most

Symbol Used to Represent State Change	Current State	Next State
1	0	1
+	1	1
−	1	0
0	0	0
X	don't-care	

FIGURE L27.1

significant bit and D the least significant bit. On your map, identify the square associated with count 3 ($ABCD$ = 0011) and place a 0 (zero) in this square (the state representing count 3 will be referred to as state 0). Next locate the square associated with count 4 ($ABCD$ = 0100) and place a 1 in this square (this will be called state 1). Continue this procedure, placing a 2 in the square associated with count 6, a 3 in the square associated with count 8, and so on, until all the counts in the sequence have been accounted for. This map will allow you to see how each count bit (A, B, C, and D) changes as the count moves from state 0 to state 1 to state 2, and so on. Your map should resemble Figure 8.22 except that your map will be a four-variable map instead of a three-variable map.

2. Draw four more Karnaugh maps, one for each $J-K$ flip-flop of your counter. Referring to your map from part 1, notice that several squares are not part of the count sequence. These states never occur in the count and are therefore don't-care states. On your four maps, place an X in the don't-care squares (the same squares on all four maps). Now label one of your four maps as A (referring to flip-flop A). Look at the state of A in the square labeled 0 and the square labeled 1. How does A change between state 0 and state 1. Refer to Figure L27.1 to determine what symbol should therefore be placed in the square representing state 0. For example in state 0, A is low (0) and in state 1, A is also low (0). Thus the table in Figure L27.1 shows that a 0 should be placed in the square representing state 0 on the map for flip-flop A. Proceed through the entire sequence of states, filling in the map for flip-flop A [with the symbols 0, 1, +, and − as appropriate]. When you have finished, do the same for flip-flops B, C, and D.

3. Use the rules in Figure L27.2 to group the symbols on the map for flip-flop A, and thus determine what Boolean function should be input to the J and K inputs of flip-flop A. Do the same thing with the maps for flip-flops B, C, and D. You should now have Boolean functions representing the J and K inputs to all four flip-flops. From these Boolean functions, draw a circuit that will implement your counter. Remember, the clock input is connected to all four $J-K$ flip-flops (such as is shown in Figure 8.25).

4. Construct your circuit and verify its operation. The pin-outs of the 74LS76A dual $J-K$ flip-flop are shown in Figure L19.1, and the pin-outs of the 74LS08 quad 2-input AND gate are shown in Figure L9.1. Monitor the four flip-flop

J input equation

1. Each square with a 1 must be used in a group.
2. Each square with a 0 may not be used in a group.
3. All other symbols are optional.

K input equation

1. Each square with a – must be used in a group.
2. Each square with a + may not be used in a group.
3. All other symbols are optional.

FIGURE L27.2

set outputs (*A*, *B*, *C*, and *D*) with your logic indicators, and use your debounced switch as a clock input. Does your counter indeed cycle through the states 3, 4, 6, 8, 9, 12, and 15?

Discussion

1. In part 1 you labeled the desired count sequence on a four-variable Karnaugh map. Your result should have looked like Figure L27.3.
2. In part 2 you used the Karnaugh map from part 1 to construct individual maps for flip-flops *A*, *B*, *C*, and *D*. These maps should have been as shown in Figure L27.4. Notice that considerable reduction is possible, so the *J* and *K* input equations for each flip-flop turn out to be quite simple. You should have obtained the Boolean functions shown in Figure L27.4.
3. With the input equations from part 2, you designed a circuit to implement your counter. The design should have been as shown in Figure L27.5. Note that

	$\bar{A}$	$\bar{A}$	A	A	
$\bar{B}$	$\bar{A}\bar{B}\bar{C}\bar{D}$	$\bar{A}\bar{B}C\bar{D}$	$A\bar{B}C\bar{D}$	3 $A\bar{B}\bar{C}\bar{D}$	$\bar{D}$
$\bar{B}$	$\bar{A}\bar{B}\bar{C}D$	0 $\bar{A}\bar{B}CD$	$A\bar{B}CD$	4 $A\bar{B}\bar{C}D$	D
B	$\bar{A}B\bar{C}D$	$\bar{A}BCD$	6 $ABCD$	$AB\bar{C}D$	D
B	1 $\bar{A}B\bar{C}\bar{D}$	2 $\bar{A}BC\bar{D}$	$ABC\bar{D}$	5 $AB\bar{C}\bar{D}$	$\bar{D}$
	$\bar{C}$	C	C	$\bar{C}$	

FIGURE L27.3

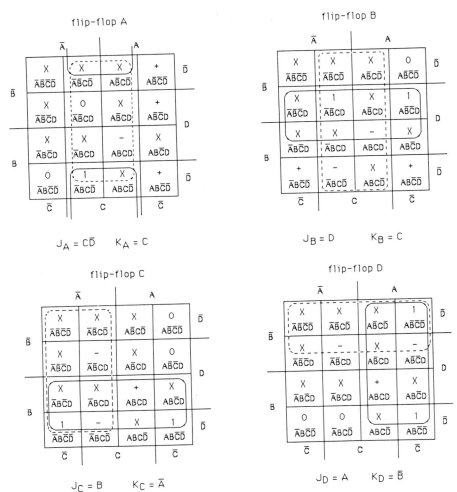

FIGURE L27.4

inverters are not necessary since the *J–K* flip-flops have both set and reset outputs. When you constructed and tested your circuit, you should have obtained the desired count sequence. You may have noticed that when power was first applied to your circuit an illegal count was present. Normally, toggling the system clock a few times cycles the counter to a count within the desired sequence. From then on, only the desired count sequence is obtained as the clock is toggled. If you wish, however, you may use the preset and reset inputs to start the count at the beginning of the desired sequence before any clock pulses are input to the circuit.

To logic indicators

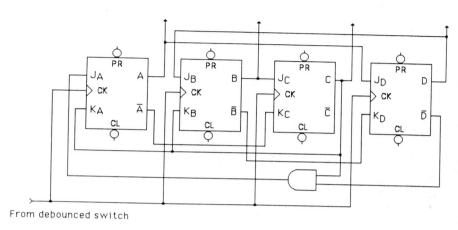

From debounced switch

FIGURE L27.5

LABORATORY 28   FOUR-BIT D/A CONVERTER

Objectives   To construct and test a simple 4-bit, digital-to-analog converter using an *R–2R* ladder.

Materials
Required   +5-V power supply

1   digital voltmeter (DVM)

1   74LS193 counter

5   50-kΩ resistors (1/4 W)

3   25-kΩ resistors (1/4 W)

4   logic indicators

1   debounced switch

Procedure   1. Construct the circuit shown in Figure L28.1. The detailed pin-outs of the 74LS193 counter are shown in Figure 6.8 should you wish to review them in detail. Apply power to your circuit and briefly disconnect the CLEAR input so that the count is reset to 0000. Use your digital voltmeter (DVM) to measure the voltage obtained at the point labeled Output. What does it read when the count is 0000? Now toggle the debounced switch so that the count becomes 0001. Record the

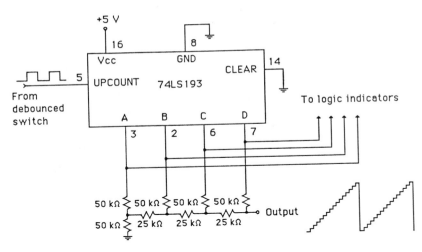

FIGURE L28.1

DVM reading. Continue to toggle the count and record the output voltage at each count. How uniform are the voltage steps as the count progresses from one count to the next? What is the average size of a voltage step from one count to the next? Write a simple formula that gives the voltage output in terms of the count. Can you visualize how an 8-bit D/A converter could be constructed by using more resistors and a second 74LS193 counter?

2. If you have an oscilloscope available and a TTL square-wave generator with about a 1-kHz frequency, replace your debounced switch with the output from the square-wave generator and monitor the output with your scope. Does the output look like what is shown on Figure L28.1?

Discussion 1. In part 1 you constructed a simple 4-bit D/A converter using a binary counter and an $R-2R$ ladder. The voltage obtained from the output was simply the output count multiplied by a fixed voltage per count. Assuming that the counter outputs actually produced $+5$ V when they were high and 0 V when they were low (which they probably do not), the voltage step between consecutive counts would be $+5$ V/16 = 0.3125 V/step. It is unlikely that your steps were this size or that they were uniform. In the first place, the outputs from the counter are simply TTL outputs and are not guaranteed to be exactly 0 and $+5$ V. In the second place, the resistors that you used were probably not exactly 50 and 25 kΩ. However, you should have obtained distinct, more or less uniform voltage steps as you toggled the count upward from 0000 to 1111. Note that the 74LS193 also can function as a data latch. That is, it has four parallel inputs and a LOAD input. Thus any digital number in the range 0000 to 1111 can be loaded into the counter at any time. It is not necessary to cycle through several intermediate counts to generate the desired output voltage.

2. In part 2 you were able to observe the sawtooth output generated as the counter cycled repeatedly from 0000 to 1111. Because the counts came in at 1000/s, the signal could be conveniently displayed on your oscilloscope. This gave a better visual picture of the voltage steps. The output on the scope face should have resembled the staircase shown in Figure L28.1. Depending on how close the resistors were to their numbered values, the steps of the staircase may have been slightly nonuniform.

EQUIPMENT SUPPLIERS

There are many good suppliers of electronic components. Among the frequently used mail-order suppliers are:

Jameco Electronics
1355 Shoreway Road
Belmont, CA 94002
(415) 592-8097

JDR Microdevices
1224 S. Bascom Ave.
San Jose, CA 95128
(800) 538-5000

Digi-key Corporation
701 Brooks Ave. South
P.O. Box 677
Thief River Falls, MN 56701
(800) 344-4539

There are many good suppliers of digital trainers suitable for conducting the laboratories in this appendix. Among these are (in addition to Jameco listed above):

Heath Company
Benton Harbor, MI 49022
(800) 253-0570

Priority One Electronics
21622 Plummer St.
Chatsworth, CA 91311-4194
(800) 423-5922

INDEX

HBJ Harcourt Brace Jovanovich, Inc.
COLLEGE DEPARTMENT
MIDWEST REGION
7555 CALDWELL AVENUE
CHICAGO, IL 60648

RAVI TAMMANA
DEPT OF TECHNOLOGY
SOUTHERN ILLINOIS UNIV
CARBONDALE   IL 62901

254

SHIP VIA				REGION				WHSE.	PAGE
LIBRARY RATE				3 — MIDWEST				22	1 OF 1

SALES CALL DATE	ENTRY DATE	PROCESS DATE	SLSPR CODE	CNTY	ST.	SAMPLE ACCT. NO. 002249	LABEL NUMBER
		08/20/90	314			3192-9995	

QUANTITY	TITLE CODE	TITLE
1	05-17636 DIGIT - -	U0205DA D0156AA DIGITAL ELECTRONICS EFF DATE:082090 (DESK)
1	1.4060	TOTAL ORDER QUANTITY/WEIGHT

PACKING SLIP